THE LIBRARY
ST. MARY'S COLLEGE OF MARYLAND
ST. MARY'S CITY, MARYLAND 20686

D1559089

ORBITAL INTERACTION THEORY OF ORGANIC CHEMISTRY

ORBITAL INTERACTION THEORY OF ORGANIC CHEMISTRY

ARVI RAUK

Department of Chemistry
The University of Calgary
Calgary, AB Canada

A Wiley-Interscience Publication
JOHN WILEY & SONS, INC.
New York   Chichester   Brisbane   Toronto   Singapore

This text is printed on acid-free paper.

Copyright © 1994 by John Wiley & Sons, Inc.

All rights reserved. Published simultaneously in Canada.

Reproduction or translation of any part of this work beyond that permitted by Section 107 or 108 of the 1976 United States Copyright Act without the permission of the copyright owner is unlawful. Requests for permission or further information should be addressed to the Permissions Department, John Wiley & Sons, Inc., 605 Third Avenue, New York, NY 10158-0012.

Library of Congress Cataloging in Publication Data:

Rauk, Arvi, 1942–
    Orbital interaction theory of organic chemistry / by Arvi Rauk.
      p. cm.
    "A Wiley-Interscience publication."
    Includes bibliographical references and index.
    ISBN 0-471-59389-3 (acid-free)
    1. Molecular orbitals.   2. Chemistry, Physical organic.
 I. Title.
QD461.R33 1994
547.1′28—dc20   93-27021

Printed in the United States of America

10 9 8 7 6 5 4 3 2 1

CONTENTS

PREFACE xiii

1 SYMMETRY AND STEREOCHEMISTRY 7

Purpose / 1
Definition of a Group / 2
Molecular Point Groups / 2
Schoenflies Notation / 3
Interrelations of Symmetry Elements / 3
Type Classification / 3
Isomerism and Measurements / 6
Stereoisomerism of Molecules / 7
Stereotopic Relationships of Groups in Molecules / 8
NMR and Stereochemistry / 11
Symmetry and Structural Parameters / 12
Note on Hybridization / 12
Symmetry and Orbitals / 14
In What Combination? / 15

2 MOLECULAR ORBITAL THEORY 18

Introduction / 18
The Electronic Hamiltonian Operator / 19
The Electronic Schrödinger Equation / 20

Expectation Values / 21
The Many-Electron Wavefunction / 21
The Electronic Hartree–Fock Energy / 23
Variation of E_{HF} / 27
LCAO Solution of the Fock Equations / 31
Integrals / 33
The Basis Set (STO-3G, 6-31G*, and All That) / 34
Interpretation of the Solutions of the HF Equations / 37
  Orbital Energies and the Total Electronic Energy / 37
Restricted Hartree–Fock (RHF) Theory / 38
  Mulliken Population Analysis / 40
  Dipole Moments / 40
  Total Energies / 41
  Orbital Energies and Orbitals / 42
  Configuration Energies / 43
  Energy Index / 44
  Bond Polarity Index / 45
  Bond Polarity Index and Orbital Interaction Diagrams / 46
Geometry Optimization / 46
Normal Coordinates and Harmonic Frequency Analysis / 47
Successes and Failures of Hartree–Fock Theory / 47
Post-Hartree–Fock Methods / 49
Configuration Interaction (CI) Theory / 49
Excited States from CI Calculations / 51
Many Body Perturbation Theory / 52
  Rayleigh–Schrödinger Perturbation Theory / 52
  Møller–Plesset Perturbation Theory (MPPT) / 55

3 ORBITAL INTERACTION THEORY   57

Relationship to the Hartree–Fock Equations / 57
The Hückel Approximation / 57
  Orbital Energies and the Total Electronic Energy / 57
Case Study of a Two-Orbital Interaction / 58
  Case 1: $\epsilon_A = \epsilon_B$, $S_{AB} = 0$ / 62
  Case 2: $\epsilon_A = \epsilon_B$, $S_{AB} > 0$, $S_{AB} \ll 1$ / 63
  Case 3: $\epsilon_A > \epsilon_B$, $S_{AB} = 0$ / 64
  Case 4: $\epsilon_A > \epsilon_B$, $S_{AB} > 0$ / 66
The Effect of Overlap / 68
The Relationship to Perturbation Theory / 69

Generalizations for Intermolecular Interactions / 70
  Generalization 1: $\Delta\epsilon_L \approx \Delta\epsilon_U$ / 70
  Generalization 2: $|\Delta\epsilon_L| < |\Delta\epsilon_U|$ / 70
  Generalization 3: $(|\Delta\epsilon_U| - |\Delta\epsilon_L|) < |\Delta\epsilon_L|, |\Delta\epsilon_U|$ / 70
  Generalization 4: $h_{AB} \approx k(\epsilon_A + \epsilon_B) S_{AB}$, ($k$ a Positive Constant, S_{AB} Assumed Positive) / 70
  Generalization 5: $\Delta\epsilon_L \approx \Delta\epsilon_U \approx \dfrac{h_{AB}^2}{\epsilon_A - \epsilon_B}$ / 71
  Generalization 6: $\phi_L \approx \chi_B + d\chi_A$; $\phi_U \approx \chi_A - d\chi_B$
  $d = -\dfrac{h_{AB}}{\epsilon_A - \epsilon_B}$ $d < 1$ / 71
Energy and Charge Distribution Changes from Orbital Interaction / 71
  The Four-Electron–Two-Orbital Interaction / 72
  The Three-Electron–Two-Orbital Interaction / 73
  The Two-Electron–Two-Orbital Interaction / 74
  The One-Electron–Two-Orbital Interaction / 76
  The Zero-Electron–Two-Orbital Interaction / 76
  Interactions between Molecules: Many Electrons, Many Orbitals / 77
    General Principles Governing the Magnitude of h_{AB} and S_{AB} / 77
  Interactions of MOs / 78
Electrostatic Effects / 80
Group Orbitals / 81
Assumptions for Application of Qualitative MO Theory / 85
An Example: The Carbonyl Group / 86
  Construction of the Interaction Diagram / 86
  Interpretation of the Interaction Diagram / 89
Why Does it Work, and When Might it Not? / 91

4 SIGMA BONDS AND ORBITAL INTERACTION THEORY   94

The C—X σ Bonds: X = C, N, O, F and X = F, Cl, Br, I / 94
Case Study 1: σ Bonds: Homolytic Versus Heterolytic Cleavage / 96
  Heterolytic Cleavage of σ Bonds / 96
  Homolytic Cleavage of σ Bonds / 97
Bonding in Cyclopropane / 98
Interactions of σ Bonds / 99
σ Bonds as Electron Donors or Acceptors / 101
σ Bonds as Electron Acceptors / 101
  As a σ Acceptor / 101

As a π Acceptor / 102
σ Bonds as Electron Donors / 103
    As a σ Donor / 103
    As a π Donor / 103

5 SIMPLE HÜCKEL MOLECULAR ORBITAL THEORY 106

The Simple Hückel Assumptions / 106
The Charge and Bond Order in Simple Hückel MO Theory:
($S_{AB} = 0$, One Orbital Per Atom) / 112
    P_A, the Electron Population of Center A; C_A, the Net Charge of Center A / 112
    B_{AB}, the Bond Order Between Centers A and B / 113
Factors Governing the Energies of MOs: Simple Hückel MO Theory / 113
    The Reference Energy, α, and Energy Scale, $|\beta|$ / 113
    Heteroatoms in SHMO Theory / 114
    The Effect of Coordination Number on α and β / 114
    Hybridization at C in Terms of α and β / 117
Gross Classification of Molecules on the Basis of MO Energies / 117

6 OLEFIN REACTIONS AND PROPERTIES 120

Case Studies 2: Reactions of Olefins / 120
    Effect of X: Substituents / 121
    Effect of Z Substituents / 123
    Effect of "C" Substituents / 123
    Effect of Distortion of the Molecular Skeleton / 124
    Alkynes / 125

7 REACTIVE INTERMEDIATES 126

Reactive Intermediates, $[CH_3]^+$, $[CH_3]^-$, $[CH_3]^{\cdot}$, and $[:CH_2]$ / 126
    Carbocations / 126
    Intermolecular Reactions of Carbocations / 128
    Intramolecular Reactions of Carbocations / 128
    Carbanions / 129
    Carbon Free Radicals / 131
    Carbenes / 134
Nitrenes, [NH], and Nitrenium Ions, $[NH_2]^+$ / 136
    Nitrenes / 136
    Nitrenium Ions / 139

8 CARBONYL COMPOUNDS 142

Case Study 3: Reactions of Carbonyl Compounds / 142
    Electrophilic Attack on a Carbonyl Group / 142
    Basicity and Nucleophilicity of the Oxygen Atom / 144
    Nucleophilic Attack on a Carbonyl Group / 148
    The Amide Group / 148
Thermodynamic Stability of Substituted Carbonyl Groups / 149

9 NUCLEOPHILIC SUBSTITUTION REACTIONS 151

Case Study 4: Nucleophilic Substitution at Saturated Carbon / 151
    Unimolecular Nucleophilic Substitution S_N1 / 151
    Bimolecular Nucleophilic Substitution S_N2 / 153
    Another Description of the S_N2 Reaction: The VBCM Model / 157

10 BONDS TO HYDROGEN 150

Hydrogen Bonds and Proton Abstraction Reactions / 160
    Hydrogen Bonds / 160
    Proton Abstraction Reactions / 162
    The E2 Elimination Reaction / 164
    The E1cB Mechanism Reaction / 165
    The E1 Elimination Reaction / 165
Reaction with Electrophiles: Hydride Abstraction and Hydride Bridging / 166
    Activation by π Donors (X: and C Substituents) / 166
    Hydride Abstraction / 166
    Hydride Bridges / 168
Reaction with Free Radicals: Hydrogen Atom Abstraction and One- or Three-Electron Bonding / 169
    Hydrogen Bridged Radicals / 169

11 AROMATIC COMPOUNDS 170

Case Study 4: Reactions of Aromatic Compounds / 170
    The Cyclic π Systems by Simple Hückel MO Theory / 170
    Aromaticity in σ-Bonded Arrays? / 171
Reactions of Substituted Benzenes / 172
Electrophilic Substitutions / 172
    Effect of Substituents on Substrate Reactivity / 173
    Electrophilic Attack on X:-Substituted Benzenes / 173

Electrophilic Attack on Z:-Substituted Benzenes / 174
Electrophilic Attack on C-Substituted Benzenes / 175
Electrophilic Attack on N Aromatics: Pyrrole and Pyridine / 176
Nucleophilic Substitutions / 177
Effect on Substituents on Substrate Reactivity / 178
Nucleophilic Attack on Z:-Substituted Benzenes / 178
Nucleophilic Attack on N Aromatics: Pyrrole and Pyridine / 179
Case Study 5: Nucleophilic Substitution by Proton Abstraction / 179

12 PERICYCLIC REACTIONS 182

General Considerations / 182
Cycloadditions and Cycloreversions / 183
Stereochemical Considerations / 185
Electrocyclic Reactions / 185
Stereochemical Considerations / 187
Cheletropic Reactions / 187
Stereochemical Considerations / 187
Sigmatropic Rearrangements / 187
Stereochemical Considerations / 188
Component Analysis (*Allowed* or *Forbidden*?) / 188
The Rule for Component Analysis / 190
The Diels–Alder Reaction / 190
The Cope Rearrangement / 192
1,3-Dipolar Cycloaddition Reactions / 194

13 ORBITAL AND STATE CORRELATION DIAGRAMS 198

General Principles / 198
Woodward–Hoffmann Orbital Correlation Diagrams / 199
Cycloaddition Reactions / 199
Electrocyclic Reactions / 201
Cheletropic Reactions / 202
Photochemistry from Orbital Correlation Diagrams / 202
Limitations of Orbital Correlation Diagrams / 206
State Correlation Diagrams / 206
Electronic States from MOs / 208
Rules for Correlation of Electronic States / 209
Example: Carbene Addition to an Olefin / 210

14 PHOTOCHEMISTRY   213

Photoexcitation / 213
The Jablonski Diagram / 215
Fate of the Excited Molecule in Solution / 216
The Dauben–Salem–Turro Analysis / 217
Norrish Type II Reaction of Carbonyl Compounds / 218
Norrish Type I Cleavage Reaction of Carbonyl Compounds / 221

APPENDIX A AN INPUT DESCRIPTION FOR GAUSSIAN XX   223

Molecule Specification Input Section / 224
Active and Frozen Variable Specifications / 226

APEPNDIX B THE INTERACTIVE PROGRAM, SHMO   228

Interactive Structure Input / 228
The *Options* Menu / 232
Known Bugs and Caveats / 233

APPENDIX C EXERCISES GROUPED BY CHAPTER   234

Exercises from Chapter 1 / 234
Exercises from Chapter 2 / 237
Exercises from Chapter 3 / 239
Exercises from Chapter 4 / 243
Exercises from Chapter 5 / 245
Exercises from Chapter 6 / 252
Exercises from Chapter 7 / 253
Exercises from Chapter 8 / 258
Exercises from Chapter 9 / 262
Exercises from Chapter 10 / 262
Exercises from Chapter 11 / 264
Exercises from Chapter 12 / 265
Exercises from Chapter 13 / 270
Exercises from Chapter 14 / 271
Miscellaneous Exercises / 274

REFERENCES   285

INDEX   295

PREFACE

The premise on which this text is based is that the vast majority of chemical phenomena may be qualitatively understood by the judicious use of simple orbital interaction diagrams. The material borrows heavily from the pioneering work of Fukui [1], [2], Woodward and Hoffmann [3], Klopman [4], Salem [5], Hoffmann [6], and many others whose work will be acknowledged throughout. Parts of the text are modeled most closely on the excellent book by Fleming: *Frontier Orbitals and Organic Chemical Reactions* [7], from which a number of illustrative examples are extracted. If there is uniqueness to the present approach, it lies in the introduction of the α and β of Simple Hückel Molecular Orbital theory as reference energy and energy scale on which to draw the interaction diagrams, mixing σ and σ^* orbitals and nonbonded orbitals with the usual π orbitals of SHMO theory on the same energy scale. This approach is difficult to justify theoretically but it provides a platform on which students can construct their interaction diagrams and is very useful in practice. Numerous illustrations from the recent literature are provided.

The book is intended for students of organic chemistry at the senior undergraduate and postgraduate levels. All reactions of organic compounds are treated within the framework of generalized Lewis acid–Lewis base theory, their reactivity being governed by the characteristics of the frontier orbitals of the two reactants. All compounds have occupied molecular obitals and so can donate electrons, i.e., act as bases in the Lewis sense. All compounds have empty molecular orbitals and so can accept electrons, i.e., act as acids in the Lewis sense. The "basicity" of a compound depends on its ability to donate the electron pair. This depends on the energy of the electrons, the distribution of the electrons (shape of the MO) and also on the ability of the substrate to receive the electrons (on the shape and energy of its empty orbital). The basicity of a compound toward different substrates will be different, hence a distinction between Lowry–Bronsted basicity and nucleophilicity. A parallel definition applies for the "acidity" of the compound. The structures of compounds are determined by the energetics of the occupied orbitals. Fine distinctions, such as conformational preferences, can be made on the basis of minimization of repulsive interactions and/or maximization of attractive interactions between the frontier localized group orbitals of a compound. All aspects are examined from the point of view of orbital interaction diagrams from which gross features of reactivity and structure flow naturally. The approach is qualitatively different from, and simpler than, a number of alternative approaches, such as the VBCM (*valence bond configuration mixing*) model [8], and OCAMS (*orbital correlation analysis using maximum symmetry*) approach [9], [10].

The organization of the text follows a logical pedagogical sequence. The first chapter is not primarily about "orbitals" at all, but introduces (or recalls) to the student elements of symmetry and stereochemical relationships among molecules and among groups within a molecule. Many of the reactions of organic chemistry follow stereochemically well-defined paths, dictated, it will be argued, by the interactions of the frontier orbitals. The conceptual leap to orbitals as objects anchored to the molecular framework, which have well defined spacial relationships to each other is easier to make as a consequence. Whether or not orbitals interact can often be decided on grounds of symmetry. The chapter concludes with the examination of the symmetry properties of a few orbitals which are familiar to the student.

The second chapter introduces the student to "orbitals" proper, and offers a simplified rationalization for why Orbital Interaction theory may be expected to work. It does so by means of a detailed derivation of Hartree–Fock theory making only the simplifying concession that all wavefunctions are real. Some connection is made to the results of *ab initio* quantum chemical calculations. A brief description of how to carry out these calculations using the GAUSSIAN system of codes is provided in Appendix A. Postgraduate students can benefit from carrying out a project based on such calculations on a system related to their own research interests. A few exercises are provided to direct the student. For the purpose of undergraduate instruction, this chapter may be skipped and the essential arguments and conclusions provided to the students in a single lecture as an introduction to Chapter 3.

Orbital Interaction theory proper is introduced in Chapter 3. The independent electron (Hückel) approximation is invoked and the effective one-electron Schrödinger equation is solved for the two-orbital case. The solutions provide the basis for the orbital interaction diagram. The effect of overlap and energy separation on the energies and polarizations of the resulting molecular orbitals are explicitly demonstrated. The consequences of 0–4 electrons are examined and applications are hinted at. Group orbitals are provided as building blocks from which the student may begin to assemble more complex orbital systems.

Chapter 4 provides a brief interlude in the theoretical derivations by examining a specific application of the two-orbital interaction diagrams to the description of σ bonds and their reactions.

In Chapter 5, conventional Simple Hückel Molecular Orbital (SHMO) theory is introduced. The Hückel "α" is suggested as a reference energy, and use of "$|\beta|$" as a unit of energy is advocated. Parameters for heteroatoms and hybridized orbitals are given. Finally, the interactive computer program, SHMO, a copy of which is provided with this book, is described in Appendix B.

Chapters 6–11 describe applications or orbital interaction theory to various chemical systems in order to show how familiar concepts such as acid and base strengths, nucleo- and electrophilicity, stabilization and destabilization, thermodynamic stability, and chemical reactivity may be understood.

Pericyclic reactions are described in Chapter 12 as a special case of frontier orbital interactions, i.e., following Fukui [1]. However, the stereochemical nomenclature, *suprafacial* and *antarafacial*, and the very useful general component analysis of Woodward and Hoffmann [3] are also introduced here.

Chapter 13 deals with orbital correlation diagrams following Woodward and Hoffmann [3]. State wave functions and properties of electronic states are deduced from the orbital picture, and rules for state correlation diagrams are reviewed, as a prelude to an introduction to the field of photochemistry in Chapter 14.

In Chapter 14, the state correlation diagram approach of the previous chapter is applied to a brief discussion of photochemistry in the manner of Dauben, Salem and Turro [11]. A more comprehensive approach to this subject may be found in the text by Michl and Bonacic-Koutecky [12], Turro [13], or Gilbert and Baggott [14].

Sample problems and quizzes, grouped by chapter, are presented in Appendix C. Many are based on examples from the recent literature and references are provided. Detailed answers are worked out for a few. These serve as further examples to the reader of the application of the principles of orbital interaction theory.

ARVI RAUK

Calgary, Alberta, Canada
December, 1993

ORBITAL INTERACTION THEORY OF ORGANIC CHEMISTRY

CHAPTER 1

SYMMETRY AND STEREOCHEMISTRY

PURPOSE

Symmetry is a concept that we all make use of in an unconscious fashion. We notice it every time we look in our bathroom mirror. We ourselves are (approximately) bilaterally symmetric. A reflected right hand looks like a left hand, a reflected right ear like a left ear, but the mirror image of the face as a whole, or of the toothbrush does not look different from the original. The hand, a *chiral* object, is distinguishable from its mirror image; the toothbrush is not. The toothbrush is *achiral*, and possesses a mirror plane of symmetry which bisects it. It would not surprise us if we were to inspect the two sides of the toothbrush and find them identical in many respects. It *may* surprise us to note that the two sides are distinguishable when held in the hand, i.e., in a chiral environment (the fingers hold one side and the thumb the other). However, the achiral toothbrush fits equally comfortably into either the right or the left hand. Chiral objects do not. They interact differently with other chiral objects and often the different interactions are known by separate words. When you hold someone's right hand in your right hand, you are *shaking* hands; when it is the other person's left hand in your right, you are *holding* hands. Similar properties and interactions exist in the case of molecules as well.

In this chapter we will familiarize ourselves with basic concepts in *molecular* symmetry [15]. The presence or absence of symmetry has consequences on the appearance of spectra, the relative reactivity of groups, and many other aspects of chemistry, including the way we will make use of orbitals and their interactions. We will see that the orbitals that make up the primary description of the electronic structure of molecules or groups within a molecule have a definite relationship to the three-dimensional structure of the molecule as defined by the positions of the nuclei. The orientations of the nuclear framework will determine the orientations of the orbitals. The relationships between structural units (groups) of a molecule to each other can often be classified in terms of the symmetry that the molecule as a whole possesses. We will begin by introducing the basic terminology of molecular symmetry. Finally we will apply simple symmetry classification: to local group orbitals to decide whether or not interaction is allowed in the construction of molecular orbitals, to molecular orbitals in order to determine the stereochemical course of electrocyclic reactions and to help determine the principal interactions in bimolecular reactions, and to electronic states, in order to construct state correlation diagrams.

We begin by introducing molecular point groups according to the Schoenflies notation and assigning molecular and group symmetry following Jaffe and Orchin [16] where greater detail may be found.

DEFINITION OF A *GROUP*

A group, $G = \{.\,,\,.\,,\,.\,,\,.\,g_i\,,\,.\,,\,.\,,\}$, is a set of elements related by an operation which we will call *group multiply* for convenience, and has the following properties:

1. The product of any two elements is in the set, i.e., the set is closed under group multiplication.
2. The associative law holds, e.g., $g_i(g_j g_k) = (g_i g_j) g_k$
3. There is a *unit* element, e, such that $eg_i = g_i e = g_i$.
4. There is an *inverse*, g_i^{-1}, to each element, such that $(g_i^{-1}) g_i = g_i (g_i^{-1}) = e$. An element may be its own inverse.

MOLECULAR POINT GROUPS

A molecular point group is a set of symmetry elements. Each symmetry element describes an operation which when carried out on the molecular skeleton, leaves the molecular skeleton unchanged. Elements of point groups may represent any of the following operations:

1. Rotations about axes through the origin.

    C_n = rotation through $2\pi/n$ radians (in solids, n = 1, 2, 3, 4, and 6).

2. Reflections in a plane containing the origin (center of mass)

    σ = reflection in a plane

3. Improper rotations—a rotation about an axis through the origin followed

by a reflection in a plane containing the origin and perpendicular to the axis of rotation.

S_n = rotation through $2\pi/n$ radians followed by σ_h.

SCHOENFLIES NOTATION

E = identity
C_n = rotation about an axis through $2\pi/n$ radians. The *principal* axis is the axis of highest n.
σ_h = reflection in a horizontal plane, i.e., the plane through the origin perpendicular to the axis of highest n.
σ_v = reflection in a vertical plane, i.e., the plane containing the axis of highest n.
σ_d = reflection in a diagonal plane, i.e., the plane containing the axis of highest n and bisecting the angle between the two-fold axes perpendicular to the principal axis. This is a special case of σ_v.
S_n = improper rotation through $2\pi/n$, i.e., C_n followed by σ_h.
$i = S_2$ = inversion through the center of mass, i.e., $\mathbf{r} \rightarrow -\mathbf{r}$.

INTERRELATIONS OF SYMMETRY ELEMENTS

1. (a) The intersection of two reflection planes must be a symmetry axis. If the angle, ϕ, between the planes is π/n, the axis is n-fold.
1. (b) If a reflection plane contains an n-fold axis, there must be n-1 other reflection planes at angles of π/n.
2. (a) Two 2-fold axes separated by an angle π/n require a perpendicular n-fold axis.
2. (b) A 2-fold axis and an n-fold axis perpendicular to it require n-1 additional 2-fold axes separated by angles of π/n.
3. An even-fold axis, a reflection plane perpendicular to it, and an inversion center are interdependent. Any two of these implies the existence of the third.

TYPE CLASSIFICATION

The following classification by types is due to Jaffe and Orchin [16]. Representative examples are given below for a number of types. The reader is challenged to find the rest.

Type 1. No rotation axis; point groups C_1, C_s, C_i.

  (a) $C_1 = \{E\}$. This group has no symmetry elements. It is the point group of *asymmetric* compounds.

4   SYMMETRY AND STEREOCHEMISTRY

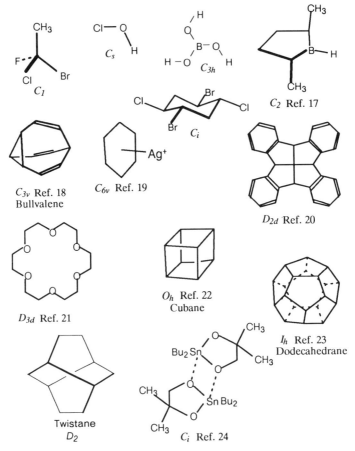

Figure 1.1. Examples of molecules belonging to various point groups.

(b) $C_s = \{E, \sigma\}$. This group has only a single plane of symmetry. Methanol (CH_3OH) is an example.

(c) $C_i = \{E, i\}$. This group has only a center of inversion. Two examples are shown in Figure 1.1.

Type 2. Only one axis of rotation; point groups C_n, S_n, C_{nv}, C_{nh}.

(a) C_n. This group has only a single rotational axis of order greater than one. These molecules are *dissymmetric* (chiral) and can be made optically active unless the enantiomeric forms are readily interconvertible.

$C_2 = \{E, C_2\}$. Hydrogen peroxide (HOOH) and *gauche* 1,2-dichloroethane are examples.

$C_3 = \{E, C_3, C_3{}^2\}.$
$C_4 = \{E, 2C_4, C_2 (= C_4{}^2)\}$

(b) S_n.

$S_4. = \{E, C_2, S_4, S_4{}^3\}$. The "$D_{2d}$" structure in Figure 1 actually belongs to S_4 since the five-membered rings are not planar.

$S_6. = \{E, C_3, C_3{}^2 \, i, S_6, S_6{}^5\}$

(c) C_{nv}. This group has symmetry elements C_n, and n σ_v.

$C_{2v} = \{E, C_2, \sigma_v, \sigma_v,\}$. Water, formaldehyde, and methylene chloride (CH_2Cl_2) are common examples.

$C_{3v} = \{E, 2C_3, 3\sigma_v\}$. Chloroform ($CHCl_3$) and ammonia are typical examples. See also bullvalene in Figure 1.1.

$C_{4v} = \{E, 2C_4, C_2, 2\sigma_v, 2\sigma_d\}$
$C_{5v} = \{E, 2C_5, C_5{}^2, 5\sigma_v\}$
$C_{6v} = \{E, 2C_6, 2C_3, C_2, 3\sigma_v, 3\sigma_d\}$

$C_{\infty v}$. HCl and CO and other linear polyatomic molecules without a center of inversion.

(d) C_{nh}. This group has the symmetry element C_n, and a horizontal mirror plane, σ_h. When n is even, a σ_h implies an i.

$C_{2h} = \{E, C_2, i, \sigma_h\}$—e.g., ($E$)-1,2-dichloroethene
$C_{3h} = \{E, 2C_3, \sigma_h, 2S_3\}$—e.g., boric acid [$B(OH)_3$, see Figure 1.1]
$C_{4h} = \{E, 2C_4, C_2, i, \sigma_h, 2S_4\}$

Type 3. One n-fold axis and n 2-fold axes; point groups D_n, D_{nh}, D_{nd}.

(a) D_n. This group has only a single rotational axis of order n greater than one, and n 2-fold axes perpendicular to the principal axis. These molecules are dissymmetric and can be made optically active unless various conformations are readily interconvertible.

$D_2 = \{E, 3C_2\}$—twisted ethylene, twistane (Figure 1.1).
$D_3 = \{E, 2C_3, 3C_2\}$—trisethylenediamine complexes of transition metals.

(b) D_{nh}. This group has only a single rotational axis of order n greater than one, n 2-fold axes perpendicular to the principal axis, and a σ_h (which also results in n σ_v).

$D_{2h} = \{E, 3C_2, 3\sigma_v, i\}$. Ethylene, diborane, and naphthalene have this symmetry.

$D_{3h} = \{E, 2C_3, 3C_2, 3\sigma_v, \sigma_h, 2S_3\}$. Cyclopropane belongs to this point group.

$D_{4h} = \{E, 2C_4, C_2, 2C_2', 2C_2'', i, 2S_4, \sigma_h, 2\sigma_v, 2\sigma_d\}$—the point group of the square or planar cyclobutane. What about cyclobutadiene?

$D_{5h} = \{E, 2C_5, 2C_5^2, 5C_2, 2S_5, 2S_5^2, \sigma_h, 5\sigma_v\}$—cyclopentadienyl anion

$D_{6h} = \{E, 2C_6, 2C_3, C_2, 3C_2', 3C_2'', i, 2S_6, 2S_3, \sigma_h, 3\sigma_v, 3\sigma_d\}$—benzene

$D_{\infty h}$. The other point group of linear molecules, e.g., carbon dioxide and acetylene.

(c) D_{nd}. This group has only a single rotational axis of order n greater than one, n 2-fold axes perpendicular to the principal axis, and n diagonal planes, σ_d which bisect the angles made by successive 2-fold axes. In general, D_{nd} contains an S_{2n}, and if n is odd, it contains i.

$D_{2d} = \{E, 3C_2, 2\sigma_d, 2S_4\}$. Allene has this symmetry, as do puckered cyclobutane and cyclooctatetraene.

$D_{3d} = \{E, 2C_3, 3C_2, i, 3\sigma_d, 2S_6\}$—e.g., cyclohexane and ethane. See also Figure 1.1.

$D_{4d} = \{E, 2C_4, C_2, 2C_2', 2C_2'', 2S_8, 2S_8^3, 4\sigma_d\}$

$D_{5d} = \{E, 2C_5, 2C_5^2, 5C_2, i, 2S_{10}, 2S_{10}^3, 5\sigma_d\}$

Type 4. More than one axis higher than 2-fold; point groups T_d, O_h, I_h, K_h (also T_h, T, O, I). Methane (T_d), cubane (O_h, Figure 1.1), dodecahedrane (I_h, Figure 1.1), and buckminsterfullerene, C_{60}, (I_h, Chapter 11). K_h is the point group of the sphere.

Exercise 1.1. A number of molecules representative of some of the point groups discussed are shown in Figure 1.1. Locate all of the elements of symmetry for each.

ISOMERISM AND MEASUREMENTS

The following assertions are made for the student to think about! [15] Isomers: molecules having the same molecular formula but differing in structure and *separated by energy barriers*. If isomers convert at immeasurably fast rates they are *not* considered isomers. Therefore, the method of measurement used to distinguish isomers must be faster than the rate of interconversion.

Table 1.1 lists minimum lifetimes for observation of separate species and the appropriate spectroscopic methods.

TABLE 1.1. Minimum Lifetimes for Observation of Separate Species

Type of Observation	Lifetime (sec)
Electron diffraction	10^{-20}
Neutron, X-ray diffraction	10^{-18}
UV, visible	10^{-15}
IR, Raman	10^{-13}
Microwave	10^{-4}–10^{-10}
ESR	10^{-4}–10^{-8}
NMR	10^{-1}–10^{-9}
Mossbauer (iron)	10^{-7}
Molecular beam	10^{-6}
Physical isolation and separation	$> 10^2$

STEREOISOMERISM OF MOLECULES

The student will recall that the stereomeric relationship between pairs of substances may be derived through the sequence of questions and answers represented by the flowchart [15] in Figure 1.2. In terms of properties, three broad categorizations arise:

1. Identical molecules—not distinguishable under any conditions, *chiral or achiral*.

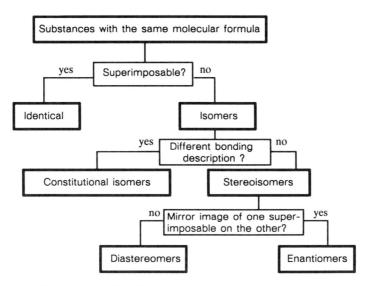

Figure 1.2. Flowchart for deciding stereomeric relationships between pairs of substances.

8 SYMMETRY AND STEREOCHEMISTRY

2. Enantiomers—same in all scalar properties, distinguishable only under chiral conditions. Only molecules whose point groups are C_n ($n > 1$), D_n ($n > 1$), T, O, or I are chiral and can exist in enantiomeric forms.
3. Constitutional isomers and diastereomers—differ in all scalar properties and are distinguishable under any conditions, chiral or achiral. Geometric isomers, which are related by the orientation of groups around a double bond, are a special case of diastereomers.

Molecules are *chiral* if their molecular point groups do *not* include any S_n ($n > 1$) symmetry elements. Otherwise they are *achiral*. An achiral molecule is not distinguishable from its own mirror image. This is often phrased as "an achiral molecule is superimposable on its own mirror image." A chiral molecule is *not* superimposable on its mirror image. A molecule which is identical to the mirror image of another molecule is the enantiomer of that molecule. According to the definitions above, an object is either chiral or it is not; it belongs to a particular point group or it does not. However, efforts have been made to define *degrees* of chirality [25] and continuous measures of symmetry [26].

The concepts of *chirality* and *isomerism* may readily be extended to pairs or larger assemblages of molecules, hence the reference to chiral and achiral *environments* above.

STEREOTOPIC RELATIONSHIPS OF GROUPS IN MOLECULES

The concepts used to describe relationships between pairs of molecules may also be extended to pairs of groups within a molecule [15]. This is particularly useful in determining the appearance of an NMR spectrum or the possibility of selective reaction at similar functional groups. Regions (such as faces of planar portions) around molecules may be similarly classified. The same relationships could also be applied to (groups of) atomic orbitals within the molecule. These are collectively referred to as "groups" for the purpose of the flowchart in Figure 1.3. As from the analysis of Figure 1.2, three broad groupings of properties emerge:

1. Homotopic groups—not distinguishable under any conditions, *chiral or achiral*. In order to have homotopic groups, a molecule must have a finite axis of rotation. Thus the only molecules which *cannot* have homotopic groups are those whose point groups are C_1, C_s, C_i, and $C_{\infty v}$.
2. Enantiotopic groups—same in all scalar properties, distinguishable only under chiral conditions.
3. Constitutionally heterotopic and diastereotopic groups—differ in all scalar properties and are distinguishable under any conditions, chiral or achiral.

STEREOTOPIC RELATIONSHIPS OF GROUPS IN MOLECULES

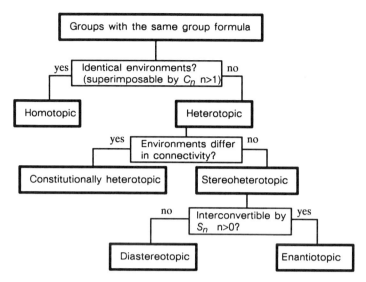

Figure 1.3. Flowchart for deciding stereotopic relationships between pairs of groups.

Groups may be compared by <u>internal comparison</u> (groups in the same molecule) or by <u>external comparison</u> (groups in different molecules).

One can also compare <u>faces</u> of a molecule.

Exercise 1.2. Verify the group designations:

Homotopic groups—(H_1, H_4), (H_2, H_3), (H_5, H_6)
Enantiotopic groups—(H_1, H_2), (H_3, H_4), (H_1, H_3), (H_2, H_4)
Constitutionally heterotopic groups—any of H_1–H_4 with H_5 or H_6
F_1 and F_2 are homotopic faces.
There are no diastereotopic groups in this molecule.

Exercise 1.3. Verify the classification of the pairs of groups in tricyclo[3.1.0.0^{2,4}]hexane.

Homotopic—(H_1, H_6), (H_2, H_5), (H_3, H_7), (H_4, H_8)
Enantiotopic—(H_3, H_4), (H_3, H_8), (H_4, H_7), (H_7, H_8)
Diastereotopic—(H_1, H_2), (H_1, H_5), (H_2, H_6), (H_5, H_6)
Constitutionally heterotopic—(H_1, H_3), (H_1, H_4), (H_2, H_8), etc.

Asymmetric molecules cannot contain homotopic or enantiotopic groups.

10 SYMMETRY AND STEREOCHEMISTRY

The first example of an asymmetric synthesis using the *chiral, optically pure* base, brucine [27]:

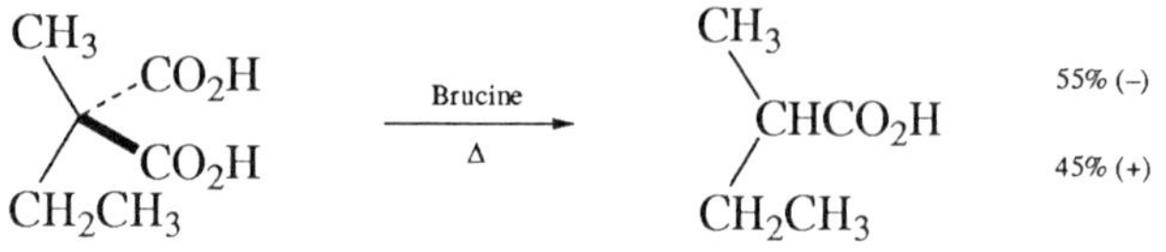

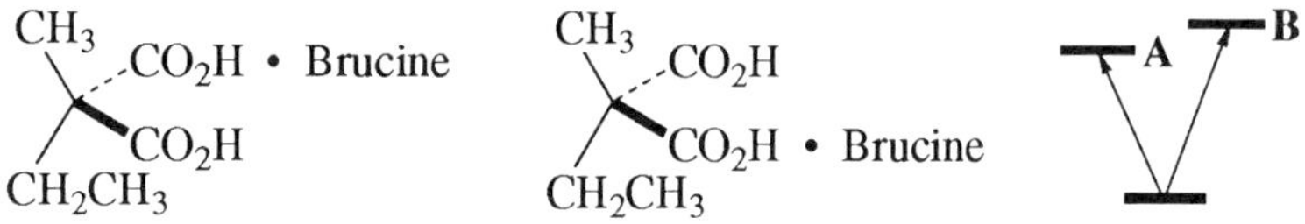

[Scheme: ethylmethylmalonic acid derivative + Brucine, Δ → monocarboxylic acid product, 55% (−), 45% (+)]

[Diastereomeric brucine salts A and B, with energy diagram showing A and B transition states]

A   B

A and **B** are diastereomeric and therefore different in all properties, including activation energy for decarboxylation. A carbon atom which contains two enantiotopic groups is **prochiral**.

Exercise 1.4. Compare all of the groups and faces of the trans-3,4-dimethylcyclopentanones below, both by internal comparison and by external comparison.

[Two trans-3,4-dimethylcyclopentanone structures with faces/groups labeled F_1, F_2 and F_3, F_4]

Exercise 1.5. Analyze the following sequence from the tricarboxylic acid cycle.

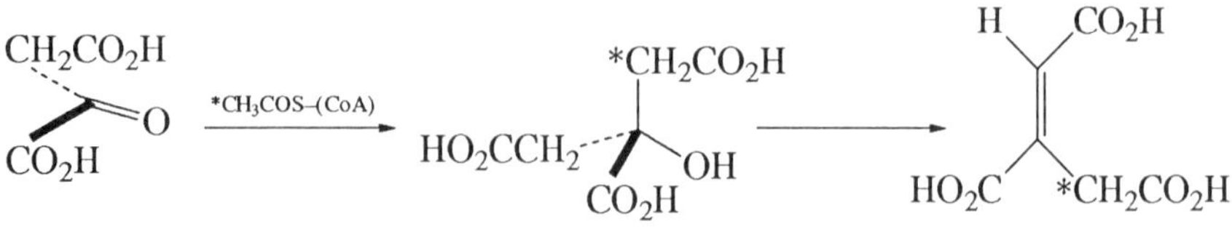

[Scheme: oxaloacetic acid + *CH_3COS-(CoA) → citric acid → aconitic acid]

oxaloacetic acid   citric acid   aconitic acid

Many of the ideas espoused here are from the work of Mislow [28]. For an alternative discussion of the concepts introduced in this section, see reference [29].

NMR AND STEREOCHEMISTRY

<u>Homotopic groups</u>—chemical shifts are indistinguishable in chiral or achiral solvents, i.e., the groups are *isochronous*.

<u>Enantiotopic groups</u>—isochronous in achiral solvents and distinguishable (*anisochronous*) in chiral solvents.

To be <u>anisochronous</u>: (1) groups may be related by symmetry taking into consideration internal motions which are rapid on the NMR time scale. (2) there must be sufficient field gradient so that the difference is observable.

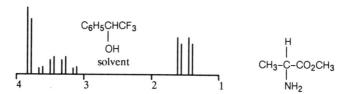

It requires a fair degree of association to make the chemical shift difference visible (Figures 1.4 and 1.5), e.g., optical purities of α-amino acids.

<u>Lanthanide chemical shifts</u>—See [31] for examples of the use of chiral lanthanide chemical shift reagents.

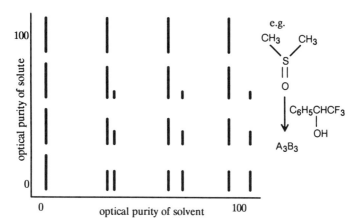

Figure 1.4. Simulated NMR spectrum of methyl alanine in a chiral solvent.

Figure 1.5. Effect of solvent and solute optical purity on the appearance of NMR signals of enantiotopic groups.

Enantiomeric purity is often determined by derivatization with an optically pure chiral agent. For alcohols and amines, α-methoxy-α-trifluoromethylphenylacetic acid (MTPA) and α-cyano-α-fluorophenylacetic acid (CFPA) [32] work well.

SYMMETRY AND STRUCTURAL PARAMETERS

Structural parameters (bond lengths, bond angles, dihedral angles) **must** be the same in a molecule when they are interconvertible by a symmetry operation, i.e., congruent. Conversely, structural parameters **cannot** be the same in a molecule when they are not congruent. If the structural parameters are not congruent, it is not possible to use symmetry arguments to predict the magnitude of the difference.

Some relationships are shown below between the bond lengths and angles of nominally tetrahedral molecules. The notation **a** and **b** denote groups which are different in some way. The point groups shown denote the *molecular* point group. In order for the relationships to hold *exactly*, the structures of **a** and **b** must be such as to preserve the overall symmetry. The relationships may be *approximately* obeyed if the denoted point groups are a fair representation of the *local* symmetry. For example, the first structure will have exactly T_d symmetry if **a** is H, Cl, or Me but not if **a** is Et, since the ethyl group does not have a 3-fold axis of symmetry.

Exercise 1.6. What point groups are available for different orientations of the ethyl groups in the "T_d" structure with **a** = Ethyl?

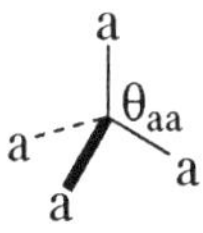

T_d: $\cos\theta_{aa} = -1/3$

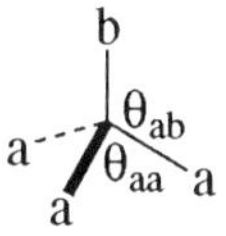

C_{3v}: $3\sin^2\theta_{ab} = 2(1 - \cos\theta_{aa})$

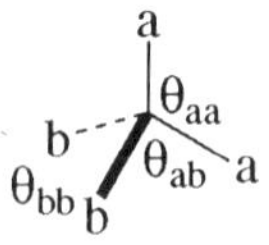

C_{2v}: $\cos\theta_{ab} = -\cos(1/2\,\theta_{aa})\cos(1/2\,\theta_{bb})$

NOTE ON HYBRIDIZATION

The concept of hybridization was introduced in order to provide a mechanism for achieving directionality in bonding, recognizing implicitly that linear combinations of the $2s$ and some of the three $2p$ orbitals point in well-defined directions relative to other such combinations. Thus if one takes a 1:1 linear combination of the $2s$ orbitals and *one* of the $2p$ orbitals (leaving the other two

2p orbitals alone), one obtains two *sp* hybrid orbitals which are directed at an angle of 180° to each other. As we shall see later, orbitals mix (or hybridize) so as to provide the best overlap for bonding. Mixing the 2s orbital with *two* of the three 2p orbitals yield three equivalent sp^2 hybrid orbitals which are exactly arranged at 120° relative to each other, yielding the familiar trigonal planar pattern of bonds when each sp^2 hybrid orbital forms a sigma bond to a different but identical atom or group. Likewise, four equivalent sp^3 hybrid orbitals directed toward the corners of a tetrahedron, with equal inter-orbital angles of 109.47° are obtained when the 2s and all *three* 2p orbitals are mixed. The *one*, *two*, and *three* refer to the "weights" of the 2p orbitals relative to the 2s orbital in the *sp*, sp^2, and sp^3 hybrid orbitals, respectively. The angles between two equivalent hybrid orbitals are determined by the "weights" of the 2p:2s mixture. Conversely, observation of inter-bond angles of 180°, 120°, and 109.47° between two equivalent (by symmetry) geminal C—X bonds implies that the carbon atom is using *sp*, sp^2, and sp^3 hybrid orbitals, respectively, to form those bonds. Hybridization can be inferred from the observed angles. Since the observed inter-bond angles are rarely the idealized values, 180°, 120°, and 109.47°, it follows that the orbitals are not the idealized hybrids but rather hybrids where the "weight" of 2p orbital relative to 2s orbital is a positive real number, say λ^2. In this case, a general hybrid orbital, h_i, will have the composition $s + \lambda_i p$, which is equivalent to $sp^{\lambda_i^2}$ hybridization. The "weight," λ_i^2, may range from 0 to ∞ (pure *s* to pure *p*). Normalization of the hybrid orbitals requires that the following relationships hold

$$\sum_i \frac{1}{1 + \lambda_i^2} = 1 \qquad \frac{1}{1 + \lambda_i^2} = s \text{ character of hybrid orbital } h_i$$

$$\sum_i \frac{\lambda_i^2}{1 + \lambda_i^2} = 1, 2, \text{ or } 3 \qquad \frac{\lambda_i^2}{1 + \lambda_i^2} = p \text{ character of hybrid orbital } h_i$$

where the sums run over the number of *hybridized* orbitals.

For any pair of hybrid orbitals, h_i and h_j, the following relationship exists

$$1 + \lambda_i \lambda_j \cos \theta_{ij} = 0 \quad \text{where } \theta_{ij} \text{ is the angle between two } \textit{hybridized} \text{ orbitals.}$$

Exercise 1.7. What is the hybridization of the carbon orbitals which form the C—H and C—C bonds of cyclopropane (HCH = 114°)? Verify that if the carbon hybrids which are used for the C—H bonds are exactly sp^2, then the two equivalent hybrids for the C—C bonds must be sp^5 and the interorbital angle is 101.5°!

Empirically, C^{13}—H spin–spin coupling constants are proportional to the "*s* character" of the hybrid-orbital used in the σ bond to H.

$$J_{CH}(\text{cps}) \approx \frac{500}{1 + \lambda_i^2}$$

SYMMETRY AND ORBITALS

Symmetry properties of atomic and molecular orbitals will prove useful in a variety of contexts. We will familiarize ourselves with the characteristics of the basic types of orbitals which will be used throughout the remainder of these notes. It is not proper to assign a point group label to orbitals because of the phase characteristics, but rather to the charge distribution which would result upon squaring the orbital. The orbital may then be characterized by designating the label of the *irreducible representation* according to which it transforms within the context of the local or global molecular point group. These attributes are specifically described for atomic s, p, and sp^n (hybrid) atomic orbitals, and for molecular orbitals below.

Atomic Orbitals. The symmetry characteristics of s, p, and sp^n (hybrid) atomic orbitals are illustrated in Figure 1.6. Thus the charge distribution due to an electron in an atomic s orbital is spherically symmetric (point group K_h) and the s orbital itself will transform as the totally symmetric irreducible representation. Alternatively, one may assign a label, S or A, which describes the behavior of the orbital under any relevant symmetry operations. For instance, the s orbital does not change sign (phase) upon reflection in any plane containing its center nor upon rotation through any angle about any axis of symmetry. It is *symmetric* with respect to any symmetry operation and this characteristic is conveniently assigned the label, S, for whichever symmetry operation is considered. On the other hand, the charge distribution due to an electron in an atomic p orbital is dumbbell shaped (axially symmetric with a horizontal mirror plane—point group $D_{\infty h}$). The p orbital itself will transform as the irreducible representation, Σ_u^+; that is, the p orbital does not change sign (phase) upon reflection in any plane containing its principal axis nor upon rotation through any angle about the principal axis, but *does* change sign (phase) upon reflection across the horizontal mirror plane (its own nodal plane) and rotation about any axis of symmetry (necessarily 2-fold) contained in that plane. It is *symmetric* (S) with respect to any of the first set of symmetry operations.

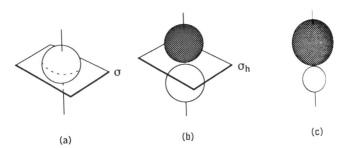

Figure 1.6. The symmetry characteristics of: **a** s; **b** p; and **c** sp^n (hybrid) atomic orbitals. The shapes of the electron distributions are similar if one ignores the phases.

It is *antisymmetric* with respect to any of the second set of symmetry operations and is assigned the label, A, for these. Hybrid atomic orbitals, sp^n, retain only the axial symmetry of the pure s and p orbitals. The node (boundary separating the two phases of the orbital) is now a curved surface and no longer a symmetry element. The charge distribution belongs to the point group, $C_{\infty v}$, and the hybrid orbital transforms as the a_1 irreducible representation of $C_{\infty v}$.

Molecular and Group Orbitals. Let us accept that molecular orbitals (MOs) and group orbitals are both described as linear combinations of atomic orbitals. Exactly how and why this is the case will be seen in Chapter 2. For the purpose of this section, *proper* molecular orbitals are those linear combinations which transform as irreducible representations of the *molecular* point group, i.e., are *symmetry adapted*. *Group* orbitals are linear combinations which are symmetric or antisymmetric with respect to any *local* symmetry operations of that part of the molecule which constitutes the group, e.g., a methyl group. At an intermediate level of description, molecular orbitals may be thought of as linear combinations of group orbitals. We shall frequently use the term *localized orbital*. This term has a formal definition in the literature of electronic structure theory, but we shall use it in a loose sense to describe a characteristic piece of a true MO or a group MO such as a sigma bond between a particular pair of atoms, or an atomic orbital describing a nonbonded pair of electrons. A localized MO may indeed be a proper MO or a group MO which happens to be concentrated in one region of the molecule. More likely, however, a proper MO or a group MO would be described as a linear combination of localized MOs. Some examples of proper MOs and group MOs are shown in Figure 1.7. Notice that the "proper" MOs of water which describe the "lone-pairs" of electrons are in- and out-of-phase combinations of the "rabbit ears" often pictured in elementary texts. The out-of-phase combination has no "s" character at all. It is a pure p orbital on the oxygen atom. The same is true of the "proper" MOs which describe the O—H bonds.

IN WHAT COMBINATION?

While it is easy to make sketches of hybrid, group, and molecular orbitals as shown in Figure 1.7, the criteria for choosing the *degree* of hybridization, or the specific *amount* of mixing of orbitals from different atoms to make MOs are not obvious. As we have seen, if the molecule has nontrivial symmetry (i.e., is not asymmetric, point group C_1), then the charge distribution must have the same symmetry as the molecular framework and "proper" MOs should also reflect the symmetry. Elements of symmetry can serve as a guide for the amount of mixing. The π bonding MO of ethylene (Figure 1.7) is partly determined by the symmetry. The $2p$ orbitals of each C must mix with equal weights. But why is the in-phase combination occupied and not the out-of-phase combination? The answer lies in the quantum mechanical theory of electronic structure (MO theory).

16 SYMMETRY AND STEREOCHEMISTRY

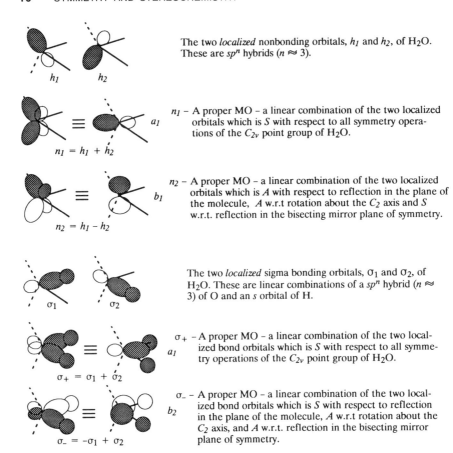

The π bond of ethylene (and other olefins) is a proper MO, highly localized to the two carbon atoms. It is the linear combination of the two 2p orbitals which is S with respect to reflection in the bisecting plane and A w.r.t. a 180° rotation about the C_2 axis which contains that plane. All 'π'-type orbitals are A w.r.t. reflection in the nodal plane of the p orbitals themselves.

The π* antibonding orbital of ethylene (and other olefins) is a also a proper MO, highly localized to the two carbon atoms. It is the linear combination of the two 2p orbitals which is A with respect to reflection in the bisecting plane and S w.r.t. a 180° rotation about the C_2 axis which contains that plane.

Figure 1.7. Examples of symmetry adapted ("proper") MOs and their constituent atomic or localized orbitals.

In Chapter 2, the physical and mathematical basis of the most familiar version of MO theory is presented. It is argued that the chemical and physical properties of molecules arise in large part from the distribution of the electrons in the molecule. It will be seen that the simplest function which correctly describes the simultaneous distribution of all the electrons in a molecule is a product of functions (MOs) which individually describe the distribution of one electron at a time. Actually one must take a linear combination of such products to allow for the fact that any electron may have any of the one-electron distributions and to recognize the *fermion* character of electrons. In other words, a many-electron wavefunction is expressed as an antisymmetrized sum of products of one-electron wavefunctions or molecular orbitals. An optimum set of MOs is derived by *minimizing the energy* of the assemblage of electrons and atoms with respect to variations in the MOs. Since the MOs are expanded in terms of atomic orbitals (strictly speaking, atomic–orbital-like functions), the process involves a variation of the amount of mixing of the atomic orbitals until a mixing combination is found which yields the lowest possible energy. Indeed, a prescription for finding this optimum combination falls out of the theory. In Chapter 2, the theory is presented in a mathematically rigorous fashion and concludes with a brief description of ways to improve the theory as well as a practical guide to one of the current computer programs, GAUSSIAN 92, which implements the theory for solving chemical problems.

The level of treatment in Chapter 2 is most suitable for graduate students. Undergraduates may skip to Chapter 3 and continue where the orbital interaction diagram is derived.

CHAPTER 2

MOLECULAR ORBITAL THEORY

INTRODUCTION

Orbital Interaction theory has its roots in molecular orbital (MO) theory. MO theory in one form or another plays a central role in the understanding of all aspects of chemical phenomena, whether it be in the form of a discussion of *hybridization* in connection with the geometry of tetrahedral carbon, *aromaticity* and the $4n + 2$ rule, or *orbital symmetry allowedness* of the Diels–Alder reaction. Many of the concepts are introduced in introductory general chemistry or organic chemistry courses. Indeed, for three quarters of a century, quantum mechanics in the form of the Schrödinger equation, has provided the underpinning for all but the most esoteric of chemical phenomena, and it is quite appropriate (and even essential) that students of chemistry be introduced to some aspects of it, even at a stage in their education when they do not have the mathematical background to follow the derivation of the necessary equations.

The material in this chapter must appear daunting to an organic chemist not used to the formalism of quantum mechanics! However, the mathematics is not beyond the capabilities of a typical graduate student or even a good third- or fourth-year undergraduate student. For undergraduate instruction, this chapter may well be skipped unless it is the intention to introduce the student to the *ab initio* MO computer programs which are increasingly at the forefront of chemical research and instruction.

The purpose of this chapter is to give the student a firm understanding of the requirements for the description of many-electron wavefunctions, the standard procedures used to obtain energies and wavefunctions, and the role of one-electron wavefunctions (orbitals) in the scheme of things. It provides in as simple a way as possible the theory behind the most straightforward applications of prevailing nonempirical quantum chemistry computer codes such as GAUSSIAN 92 [33]. At the same time, the simplifications which can be made to derive the "empirical or semiempirical MO" methods are placed in proper perspective.

THE ELECTRONIC HAMILTONIAN OPERATOR

The starting point is the classical energy expression for a molecule. A molecule, after all, is just a collection of charged particles in motion interacting through electrostatic forces (i.e., obey Newtonian mechanics and Coulomb's Law). For an **isolated system** of N_N nuclei and N_e electrons, the classical **nonrelativistic** total energy may be written:

$$E = \sum_{i=1}^{N_e} \frac{p_i^2}{2m_e} + \sum_{I=1}^{N_N} \frac{P_I^2}{2M_I} - \sum_{i=1}^{N_e} \sum_{I=1}^{N_N} \frac{Z_I e^2}{r_{iI}} + \sum_{i=1}^{N_e-1} \sum_{j=i+1}^{N_e} \frac{e^2}{r_{ij}} + \sum_{I=1}^{N_N-1} \sum_{J=I+1}^{N_N} \frac{Z_I Z_J e^2}{R_{IJ}} \quad (2.1)$$

where m_e and e are the mass and charge of an electron, M_I, Z_I, and P_I are the mass, atomic number, and momentum of the Ith nucleus, p_i is the momentum of the ith electron, r_{ij} and R_{IJ} are the separation between the ith and jth electron and the Ith and Jth nucleus, respectively, and r_{iI} is the separation between the ith electron and the Ith nucleus. The first two terms of equation (2.1) describe the kinetic energy of the system due to electrons and nuclei, respectively. The last three terms describe the potential energy, given by Coulomb's Law, of electron–nuclear attraction, electron–electron repulsion, and nuclear–nuclear repulsion, respectively. Notice that the energy is zero when the particles are infinitely far apart and not moving. Since the ratio of electron to nuclear masses is at least 1/1820, electronic velocities are much higher than nuclear velocities and it is common practice to invoke the **Born–Oppenheimer (B–O) Approximation** of stationary nuclei. This does not imply that the nuclei are not moving, but rather that the electronic distribution can respond instantaneously (adiabatically) to changes in the nuclear positions. Thus in the B–O approximation, the second term on the right-hand side of equation (2.1) is zero and the last term is a constant which one could work out on a calculator since the nuclear coordinate values are known and fixed. The total energy depends on the nuclear coordinates which we will represent collectively as **R**,

$$E(\mathbf{R}) = \sum_{i=1}^{N_e} \frac{p_i^2}{2m_e} - \sum_{i=1}^{N_e} \sum_{I=1}^{N_N} \frac{Z_I e^2}{r_{iI}} + \sum_{i=1}^{N_e-1} \sum_{j=i+1}^{N_e} \frac{e^2}{r_{ij}} + \sum_{I=1}^{N_N} \sum_{J=I+1}^{N_N} \frac{Z_I Z_J e^2}{R_{IJ}}$$

(2.2)

The electronic Hamiltonian operator, H^e, may be derived from the classical energy expression by replacing all momenta, p_i, by the derivative operator, $-i\hbar\nabla(i)$. Thus,

$$H^e(\mathbf{R}) = -\sum_{i=1}^{N_e} \frac{\hbar^2}{2m_e}\nabla(i)^2 - \sum_{i=1}^{N_e}\sum_{I=1}^{N_N} \frac{Z_I e^2}{r_{iI}} + \sum_{i=1}^{N_e-1}\sum_{j=i+1}^{N_e} \frac{e^2}{r_{ij}} \quad (2.3)$$

$$= \sum_{i=1}^{N_e} h(i) + \sum_{i=1}^{N_e-1}\sum_{j=i+1}^{N_e} \frac{e^2}{r_{ij}} \quad (2.4)$$

where the one electron Hamiltonian (core Hamiltonian) for the ith electron, $h(i)$, is given by

$$h(i) = -\frac{\hbar^2}{2m_e}\nabla(i)^2 - \sum_{I=1}^{N_N} \frac{Z_I e^2}{r_{iI}} \quad (2.5)$$

The explicit dependence on $\mathbf{R}$ is not shown for equation (2.5) and will not be given in subsequent equations, it being understood that unless otherwise stated, we are working within the B–O approximation. The Laplacian operator, $\nabla(i)^2$, in cartesian coordinates for the ith electron, is given by

$$\nabla(i)^2 = \frac{\partial^2}{\partial x_i^2} + \frac{\partial^2}{\partial y_i^2} + \frac{\partial^2}{\partial z_i^2} \quad (2.6)$$

THE ELECTRONIC SCHRÖDINGER EQUATION

The total energy of the molecule is the sum of the electronic energy, E^e, and the nuclear energy [the last term of equation (2.2)], which is constant within the B–O approximation. The electronic energy, E^e, must be obtained by solution of the electronic Schrödinger equation,

$$H^e\Psi = E^e\Psi \quad (2.7)$$

where Ψ is the **wavefunction** which describes the distribution of the electrons whose energy is E^e. Unfortunately, no *exact* solution for equation (2.7) exists, except for systems consisting of only one electron. Nevertheless, it can be shown that there is an infinite number of solutions, i.e., wavefunctions Ψ_n each corresponding to a different distribution of electrons with energy E_n^e, and that there is a *lowest* energy distribution, which is customarily denoted Ψ_O with associated energy E_0^e. Henceforth, the superscript denoting "electronic" will be omitted. Unless otherwise stated, all quantities will be electronic quantities. The sole exception is the total energy of the molecule which is obtained by adding the constant nuclear energy to the electronic energy.

EXPECTATION VALUES

It is one of the postulates of quantum mechanics that for every observable quantity, o, there is a corresponding operator, O, such that an average or **expectation** value of the observable may be obtained by evaluating the expression

$$o = \frac{\int \Psi O \Psi \, d\tau}{\int \Psi^2 \, d\tau} \qquad (2.8)$$

In equation (2.8), Ψ is the wavefunction which describes the distribution of particles in the system. It may be the exact wavefunction [the solution to equation (2.7)] or a reasonable approximate wavefunction. For most molecules, the ground electronic state wavefunction is real, and in writing the expectation value in the form of equation (2.8), we have made this simplifying (though not necessary) assumption. The electronic energy is an observable of the system and the corresponding operator is the Hamiltonian. Therefore, one may obtain an estimate for the energy even if one does not know the exact wavefunction but only an approximate one, $\Psi^\sim$, i.e.,

$$E^\sim = \frac{\int \Psi^\sim H \Psi^\sim \, d\tau}{\int |\Psi^\sim|^2 \, d\tau} \qquad (2.9)$$

It can be proven that for the ground state, $E^\sim$ is always greater than, or equal to the exact energy, E_0, and that the two are equal only if $\Psi^\sim = \Psi_0$. This fact provides a prescription for obtaining a solution to equation (2.7) which is as accurate as possible. The procedure is called the **variation method** and is as follows: 1) Construct a wavefunction with the correct form to describe the system, building in flexibility in the form of a set of parameters. 2) Differentiate $E^\sim$ [equation (2.9)] with respect to each of the parameters in turn and set the resulting equations to zero. 3) Solve the resulting set of simultaneous equations to obtain the optimum set of parameters which give the lowest energy (closest to the exact energy). The wavefunction constructed using these parameters then should be as close as possible to the exact wavefunction. The first task is the construction of the wavefunction.

THE MANY-ELECTRON WAVEFUNCTION

The minimum requirements for a many-electron wavefunction, namely, antisymmetry with respect to interchange of electrons, and indistinguishability of electrons, are satisfied by an antisymmetrized sum of products of one-electron wavefunctions (orbitals), $\phi(1)$,

$$\Phi(1, 2, \cdots, N_e) = (N_e!)^{-1/2} \sum_{p=0}^{N_e!-1} (-1)^p P^p [\phi_1(1)\phi_2(2) \cdots \phi_{N_e}(N_e)]$$

(2.10)

the term in brackets, $[\phi_1(1)\phi_2(2) \cdots \phi_{N_e}(N_e)]$, is a Hartree product. The numbers in parentheses refer to particular electrons, or more specifically, to the x, y, z, and spin coordinates of those electrons. The subscripts refer to the characteristics of the individual orbital (spatial distribution and spin). There is a different orbital for each of the N_e electrons. This is required by the Pauli exclusion principle. The Hartree product represents a particular assignment of the electrons to orbitals. Any other assignment of the electrons to the same orbital set is equally likely and must be allowed to preserve indistinguishability of the electrons. The permutation operator, P, permutes the coordinates of two electrons, i.e., the electrons swap orbitals. Successive powers of P accomplish other interchanges; even and odd powers accomplish even and odd numbers of interchanges. There are $N_e!$ possible permutations of N_e electrons among N_e orbitals; the sum over $N_e!$ terms accomplishes this. The antisymmetry requirement for electronic wavefunctions is satisfied by the factor $(-1)^p$. The orbitals form an orthonormal set, i.e., for any pair ϕ_a and ϕ_b,

$$\int \phi_a(1) \phi_b(1) \, d\tau_1 = \delta_{ab} \quad (2.11)$$

where the integration is over all possible values of the three spatial coordinates, x, y, and z, and the "spin coordinate," s, and $d\tau_1$ represent the volume element, $dx_1 \, dy_1 \, dz_1 \, ds_1$. The factor $(N_e!)^{-1/2}$ ensures that $\Phi(1, 2, , , N_e)$ is also normalized. Equation (2.10) may be expressed in determinantal form and is often referred to as a determinantal wavefunction

$$\Phi(1, 2, \cdots, N_e) = (N_e!)^{-1/2} \begin{vmatrix} \phi_1(1) & \phi_2(1) & \phi_3(1) & \cdots & \phi_{N_e}(1) \\ \phi_1(2) & \phi_2(2) & \phi_3(2) & \cdots & \phi_{N_e}(2) \\ \phi_1(3) & \phi_2(3) & \phi_3(3) & \cdots & \phi_{N_e}(3) \\ \vdots & \vdots & \vdots & & \vdots \\ \phi_1(N_e) & \phi_2(N_e) & \phi_3(N_e) & \cdots & \phi_{N_e}(N_e) \end{vmatrix}$$

(2.12)

Equation (2.10) [or (2.12)] has an inherent restriction built into it since other wavefunctions of the same form are possible if one could select any N_e orbitals from an infinite number of them rather than the N_e used in equation (2.10). One could thus generate an infinite number of determinantal wavefunctions in the form of equation (2.10) and without approximation, the *exact* wavefunction, $\Psi(1, 2, \cdots, N_e)$ could be expressed as a linear combination of them.

$$\Psi(1, 2, \cdots, N_e) = \sum_{a=0}^{\infty} d_a \Phi_a \tag{2.13}$$

THE ELECTRON HARTREE–FOCK ENERGY

Although in principle an exact solution to the Schrödinger equation can be expressed in the form of equation (2.13), the wavefunctions Φ_a and coefficients d_a cannot be determined for an infinitely large set. In the **Hartree–Fock** approximation, it is assumed that the summation in equation (2.13) may be approximated by a single term, i.e., that the correct wavefunction may be approximated by single determinantal wavefunction Φ_0, the first term of equation (2.13). The method of variations is used to determine the conditions which lead to an optimum Φ_0, which will then be designated Φ_{HF}.

$$E_0 = \frac{\int \Phi_0 H \Phi_0 \, d\tau}{\int |\Phi_0|^2 \, d\tau} = \int \Phi_0 H \Phi_0 \, d\tau \tag{2.14}$$

The last equality holds since Φ_0 is normalized and will be constrained to remain so upon variation of the orbitals.

Before we substitute equation (2.10) into equation (2.14), a simplifying observation can be made. Equation (2.10) [and (2.12)] can be expressed in terms of the **antisymmetrizer operator**, A,

$$\Phi(1, 2, \cdots, N_e) = A[\phi_1(1)\phi_2(2) \cdots \phi_{N_e}(N_e)] \tag{2.15}$$

where

$$A = (N_e!)^{-1/2} \sum_{p=0}^{N_e!-1} (-1)^p P^p \tag{2.16}$$

Then

$$E_0 = \int A[\phi_1(1)\phi_2(2) \cdots \phi_{N_e}(N_e)] HA[\phi_1(1)\phi_2(2) \cdots \phi_{N_e}(N_e)] \, d\tau$$

$$= \int \phi_1(1)\phi_2(2) \cdots \phi_{N_e}(N_e) HA^2[\phi_1(1)\phi_2(2) \cdots \phi_{N_e}(N_e)] \, d\tau$$

$$[A, H] = 0$$

24 MOLECULAR ORBITAL THEORY

$$= (N_e!)^{1/2} \int \phi_1(1)\phi_2(2) \cdots \phi_{N_e}(N_e) HA[\phi_1(1)\phi_2(2) \cdots \phi_{N_e}(N_e)] \, d\tau$$

$$A^2 = (N_e!)^{1/2} A$$

$$= \int \phi_1(1)\phi_2(2) \cdots \phi_{N_e}(N_e) H$$

$$\cdot \sum_{p=0}^{N_e!-1} (-1)^p P^p [\phi_1(1)\phi_2(2) \cdots \phi_{N_e}(N_e)] \, d\tau \qquad (2.17)$$

Finally,

$$E_0 = \int \phi_1(1)\phi_2(2) \cdots \phi_{N_e}(N_e) \sum_{i=1}^{N_e} h(i) \sum_{p=0}^{N_e!-1}$$

$$\cdot (-1)^p P^p [\phi_1(1)\phi_2(2) \cdots \phi_{N_e}(N_e)] \, d\tau$$

$$+ \int \phi_1(1)\phi_2(2) \cdots \phi_{N_e}(N_e) \sum_{i=1}^{N_e-1} \sum_{j=i+1}^{N_e} \frac{e^2}{r_{ij}} \sum_{p=0}^{N_e!-1}$$

$$\cdot (-1)^p P^p [\phi_1(1)\phi_2(2) \cdots \phi_{N_e}(N_e)] \, d\tau \qquad (2.18)$$

Consider the first term in equation (2.18). Specifically, take the term for the ith electron and the "do nothing" permutation ($p = 0$)

$$\int \phi_1(1)\phi_2(2) \cdots \phi_{N_e}(N_e) h(i) (-1)^0 P^0 [\phi_1(1)\phi_2(2) \cdots \phi_{N_e}(N_e)] \, d\tau$$

$$= \int \phi_1(1)\phi_2(2) \cdots \phi_{N_e}(N_e) h(i) \phi_1(1)\phi_2(2) \cdots$$

$$\cdot \phi_{N_e}(N_e) \, d\tau_1 \, d\tau_2 \cdots d\tau_{N_e}$$

$$= \int \phi_1(1)\phi_1(1) \, d\tau_1 \int \phi_2(2)\phi_2(2) \, d\tau_2 \cdots \int \phi_i(i) h(i) \phi_i(i) \, d\tau_i \cdots$$

$$\cdot \int \phi_{N_e}(N_e) \phi_{N_e}(N_e) \, d\tau_{N_e}$$

$$= 1 \cdot 1 \cdots \cdots \int \phi_i(i) h(i) \phi_i(i) \, d\tau_i \cdots \cdots 1$$

$$= \int \phi_i(i) h(i) \phi_i(i) \, d\tau_i$$

$$= h_i \qquad (2.19)$$

Another permutation, say P^k, may interchange electrons i and j. For that term, the integration becomes

$$\int \phi_1(1)\phi_2(2) \cdots \phi_i(i) \cdots \phi_j(j) \cdots \phi_{N_e}(N_e) h(i)(-1)^k$$
$$\cdot P^k[\phi_1(1)\phi_2(2) \cdots \phi_i(i) \cdots \phi_j(j) \cdots \phi_{N_e}(N_e)] \, d\tau$$
$$= -\int \phi_1(1)\phi_2(2) \cdots \phi_i(i) \cdots \phi_j(j) \cdots \phi_{N_e}(N_e) h(i)$$
$$\cdot \phi_1(1)\phi_2(2) \cdots \phi_i(j) \cdots \phi_j(i) \cdots \phi_{N_e}(N_e) \, d\tau_1 \, d\tau_2 \cdots d\tau_{N_e}$$
$$= -\int \phi_1(1)\phi_1(1) \, d\tau_1 \int \phi_2(2)\phi_2(2) \, d\tau_2 \cdots \int \phi_i(i) h(i) \phi_j(i) \, d\tau_1$$
$$\cdots \int \phi_j(j)\phi_i(j) \, d\tau_j \cdots \int \phi_{N_e}(N_e)\phi_{N_e}(N_e) \, d\tau_{N_e}$$
$$= -1 \cdot 1 \cdots \int \phi_i(i) h(i) \phi_j(i) \, d\tau_i \cdots 0 \cdots 1$$
$$= 0 \tag{2.20}$$

The negative sign in the second line arises because, by construction, all permutations which yield a single interchange of electrons will be generated by an odd power of the permutation operator (k is odd). The "zero" result arises from the orthogonality of the orbitals [equation (2.11)]. Indeed, all other permutations will give identically zero for the same reason. Since each electron is in a different orbital, the entire first term of equation (2.18) becomes

$$\int \phi_1(1)\phi_2(2) \cdots \phi_{N_e}(N_e) \sum_{i=1}^{N_e} h(i) \sum_{p=0}^{N_e!-1} (-1)^p P^p$$
$$\cdot [\phi_1(1)\phi_2(2) \cdots \phi_{N_e}(N_e)] \, d\tau = \sum_{a=1}^{N_e} h_a \tag{2.21}$$

where we have changed the subscript to a to indicate that the sum now extends over orbitals rather than electrons (although the number of each is the same). It is worthwhile stating h_a explicitly, using equation (2.5)

$$h_a = \int \phi_a(1) \left(-\frac{\hbar^2}{2m_e} \nabla(1)^2 - \sum_{I=1}^{N_N} \frac{Z_I e^2}{r_{1I}} \right) \phi_a(1) \, d\tau_1 \tag{2.22}$$

Equation (2.22) is the energy of a single electron with spatial distribution given by the MO, ϕ_a. Equation (2.21) is the total one-electron contribution to the total electronic energy.

The two-electron contribution is derived in the same way from the second term in equation (2.18). Consider an arbitrary pair of electrons, i and j, and two permutations, the "do nothing" permutation ($p = 0$) and the specific permutation, $P^p \equiv P_{ij}$, which interchanges just the ith and jth electron. For the second permutation, p is odd and $(-1)^p$ will yield a minus sign. Thus,

$$\int \phi_1(1)\phi_2(2) \cdots \phi_{N_e}(N_e) \frac{e^2}{r_{ij}} [1 - P_{ij}][\phi_1(1)\phi_2(2) \cdots \phi_{N_e}(N_e)] \, d\tau$$

$$= \int \phi_1(1)\phi_1(1) \, d\tau_1 \int \phi_2(2)\phi_2(2) \, d\tau_2 \cdots \iint \phi_i(i)\phi_j(j)$$

$$\cdot \frac{e^2}{r_{ij}} \phi_i(i)\phi_j(j) \, d\tau_i \, d\tau_j \cdots \int \phi_{N_e}(N_e)\phi_{N_e}(N_e) \, d\tau_{N_e}$$

$$- \int \phi_1(1)\phi_1(1) \, d\tau_1 \int \phi_2(2)\phi_2(2) \, d\tau_2 \cdots \iint \phi_i(i)\phi_j(j)$$

$$\cdot \frac{e^2}{r_{ij}} \phi_i(j)\phi_j(i) \, d\tau_i \, d\tau_j \cdots \int \phi_{N_e}(N_e)\phi_{N_e}(N_e) \, d\tau_{N_e}$$

$$= \iint \phi_i(i)\phi_j(j) \frac{e^2}{r_{ij}} \phi_i(i)\phi_j(j) \, d\tau_i \, d\tau_j - \iint \phi_i(j)\phi_j(i)$$

$$\cdot \frac{e^2}{r_{ij}} \phi_i(j)\phi_j(i) \, d\tau_i \, d\tau_j$$

$$= J_{ij} - K_{ij} \quad (2.23)$$

All other permutations give identically zero terms due to the orthogonality of the orbitals. The total two-electron contribution to the electronic energy is

$$\int \phi_1(1)\phi_2(2) \cdots \phi_{N_e}(N_e) \sum_{i=1}^{N_e-1} \sum_{j=i+1}^{N_e} \frac{e^2}{r_{ij}} \sum_{p=0}^{N_e!-1} (-1)^p P^p$$

$$\cdot [\phi_1(1)\phi_2(2) \cdots \phi_{N_e}(N_e)] \, d\tau = \sum_{a=1}^{N_e-1} \sum_{b=a+1}^{N_e} (J_{ab} - K_{ab}) \quad (2.24)$$

As above, we have changed the subscripts to indicate that the summations run over the orbitals rather than the electrons. The two electron repulsion integrals, J_{ab} and K_{ab}, are formally defined as

$$J_{ab} = \iint \phi_a(1)\phi_b(2) \frac{e^2}{r_{12}} \phi_a(1)\phi_b(2) \, d\tau_1 \, d\tau_2 \quad (2.25)$$

$$K_{ab} = \iint \phi_a(1)\phi_b(2) \frac{e^2}{r_{12}} \phi_a(2)\phi_b(1) \, d\tau_1 \, d\tau_2 \quad (2.26)$$

and are the **coulomb** and **exchange** integrals, respectively. Thus,

$$E_0 = \sum_{a=1}^{N_e} h_a + \frac{1}{2} \sum_{a=1}^{N_e} \sum_{b=1}^{N_e} (J_{ab} - K_{ab}) \qquad (2.27)$$

Notice that the restriction on the second sums can be released and the factor of $\frac{1}{2}$ introduced since $J_{aa} = K_{aa}$. Once the orbitals have been "optimized" (see below) to yield the lowest possible value of the energy [equation (2.27)], the energy will be Hartree–Fock energy, E_{HF}. We will call it that from now on.

VARIATION OF E_{HF}

Variation of E_{HF} equation (2.27), with respect to variation of the orbitals is formally carried out as

$$\delta E_{HF} = \sum_{a=1}^{N_e} \delta h_a + \frac{1}{2} \sum_{a=1}^{N_e} \sum_{b=1}^{N_e} (\delta J_{ab} - \delta K_{ab}) = 0 \qquad (2.28)$$

where

$$\delta h_a = \int \phi_a(1) h(1) \delta\phi_a(1) \, d\tau_1 + \int \delta\phi_a(1) h(1) \phi_a(1) \, d\tau_1$$

$$= 2 \int \delta\phi_a(1) h(1) \phi_a(1) \, d\tau_1 \qquad (2.29)$$

$$\delta J_{ab} = \iint \delta\phi_a(1) \phi_b(2) \frac{e^2}{r_{12}} \phi_a(1) \phi_b(2) \, d\tau_1 \, d\tau_2$$

$$+ \iint \phi_a(1) \delta\phi_b(2) \frac{e^2}{r_{12}} \phi_a(1) \phi_b(2) \, d\tau_1 \, d\tau_2$$

$$+ \iint \phi_a(1) \phi_b(2) \frac{e^2}{r_{12}} \delta\phi_a(1) \phi_b(2) \, d\tau_1 \, d\tau_2$$

$$+ \iint \phi_a(1) \phi_b(2) \frac{e^2}{r_{12}} \phi_a(1) \delta\phi_b(2) \, d\tau_1 \, d\tau_2$$

$$= 2 \iint \delta\phi_a(1) \phi_b(2) \frac{e^2}{r_{12}} \phi_a(1) \phi_b(2) \, d\tau_1 \, d\tau_2$$

$$+ 2 \iint \delta\phi_b(2) \phi_a(1) \frac{e^2}{r_{12}} \phi_a(1) \phi_b(2) \, d\tau_1 \, d\tau_2 \qquad (2.30)$$

and

$$\delta K_{ab} = \iint \delta\phi_a(1)\phi_b(2) \frac{e^2}{r_{12}} \phi_a(2)\phi_b(1)\, d\tau_1\, d\tau_2$$
$$+ \iint \phi_a(1)\, \delta\phi_b(2) \frac{e^2}{r_{12}} \phi_a(2)\phi_b(1)\, d\tau_1\, d\tau_2$$
$$+ \iint \phi_a(1)\phi_b(2) \frac{e^2}{r_{12}} \delta\phi_a(2)\phi_b(1)\, d\tau_1\, d\tau_2$$
$$+ \iint \phi_a(1)\phi_b(2) \frac{e^2}{r_{12}} \phi_a(2)\, \delta\phi_b(1)\, d\tau_1\, d\tau_2$$
$$= 2 \iint \delta\phi_a(1)\phi_b(2) \frac{e^2}{r_{12}} \phi_a(2)\phi_b(1)\, d\tau_1\, d\tau_2$$
$$+ 2 \iint \delta\phi_b(2)\phi_a(1) \frac{e^2}{r_{12}} \phi_a(2)\phi_b(1)\, d\tau_1\, d\tau_2 \quad (2.31)$$

Not all variations of the orbital set are allowed. The variations are subject to the constraint that the orbitals remain orthonormal [equation (2.11)]. Thus, for all pairs of orbitals, a and b,

$$\int \delta\phi_a(1)\phi_b(1)\, d\tau_1 + \int \phi_a(1)\, \delta\phi_b(1)\, d\tau_1 = 0 \quad (2.32)$$

Constraints may be imposed on a set of simultaneous linear equations by the method of Lagrangian multipliers. Let the Lagrangian multipliers be $-\epsilon_{ab}$. Therefore, add to equation (2.28), the quantity

$$\sum_{a=1}^{N_e} \sum_{b=1}^{N_e} (-\epsilon_{ab}) \left[\int \delta\phi_a(1)\phi_b(1)\, d\tau_1 + \int \delta\phi_b(1)\phi_a(1)\, d\tau_1 \right] \quad (2.33)$$

Thus, the complete set of simultaneous equations for the variation are

$$0 = 2 \sum_{a=1}^{N_e} \int \delta\phi_a(1) h(1) \phi_a(1)\, d\tau_1$$
$$+ \sum_{a=1}^{N_e} \sum_{b=1}^{N_e} \iint \delta\phi_a(1)\phi_b(2) \frac{e^2}{r_{12}} \phi_a(1)\phi_b(2)\, d\tau_1\, d\tau_2$$
$$+ \sum_{a=1}^{N_e} \sum_{b=1}^{N_e} \iint \delta\phi_b(2)\phi_a(1) \frac{e^2}{r_{12}} \phi_a(1)\phi_b(2)\, d\tau_1\, d\tau_2$$

$$-\sum_{a=1}^{N_e}\sum_{b=1}^{N_e}\iint \delta\phi_a(1)\phi_b(2)\frac{e^2}{r_{12}}\phi_a(2)\phi_b(1)\,d\tau_1\,d\tau_2$$

$$-\sum_{a=1}^{N_e}\sum_{b=1}^{N_e}\iint \delta\phi_b(2)\phi_a(1)\frac{e^2}{r_{12}}\phi_a(2)\phi_b(1)\,d\tau_1\,d\tau_2$$

$$-\sum_{a=1}^{N_e}\sum_{b=1}^{N_e}\epsilon_{ab}\int \delta\phi_a(1)\phi_b(1)\,d\tau_1 - \sum_{a=1}^{N_e}\sum_{b=1}^{N_e}\epsilon_{ab}\int \delta\phi_b(1)\phi_a(1)\,d\tau_1$$

(2.34)

or

$$0 = 2\sum_{a=1}^{N_e}\int \delta\phi_a(1)h(1)\phi_a(1)\,d\tau_1$$

$$+ 2\sum_{a=1}^{N_e}\sum_{b=1}^{N_e}\iint \delta\phi_a(1)\phi_b(2)\frac{e^2}{r_{12}}\phi_a(1)\phi_b(2)\,d\tau_1\,d\tau_2$$

$$- 2\sum_{a=1}^{N_e}\sum_{b=1}^{N_e}\iint \delta\phi_a(1)\phi_b(2)\frac{e^2}{r_{12}}\phi_a(2)\phi_b(1)\,d\tau_1\,d\tau_2$$

$$- 2\sum_{a=1}^{N_e}\sum_{b=1}^{N_e}\epsilon_{ab}\int \delta\phi_a(1)\phi_b(1)\,d\tau_1 \qquad (2.35)$$

In deriving equation (2.35) from equation (2.34) we have made use of the fact that the indexes of the sums are arbitrary and have switched a and b in the terms in the third, fifth, and seventh terms of equation (2.34). We have also adopted without proof, the hermiticity of the Lagrangian multipliers, i.e., $\epsilon_{ab} = \epsilon_{ba}$. Canceling the 2's and collecting terms yields the result

$$0 = \sum_{a=1}^{N_e}\int d\tau_1\,\delta\phi_a(1)\left[\left\{h(1) + \sum_{b=1}^{N_e}[J_b(1) - K_b(1)]\right\}\phi_a(1)\right.$$

$$\left. - \sum_{b=1}^{N_e}\epsilon_{ab}\phi_b(1)\right] \qquad (2.36)$$

where we have introduced the coulomb and exchange one-electron operators $J_b(1)$ and $K_b(1)$ which are defined by their action, namely

$$\int \phi_a(1)J_b(1)\phi_a(1)\,d\tau_1 = \int \phi_a(1)\left(\int \frac{\phi_b(2)\phi_b(2)e^2}{r_{12}}\,d\tau_2\right)\phi_a(1)\,d\tau_1 = J_{ab}$$

(2.37)

30   MOLECULAR ORBITAL THEORY

$$\int \phi_a(1) K_b(1) \phi_a(1) \, d\tau_1 = \int \phi_a(1) \left(\int \frac{\phi_b(2) \phi_a(2) e^2}{r_{12}} \, d\tau_2 \right) \phi_b(1) \, d\tau_1 = K_{ab}$$

(2.38)

Exercise 2.1. Verify by direct substitution of equations (2.37) and (2.38) into equation (2.36) that equations (2.36) and (2.35) are equivalent.

Since the individual variations of the orbitals are linearly independent, equation (2.36) can only be true if the quantity inside the large brackets is zero for every value of a, namely

$$\left\{ h(1) + \sum_{b=1}^{N_e} [J_b(1) - K_b(1)] \right\} \phi_a(1) - \sum_{b=1}^{N_e} \epsilon_{ab} \phi_b(1) = 0 \quad (2.39)$$

Without loss of generality, the set of orbitals may be rotated so that the ϵ matrix becomes diagonal, i.e.,

$$\left\{ h(1) + \sum_{b=1}^{N_e} [J_b(1) - K_b(1)] \right\} \phi_a(1) - \epsilon_a \phi_a(1) = 0 \quad (2.40)$$

The quantity in large curly brackets is the Fock operator, $F(1)$,

$$F(1) = h(1) + \sum_{b=1}^{N_e} [J_b(1) - K_b(1)] \quad (2.41)$$

Therefore, the condition that the orbitals yield a stationary point (hopefully a minimum) on the energy hypersurface with respect to variations is that the orbitals are eigenfunctions of the Fock operator, with associated orbital energy, ϵ,

$$F(1)\phi_a(1) = \epsilon_a \phi_a(1) \quad (2.42)$$

In summary, in order to obtain a many-electron wavefunction of the single determinantal form [equation (2.12)] which will give the lowest electronic energy [equation (2.14) or (2.27)], one must use one-electron wavefunctions (orbitals) which are eigenfunctions of the one-electron Fock operator according to equation (2.42). There are many, possibly an infinite number of solutions to equation (2.42). We need the lowest N_e of them, one for each electron, for equation (2.12) [or (2.27)]. When the N_e MOs of lowest energy satisfy equation (2.42), then $E_0 = E_{HF}$ [equation (2.27)] and $\phi_0 \equiv \Phi_{HF}$ [equation (2.12)].

LCAO SOLUTION OF THE FOCK EQUATIONS

We must now bite the bullet and specify what form the molecular orbitals (MO) must have. We expand the MOs as a linear combination of a number of linearly independent functions.

$$\phi_a(1) = \sum_{i=1}^{n} \chi_i(1) c_{ia} \qquad \phi = \chi c \text{ (in matrix form)} \tag{2.43}$$

Such an expansion can always be made *without approximation* if the set of functions is mathematically *complete*. We must necessarily use a finite (and therefore incomplete) set. We will discuss the characteristics of the **basis set** below. For now let us take the χ_i as known and proceed to determining the expansion **coefficients**, c_{ia}. Substitution of equation (2.43) into equation (2.42) yields

$$F(1) \sum_{i=1}^{n} \chi_i(1) c_{ia} = \epsilon_a \sum_{i=1}^{n} \chi_i(1) c_{ia} \tag{2.44}$$

Multiplication on the left by χ_j and integration over the range of the coordinates of the electron gives

$$\sum_{i=1}^{n} \int \chi_j(1) F(1) \chi_i(1) \, d\tau_1 \, c_{ia} = \epsilon_a \sum_{i=1}^{n} \int \chi_j(1) \chi_i(1) \, d\tau_1 \, c_{ia} \tag{2.45}$$

or

$$\sum_{i=1}^{n} F_{ji} c_{ia} = \sum_{i=1}^{n} S_{ji} c_{ia} \epsilon_a \tag{2.46}$$

Equation (2.46) may be cast as a matrix equation

$$\mathbf{Fc} = \mathbf{Sc}\epsilon \tag{2.47}$$

The **overlap matrix**, S, is defined as

$$S_{ij} = \int \chi_i(1) \chi_j(1) \, d\tau_1 \tag{2.48}$$

The basis functions are normalized so that $S_{ii} = 1$, but not orthogonal, i.e., $S_{ij} \neq 0$ in general.

MOLECULAR ORBITAL THEORY

The **Fock matrix**, **F**, is

$$\begin{aligned}
\mathbf{F}_{ij} &= \int \chi_i(1) F(1) \chi_j(1) \, d\tau_1 \\
&= \int \chi_i(1) \left[h(1) + \sum_{b=1}^{N_e} [J_b(1) - K_b(1)] \right] \chi_j(1) \, d\tau_1 \\
&= \int \chi_i(1) h(1) \chi_j(1) \, d\tau_1 + \sum_{b=1}^{N_e} \left[\int \chi_i(1) J_b(1) \chi_j(1) \, d\tau_1 \right. \\
&\quad \left. - \int \chi_i(1) K_b(1) \chi_j(1) \, d\tau_1 \right] \\
&= \int \chi_i(1) h(1) \chi_j(1) \, d\tau_1 + \sum_{b=1}^{N_e} \left[\iint \chi_i(1) \phi_b(2) \frac{e^2}{r_{12}} \phi_b(2) \chi_j(1) \, d\tau_2 \, d\tau_1 \right. \\
&\quad \left. - \iint \chi_i(1) \phi_b(2) \frac{e^2}{r_{12}} \chi_j(2) \phi_b(1) \, d\tau_2 \, d\tau_1 \right]
\end{aligned} \qquad (2.49)$$

In order to construct the Fock matrix, one must already know the molecular orbitals (!) since the electron repulsion integrals require them. For this reason, the Fock equation, equation (2.47) must be solved iteratively. One makes an initial guess at the molecular orbitals and uses this guess to construct an approximate Fock matrix. Solution of the Fock equations will produce a set of MOs from which a better Fock matrix can be constructed. After repeating this operation a number of times, if everything goes well, a point will be reached where the MOs obtained from solution of the Fock equations are the same as were obtained from the previous cycle and used to make up the Fock matrix. When this point is reached, one is said to have reached *self-consistency,* or to have reached a **self-consistent field (SCF)**. In practice, solution of the Fock equations proceeds as follows. First transform the basis set, $\{\chi\}$, into an orthonormal set, $\{\lambda\}$, by means of a unitary transformation (a *rotation* in n dimensions),

$$\lambda_j = \sum_{i=1}^{n} \chi_i u_{ij} \quad S_{ij}^{\lambda} = \int \lambda_i(1) \lambda_j(1) \, d\tau_1 = \sum_{k=1}^{n} \sum_{l=1}^{n} u_{ki} S_{kl}^{\chi} u_{lj} = \delta_{ij}$$
$$\mathbf{S}^{\lambda} = \mathbf{u}^T \mathbf{S}^{\chi} \mathbf{u} = \mathbf{I} \qquad (2.50)$$

The inverse transformation is given by

$$\chi_j = \sum_{i=1}^{n} \lambda_i u_{ij}^{-1} \quad S_{ij}^{\chi} = \int \chi_i(1) \chi_j(1) \, d\tau_1 = \sum_{k=1}^{n} \sum_{l=1}^{n} u_{ki}^{-1} S_{kl}^{\lambda} u_{lj}^{-1}$$
$$= \sum_{k=1}^{n} (u^{-1})_{ik}^{T} (u^{-1})_{kj} \quad \mathbf{S}^{\chi} = (\mathbf{u}^{-1})^T \mathbf{u}^{-1} \qquad (2.51)$$

Substitution of the reverse transformation into the definition for the Fock matrix yields

$$F_{ij}^{\chi} = \int \chi_j(1) F(1) \chi_i(1) \, d\tau_1$$

$$= \sum_{k=1}^{n} \sum_{l=1}^{n} (u^{-1})_{ik}^{T} \int \lambda_k(1) F(1) \lambda_l(1) \, d\tau_1 \, u_{lj}^{-1} \quad \mathbf{F}^{\chi} = (\mathbf{u}^{-1})^{T} \mathbf{F}^{\lambda} \mathbf{u}^{-1} \quad (2.52)$$

Substitution of equations (2.50) and (2.51) into equation (2.46) and multiplication on the left by $\mathbf{c}^T$, the transpose of the coefficient matrix, yields

$$\mathbf{F}^{\chi} \mathbf{c} = \mathbf{S}^{\chi} \mathbf{c} \mathbf{e} \quad (2.47)$$

$$(\mathbf{u}^{-1})^{T} \mathbf{F}^{\lambda} \mathbf{u}^{-1} \mathbf{c} = (\mathbf{u}^{-1})^{T} \mathbf{u}^{-1} \mathbf{c} \mathbf{e} \quad (2.53)$$

$$\mathbf{c}^{T}(\mathbf{u}^{-1})^{T} \mathbf{F}^{\lambda} \mathbf{u}^{-1} \mathbf{c} = \mathbf{c}^{T}(\mathbf{u}^{-1})^{T} \mathbf{u}^{-1} \mathbf{c} \mathbf{e} \quad (2.54)$$

$$\mathbf{V}^{T} \mathbf{F}^{\lambda} \mathbf{V} = \mathbf{V}^{T} \mathbf{V} \mathbf{e} = \mathbf{e} \quad \mathbf{V} = \mathbf{u}^{-1} \mathbf{c} \quad (2.55)$$

Thus the Fock matrix in the λ basis is diagonalized by standard methods to yield the MO energies, ϵ, and the matrix $\mathbf{V}$ from which the coefficients matrix $\mathbf{c}$ may be obtained by $\mathbf{c} = \mathbf{u} \mathbf{V}$. There are several ways in which the matrix $\mathbf{u}$ and its inverse may be determined. The most commonly used is the **symmetric orthogonalization** due to Löwdin, which involves diagonalization of the overlap matrix. We will not discuss this further.

INTEGRALS

Solution of the Fock equations requires integrals involving the basis functions, either in pairs or four at a time. Some of these we have already seen. The simplest are the **overlap** integrals, stored in the form of the **overlap matrix**, $\mathbf{S}$, whose elements are given by equation (2.48),

$$\mathbf{S}_{ij} = \int \chi_i(1) \chi_j(1) \, d\tau_1 \quad (2.48)$$

The Fock integrals first encountered in equation (2.45) are constructed from **kinetic energy** integrals, **nuclear–electron attraction** integrals, and **two-electron repulsion** integrals, as follows, continuing from equation (2.49)

$$\mathbf{F}_{ij} = \int \chi_i(1) \left[\frac{-\hbar^2}{2m} \nabla^2(1) \right] \chi_j(1) \, d\tau_1 + \int \chi_i(1) \left[\sum_{I=1}^{N_N} \frac{-Z_I e^2}{r_{1I}} \right] \chi_j(1) \, d\tau_1$$

$$+ \sum_{k=1}^{n} \sum_{l=1}^{n} \sum_{b=1}^{N_e} c_{kb} c_{lb} \left[\iint \chi_i(1) \chi_k(2) \frac{e^2}{r_{12}} \chi_l(2) \chi_j(1) \, d\tau_2 \, d\tau_1 \right.$$

$$\left. - \iint \chi_i(1) \chi_k(2) \frac{e^2}{r_{12}} \chi_j(2) \chi_l(1) \, d\tau_2 \, d\tau_1 \right] \quad (2.56)$$

$$= \mathbf{T}_{ij} + \mathbf{V}_{ij}^{ne} + \sum_{k=1}^{n} \sum_{l=1}^{n} \mathbf{P}_{kl} \mathbf{G}_{ijkl} \tag{2.57}$$

The kinetic energy integrals are collected as the matrix, **T**, whose elements are defined by

$$\mathbf{T}_{ij} = \frac{-\hbar^2}{2m} \int \chi_i(1) \nabla^2(1) \chi_j(1) \, d\tau_1 \tag{2.58}$$

The nuclear–electron attraction integrals are collected as the matrix, $\mathbf{V}^{ne}$, whose elements are defined by

$$\mathbf{V}_{ij}^{ne} = -\sum_{I=1}^{N_N} Z_I e^2 \int \chi_i(1) \frac{1}{r_{1I}} \chi_j(1) \, d\tau_1 \tag{2.59}$$

The supermatrix, **G**, which contains the two-electron repulsion integrals, has elements defined by

$$\mathbf{G}_{ijkl} = \iint \chi_i(1) \chi_k(2) \frac{e^2}{r_{12}} \chi_l(2) \chi_j(1) \, d\tau_2 \, d\tau_1$$

$$- \iint \chi_i(1) \chi_k(2) \frac{e^2}{r_{12}} \chi_j(2) \chi_l(1) \, d\tau_2 \, d\tau_1 \tag{2.60}$$

In equation (2.57) we also introduced a useful matrix, the **density matrix**, **P**, whose elements are defined by

$$\mathbf{P}_{ij} = \sum_{a=1}^{N_e} c_{ia} c_{ja}, \quad \mathbf{P} = \mathbf{c} \cdot \mathbf{c}^T \tag{2.61}$$

where the sum runs over all of the occupied MOs. One of the limiting factors in *ab initio* MO calculations is the computation, storage and reading of the two-electron integrals. Whether they are stored as the supermatrix, **G**, or computed as needed, their number is approximately proportional to n^4, where n is the size of the basis set. It is highly desirable to keep n as small as possible! Much care must be taken in the choice of basis set. The choice of basis set has been called the *original sin* of computational quantum chemistry.

THE BASIS SET (STO-3G, 6-31G*, AND ALL THAT)

The basis functions we choose should describe as closely as possible the correct distribution of electrons in the vicinity of nuclei and yet be simple enough that integrations of the type required in equations (2.22), (2.25), and (2.26)

THE BASIS SET (STO-3G, 6-31G*, AND ALL THAT)

can actually be carried out efficiently. The first requirement is easily satisfied by choosing hydrogen-like-atom wavefunctions, η, the solutions to the Schrödinger equation for one-electron atoms for which exact solutions are available.

$$\eta_j(1) = N_j^\eta f(r_{1I}, \theta, \phi) e^{-\zeta_j r_{1I}} \tag{2.62}$$

Unfortunately, the exponential radial dependence of the hydrogenic functions (Figure 2.1(a)) makes the evaluation of the necessary integrals exceedingly difficult and time consuming for general computation and so another set of functions with approximately the same behavior is now universally adopted. These are **cartesian Gaussian functions**, centered on nuclei. Thus, $g_j(1)$ is a function centered on atom I:

$$g_j(1) = N_j(x_1 - X_I)^{n_x}(y_1 - Y_I)^{n_y}(z_1 - Z_I)^{n_z} e^{-a_j r_{1I}^2} \tag{2.63}$$

The superscripts, n_x, n_y, and n_z, are simple positive integers or 0. Their values determine whether the function is s-type ($n_x = n_y = n_z = 0$), p-type ($n_x + n_y + n_z = 1$ in three ways), d-type ($n_x + n_y + n_z = 2$ in six ways), etc. Specifically,

$$g_{1s}(1) = \left(\frac{8a^3}{\pi^3}\right)^{1/4} e^{-ar_1^2}, \quad g_{2p_x}(1) = \left(\frac{128a^5}{\pi^3}\right)^{1/4} x_1 e^{-ar_1^2},$$

$$g_{3d_{xy}}(1) = \left(\frac{2048a^7}{\pi^3}\right)^{1/4} x_1 y_1 e^{-ar_1^2}$$

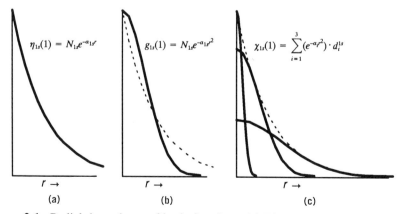

Figure 2.1. Radial dependence of basis functions: (a) The correct exponential decay (a Slater-type orbital or STO); (b) a primitive Gaussian-type function (solid line) versus an STO (dashed line); (c) a least squares expansion of the STO in terms of three Gaussian-type orbitals—STO-3G.

36 MOLECULAR ORBITAL THEORY

The correct radial behavior of the hydrogen-like atom orbital is as a simple exponential. Orbitals based on this radial dependence are called Slater-Type Orbitals or STOs. Gaussian functions are rounded at the nucleus and decrease faster than desirable (Figure 2.1(b)). Therefore, the actual basis functions are constructed by taking fixed linear combinations of the primitive Gaussian functions in such a way as to mimic exponential behavior, i.e., resemble atomic orbitals. Thus

$$\chi_i(1) = \sum_{j=1}^{n_g} g_j(1) \, d_{ji} \qquad (2.64)$$

where all of the primitive Gaussian functions are of the same type and the coefficients, d_{ji}, are chosen in such a way that χ resembles η, i.e., has approximate exponential radial dependence (Figure 2.1C).

The STO-nG basis sets are made up this way. In Table 2.1 are given the STO-3G expansions of STOs of $1s$, $2s$, and $2p$ type, with exponents of unity. In order to obtain other STOs with other exponents, ξ, one needs only to multiply the exponents of the primitive Gaussians given in Table 2.1 by the square of ξ.

A similar philosophy of *contraction* is applied to the "split valence" basis sets, for example 4-31G: the core $1s$ atomic orbital is expanded as four Gaussian functions, the valence $2s$ and $2p$ orbitals are described by two basis functions, one of which is made up of three Gaussian functions (the inner part of the valence shell), and the other is a single more diffuse Gaussian used to provide flexibility. Commonly used split valence basis sets are designated 3-21G, 4-31G, and 6-31G. These differ mainly in the quality of the description of the core electrons. Hydrogen atoms are not considered to have a core so only the "split valence" part of the designation applies to H, i.e., the split valence basis of H consists of two $1s$ functions only, while the corresponding basis set for C (for example) would consist of a single contracted $1s$ orbital and two each of $2s$, $2p_x$, $2p_y$, and $2p_z$ functions, for a total of *nine*.

The next level of improvement of the basis set involves the addition of *polarization* functions to the split valence basis set, usually the 6-31G basis. These sets are designated 6-31G* and 6-31G**. The first asterisk denotes the addition to the basis sets of heavier atoms, a set of six cartesian d-type functions to act as polarization for the s and p valence functions. The 6-31G* basis sets of Be, B, C, N, O, and F consists of *fifteen* functions. The second asterisk

TABLE 2.1. The STO-3G Basis Set Corresponding to an STO Exponent of Unity

α_{1s}	d_{1s}	α_{2sp}	d_{2s}	d_{2p}
0.109818	0.444635	0.0751386	0.700115	0.391957
0.405771	0.535328	0.231031	0.399513	0.607684
2.22766	0.154329	0.994203	−0.999672	0.155916

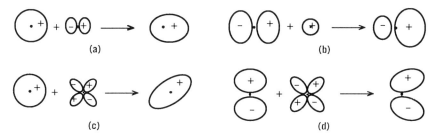

Figure 2.2. The action of polarization functions: (a) s polarized by p; (b) p polarized by s; (c) s polarized by d; (d) p polarized by d.

(if there is one) indicates the addition of a set of 2p-like Gaussian functions to the H basis set raising the number of H basis functions to *five*. Mixing of functions of differing angular momentum quantum numbers allows the formation of functions of lower symmetry and therefore a better description of the electron distribution in the molecular environment. The process is analogous to hybridization. Some examples of polarization of the s, p basis functions are illustrated in Figure 2.2.

INTERPRETATION OF THE SOLUTIONS OF THE HF EQUATIONS

Orbital Energies and the Total Electronic Energy

Solution of the HF equations yields MOs and their associated energies. The energy of ϕ_a is

$$\epsilon_a = \int \phi_a(1) F(1) \phi_a(1) \, d\tau_1 = h_a + \sum_{b=1}^{N_e} (J_{ab} - K_{ab}) \qquad (2.65)$$

where the integrals h_a, J_{ab} and K_{ab} were defined in equations (2.19), (2.25), and (2.26), respectively. The orbital energy is the kinetic energy of a single electron with the distribution specified by the MO, its attraction to all of the nuclei, and its repulsion in an averaged way with all of the other electrons in the molecule. The total electronic energy in terms of the same integrals was defined in equation (2.27).

$$E_{HF} = \sum_{a=1}^{N_e} h_a + \tfrac{1}{2} \sum_{a=1}^{N_e} \sum_{b=1}^{N_e} (J_{ab} - K_{ab}) \qquad (2.27)$$

It is clear that

$$E_{HF} = \sum_{a=1}^{N_e} \epsilon_a - \tfrac{1}{2} \sum_{a=1}^{N_e} \sum_{b=1}^{N_e} (J_{ab} - K_{ab}) \qquad (2.66)$$

The total electronic energy is not simply the sum of the orbital energies, which by themselves would over count the electron–electron repulsion.

RESTRICTED HARTREE–FOCK (RHF) THEORY

The version of the HF theory we have been studying is called *Unrestricted Hartree–Fock* (UHF) theory. It is appropriate to all molecules, regardless of the number of electrons and the distribution of electron spins (which specify the electronic state of the molecule). The spin must be taken into account when the exchange integrals are being evaluated, since if the two spin orbitals involved in this integral did have the same spin function, α or β, the integral value is zero by virtue of the orthonormality of the electron spin functions

$$\int \alpha(1)\alpha(1)\, d\sigma_1 = \int \beta(1)\beta(1)\, d\sigma_1 = 1$$

$$\int \alpha(1)\beta(1)\, d\sigma_1 = \int \beta(1)\alpha(1)\, d\sigma_1 = 0 \quad (2.67)$$

As it happens, if a molecule has the same number of electrons with spin up (α) as with spin down (β), the solution of the HF equations in the vicinity of the equilibrium geometry and for the ground electronic state usually yields the result that the spatial part of the MOs describing α and β electrons are equal in pairs. In other words, for the vast majority of molecules (F_2 is an exception), the HF determinantal wavefunction may be written as

$$\Phi_{RHF}(1, 2, 3, \cdots, N_e) = (N_e!)^{-1/2} |\phi_1'(1)\alpha(1)\ \phi_1'(2)\beta(2)$$
$$\phi_2'(3)\alpha(3) \cdots \phi_M'(N_e)\beta(N_e)| \quad (2.68)$$

which yields the familiar picture of MOs "occupied" by two electrons of opposite spins. Here M is the number of doubly occupied MOs, i.e., $M = \frac{1}{2}N_e$. If one reformulates the HF equations and total energy expression for a wavefunction which must have the form of equation (2.68), then one is doing *Restricted* HF (RHF) theory. There are considerable computational advantages to RHF theory, so unless one has some reason to suspect that the RHF solution is not the lowest energy solution, RHF is the obvious starting point. The RHF electronic energy is

$$E_{RHF} = 2 \sum_{a=1}^{M} \epsilon_a - \sum_{a=1}^{M} \sum_{b=1}^{M} (2J_{ab} - K_{ab}) \quad (2.69)$$

and the MO energy is given by

$$\epsilon_a = \int \phi_a'(1) F(1) \phi_a'(1)\, d_{t1} = h_a + \sum_{b=1}^{M} (2J_{ab} - K_{ab}) \quad (2.70)$$

Alternative formulations of the total RHF electronic energy are

$$E_{RHF} = 2 \sum_{a=1}^{M} h_a + \sum_{a=1}^{M} \sum_{b=1}^{M} (2J_{ab} - K_{ab}) \qquad (2.71)$$

and

$$E_{RHF} = \sum_{a=1}^{M} (h_a + \epsilon_a) \qquad (2.72)$$

Notice that the energy of the HF determinantal wavefunction, equation (2.68), and for that matter, for any single determinantal wavefunction, can be written by inspection; each spatial orbital contributes h_a or $2h_a$ according to its occupancy, and each orbital contributes $2J - K$ in its interaction with every other molecular orbital. Thus, the energy of the determinant for the molecular ion, M^+, obtained by removing an electron from orbital o of the RHF determinant,

$$\Phi_{RHF}^{M^+}(1, \cdots, N_e - 1)$$
$$= [(N_e - 1)!]^{-1/2} |\phi_1'(1)\alpha(1) \ \phi_1'(1)\beta(1) \cdots \phi_o'(o)\alpha(o)$$
$$\cdot \phi_{o+1}'(o+1)\alpha(o+1) \cdots \phi_M'(N_e - 1)\beta(N_e - 1)| \qquad (2.73)$$

is given by

$$E_{RHF}^{M^+} = 2 \sum_{a \neq o}^{M} h_a + h_o + \sum_{a \neq o}^{M} \sum_{b \neq o}^{M} (2J_{ab} - K_{ab}) + \sum_{b \neq o}^{M} (2J_{bo} - K_{bo}) \qquad (2.74)$$

The energy of the molecule itself, equation (2.71), could have been written

$$E_{RHF}^{M} = 2 \sum_{a \neq o}^{M} h_a + 2h_o + \sum_{a \neq o}^{M} \sum_{b \neq o}^{M} (2J_{ab} - K_{ab})$$
$$+ 2 \sum_{b \neq o}^{M} (2J_{bo} - K_{bo}) + J_{oo} \qquad (2.75)$$

Then the energy difference becomes

$$E_{RHF}^{M^+} - E_{RHF}^{M} = -h_o - \sum_{b \neq o}^{M} (2J_{bo} - K_{bo}) - J_{oo}$$
$$= -h_o - \sum_{b=1}^{M} (2J_{bo} - K_{bo})$$
$$= -\epsilon_o \qquad (2.76)$$

40 MOLECULAR ORBITAL THEORY

Thus, the ionization potential corresponding to removal of the electron from occupied MO o is just the negative of that MO's energy. This observation is known as Koopmans' theorem. One can similarly show that the energy of the lowest unoccupied MO is an estimate of the electron affinity of the molecule. In fact, ionization potentials estimated by Koopmans' theorem are fairly accurate, but the electron affinities are not useful.

Mulliken Population Analysis

The Mulliken population analysis is a simple way of gaining some useful information about the distribution of the electrons in the molecule. Let us assume again a UHF wavefunction

$$N_e = \sum_{a=1}^{N_e} \int \phi_a(1)\phi_a(1) \, d\tau_1$$

$$= \sum_{i=1}^{n} \sum_{j=1}^{n} \sum_{a=1}^{N_e} c_{ia} c_{ja} \int \chi_i(1)\chi_j(1) \, d\tau_1$$

$$= \sum_{i=1}^{n} \sum_{j=1}^{n} P_{ij} S_{ij}$$

$$= \sum_{i=1}^{n} P_i \quad \text{where } P_i = \sum_{j=1}^{n} P_{ij} S_{ij} \quad (2.77)$$

$$= \sum_{I=1}^{N_N} P_I \quad \text{where } P_I = \sum_{i}{}^{I} P_i \quad (2.78)$$

Equation (2.77) defines the atomic orbital population, P_i. Summing all of the P_i that belongs to the same atom, I, yields the atomic population, P_I, equation (2.78). The net charge, q_I, on atom I is just the difference between the nuclear charge, Z_I and the atomic population,

$$q_I = Z_I - P_I \quad (2.79)$$

Dipole Moments

The quantum mechanical dipole moment operator is equivalent to the classical dipole moment due to a collection of point charges,

$$\hat{\mu} = \sum_{i=1}^{N_e} er_i + \sum_{I=1}^{N_N} Z_I e R_I \quad (2.80)$$

As seen in equation (2.80), the dipole moment operator and the distance r_i and R_I are vectors which are usually expressed in Cartesian coordinates. The mo-

RESTRICTED HARTREE–FOCK (RHF) THEORY

lecular dipole moment within the Born–Oppenheimer approximation is evaluated as an expectation value [equation (2.8)],

$$\begin{aligned}\mu &= \int \Phi_{HF}(1, 2, \cdots, N_e) \hat{\mu} \Phi_{HF}(1, 2, \cdots, N_e) \, d\tau \\ &= \int \Phi_{HF}(1, 2, \cdots, N_e) \sum_{i=1}^{N_e} er_i \Phi_{HF}(1, 2, \cdots, N_e) \, d\tau + \sum_{I=1}^{N_N} Z_I e R_I \\ &= \sum_{a=1}^{N_e} \int \phi_a(1) er_1 \phi_a(1) \, d\tau_1 + \sum_{I=1}^{N_N} Z_I e R_I \\ &= \sum_{i=1}^{n} \sum_{j=1}^{n} \mathbf{P}_{ij} \int \chi_i(1) er_1 \chi_j(1) \, d\tau_1 + \sum_{I=1}^{N_N} Z_I e R_I \end{aligned} \quad (2.81)$$

The derivation of the second line of equation (2.81) follows the same reasoning as was used to obtain the one-electron part of the electronic energy [equation (2.21)], since both μ and h are sums of single particle operators. The dipole moment integrals over basis functions in the last line of equation (2.81) are easily evaluated. Within the HF approximation, dipole moments may be calculated to about 10% accuracy provided a large basis set is used.

Total Energies

The total energy is the sum of the total electronic energy and the nuclear-nuclear repulsion,

$$E = E_{HF} + \sum_{I=1}^{N_N-1} \sum_{J=I+1}^{N_N} \frac{Z_I Z_J e^2}{R_{IJ}} \quad (2.82)$$

Since the second term is constant for a given geometry, the total energy depends on the choice of basis set through the HF energy. This dependence is illustrated in Table 2.2.

TABLE 2.2. The SCF Total Energies (in Hartrees[a]) of CH_4, NH_3, H_2O, and HF as a Function of Basis Set[b]

Basis Set	CH_4	NH_3	H_2O	HF
STO-3G	−39.727	−55.454	−74.963	−98.571
4-31G	−40.140	−56.102	−75.907	−99.887
6-31G*	−40.195	−56.184	−76.011	−100.003
6-31G**	−40.202	−56.195	−76.023	−100.011
HF-limit[c]	−40.225	−56.225	−76.065	−100.071

[a] 1 hartree = 2626 kJ mol^{-1} = 27.21eV
[b] Table 3.13 of [34].
[c] See [35].

Orbital Energies and Orbitals

Figure 2.3 shows the MO energies for the first row hydrides. Two orbitals will have special significance for Orbital Interaction theory; the highest occupied molecular orbital (HOMO) because it represents the distribution and energy of the least tightly held electrons in the molecule, and the lowest unoccupied molecular orbital (LUMO), because it describes the easiest route to the addition of more electrons to the system. Points to notice from Figure 2.3 are as follows:

1. The "core"–"valence" separation is very large ~ 10 hartrees for C.
2. The valence shell breaks into two groups, a single orbital which is rapidly stabilized along the series C–F, and a cluster of three orbitals which remain about the same energy
3. Methane has **two** valence shell ionization potentials, ammonia has **three,** water has **four,** and HF has **three.**
4. Ammonia has the highest HOMO, methane and water have HOMOs of similar energy, and HF has the lowest HOMO of the group.
5. The LUMO energies decrease along the series CH_4–HF.

The orbital degeneracies noted in point 3) above and seen in Figure 2.3 reflect the symmetry of the molecule. The MOs of ammonia are plotted in Figure 2.4. Notice that MOs 1 and 2 are essentially molecular $1s$ and $2s$ orbitals, respectively, and that the other occupied MOs [3, 4, and 5 (HOMO)]

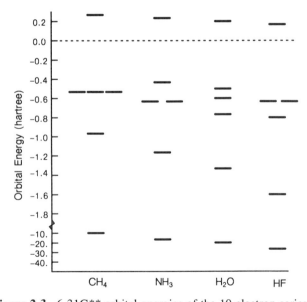

Figure 2.3. 6-31G** orbital energies of the 10-electron series.

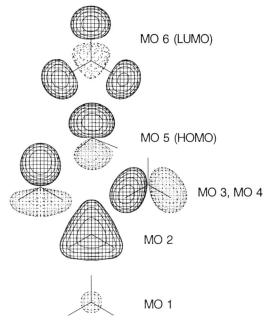

Figure 2.4. 6-31G* MOs of NH_3: MOs 1-5 are drawn at contour level $|0.1|$ while MO 6 is drawn at $|0.05|$.

resemble $2p$ orbitals. The degenerate pair together describes the three-fold symmetry of the molecule. The orientation of the node is arbitrary. As it happens the nodal plane of MO 4 as seen in Figure 2.4 does coincide with the plane containing one of the NH bonds. The HOMO would be described as the "lone pair" on N, but it is a true molecular orbital. The LUMO resembles a $3s$ orbital and is made up of the out-of-phase combination of the $2s$ of N and the $1s$ orbitals of the H atoms.

Configuration Energies

L. C. Allen has suggested that the familiar two-dimensional periodic table of elements has a missing third dimension, with units of energy [36], [37], [38]. In part, he reasons that the elements of the periodic table are grouped according to valence electron configurations by quantum numbers, n and l, which indicate orbital size and shape, but whose primary purpose is to specify energy. It is proposed that the missing third dimension is the configuration energy (CE) (also previously called *spectroscopic electronegativity* [37]), the average one-electron valence shell energy of a ground state free atom, which may be defined as follows:

$$CE = \frac{a\epsilon_s + b\epsilon_p}{a + b} \tag{2.83}$$

44   MOLECULAR ORBITAL THEORY

where a and b are the occupancies of the valence shell s and p orbitals, respectively, and ϵ_s and ϵ_p are the multiplet-averaged s and p shell ionization potentials. The latter may be measured spectroscopically or identified by Koopmans' theorem with the atomic orbital energies. For the d-block transition elements, a parallel definition applies, namely,

$$CE = \frac{a\epsilon_s + b\epsilon_d}{a + b} \qquad (2.84)$$

although the occupancy of the d shell may be difficult to assign. Values of CE closely parallel the established electronegativity scales of Pauling [39] and Allred and Rochow [40]. A comparison of the three electronegativity scales for selected main group elements is presented in Table 2.3 [41].

Energy Index

Allen and co-workers [42] have extended the idea of average energy of an isolated atom, as exemplified by the configuration energy, CE, to a definition of the *in situ* energy of an atom embedded in a molecule or molecular frag-

TABLE 2.3. A Comparison of Configuration Energy (CE) with the Electronegativity Scales of Pauling (χ_P) and Allred and Rochow ($\chi_{A\&R}$)[a]

	H							
CE	2.300							
χ_P	2.20							
$\chi_{A\&R}$	2.20							
	Li	Be	B	C	N	O	F	Ne
CE	0.912	1.576	2.051	2.544	3.066	3.610	4.193	4.787
χ_P	0.98	1.57	2.04	2.55	3.04	3.44	3.98	
$\chi_{A\&R}$	0.97	1.47	2.01	2.50	3.07	3.50	4.10	
	Na	Mg	Al	Si	P	S	Cl	Ar
CE	0.869	1.293	1.613	1.916	2.253	2.589	2.869	3.242
χ_P	0.93	1.31	1.61	1.90	2.19	2.58	3.16	
$\chi_{A\&R}$	1.01	1.23	1.47	1.74	2.06	2.44	2.83	
	K	Ca	Ga	Ge	As	Se	Br	Kr
CE	0.734	1.034	1.756	1.994	2.211	2.424	2.685	2.966
χ_P	0.82	1.00	1.81	2.01	2.18	2.55	2.96	
$\chi_{A\&R}$	0.91	1.04	1.82	2.02	2.20	2.48	2.74	
	Rb	Sr	In	Sn	Sb	Te	I	Xe
CE	0.706	0.963	1.656	1.824	1.984	2.158	2.359	2.582
χ_P	0.82	0.95	1.78	1.96	2.05	2.10	2.66	
$\chi_{A\&R}$	0.89	0.99	1.49	1.72	1.82	2.01	2.21	

[a]See [37].

ment, the *Energy Index* of atom I, EI_I. The derivation of an expression for the energy index is as follows. A partitioning of the Mulliken population analysis (see above) yields an expression for the total fractional population, f_I^a, of the basis functions on atom I in the a'th MO, namely

$$f_I^a = \sum_j^I \sum_i^n S_{ij} c_{ia} c_{ja} \qquad (2.85)$$

where the first summation runs over the basis functions of atom I, and the other quantities are as previously defined. The one electron energy of atom I in the a'th MO, EI_I^a, may be obtained by weighting the MO energy, ϵ_a, by the fraction of the total orbital occupancy (n_a) which can be assigned to that atom, i.e., $n_a f_I^a$. Thus,

$$EI_I^a = n_a \epsilon_a \sum_j^I \sum_i^n S_{ij} c_{ia} c_{ja} \qquad (2.86)$$

Summation of the quantity [equation (2.86)] over all of the MOs, averaged over the sum of the fractional populations, yields the average one electron energy of the valence electrons of atom I in the molecule, the *Energy Index*, EI_I

$$EI_I = \sum_a^M n_a \epsilon_a \sum_j^I \sum_i^n S_{ij} c_{ia} c_{ja} \Big/ \sum_a^M n_a \sum_j^I \sum_j^n S_{ij} c_{ia} c_{ja} \qquad (2.87)$$

It was hoped that EI_I would serve as an *a priori* definition for the electronegativity of the atom in the molecular environment, and by extension, also function as a measure of the electronegativity of the group of atoms attached via atom I to other parts of the molecule [42]. It transpired that another quantity, the *Bond Polarity Index*, *BPI*, is better correlated with experimental data when used as a measure of group electronegativity where the group is in competition with a reference group for the electrons in the bond(s) connecting them [38].

Bond Polarity Index

The bond polarity index of the bond(s) connecting groups A and B, BPI_{AB}, which are attached to each other through connecting atoms I and J, respectively, is defined in terms of the corresponding Energy Indices, EI_I and EI_J. Thus

$$BPI_{AB} = (EI_I - EI_I^{\text{ref}}) - (EI_J - EI_J^{\text{ref}}) \qquad (2.88)$$

where EI_I and EI_J are calculated for the molecule $A-B$ directly, and the reference Energy Indices, EI_I^{ref} and EI_J^{ref}, are calculated for the corresponding homopolar molecules, $A-A$ and $B-B$, where the bond(s) involve atoms $I(I-I)$ and $J(J-J)$, respectively. Specifically, the bond polarity index of the C—N bond to the CH_3 group in CH_3NH_2 involves the calculation of EI of the methyl and amino groups of methylamine and the reference EI of the groups in CH_3CH_3 and NH_2NH_2.

Bond Polarity Index and Orbital Interaction Diagrams

In the construction of orbital interaction diagrams, which forms the main emphasis of this text, we will be concerned with the placement of atom or group orbitals correctly relative to each other on a common energy scale, and the evaluation of the results of the orbital interactions. The Bond Polarity Index provides a direct nonempirical measure of the result of the interaction and may therefore serve as a guide to the construction of the diagrams.

GEOMETRY OPTIMIZATION

Differentiation of the total energy [equation (2.82)]

$$E = E_{HF} + \sum_{I=1}^{N_N-1} \sum_{J=I+1}^{N_N} \frac{Z_I Z_j e^2}{R_{IJ}} \qquad (2.82)$$

with respect to displacement of nucleus K in the direction α (where α is x, y, or z), yields the force acting on that nucleus at the chosen geometry,

$$F_{K\alpha} = -\frac{\partial E}{\partial R_{K\alpha}} = -\frac{\partial E_{HF}}{\partial R_{K\alpha}} + \sum_{I=1}^{N_N-1} \sum_{J=I+1}^{N_N} \frac{Z_I Z_j e^2}{R_{IJ}^2} \cdot \sum_\beta \frac{\partial R_{IJ\beta}}{\partial R_{K\alpha}} \qquad (2.89)$$

The forces acting on the nuclei are evaluated analytically and the nuclei can be displaced in the direction which minimizes the forces. Points on the B–O potential hypersurface [E, equation (2.82), as a function of all possible nuclear positions] at which the forces that are all zero are called **stationary points,** and correspond to local minima (stable structures), saddle points (transition structures), tops of "hills," etc. Derivatives of the forces (second derivatives of the energy) constitute the **Hessian matrix,** which provides information about the curvature of the surface. The eigenvalues of the Hessian matrix serve to characterize the nature of the stationary points. A local minimum has only positive eigenvalues; a transition structure has exactly one negative eigenvalue of the Hessian. Hessians are evaluated numerically during geometry optimizations by an algorithm due to Schlegel [43], or analytically. In Table 2.4 are presented some typical geometry optimization data as a function of basis set. The results in Table 2.4 are reasonably representative. HF calculations

TABLE 2.4. Bond Lengths and Bond Angles for the 10-Electron Hydrides[a]

Basis set	CH_4 C—H (Å)	NH_3 N—H (Å)	HNH (°)	H_2O O—H (Å)	HOH (°)	HF H—F (Å)
STO-3G	1.083	1.033	104.2	0.990	100.0	0.956
4-31G	1.081	0.991	115.8	0.951	111.2	0.922
6-31G*	1.084	1.004	107.5	0.948	105.5	0.911
6-31G**	1.084	1.004	107.6	0.943	106.0	0.901
Near HF-limit	1.084	1.000	107.2	0.940	106.1	0.897
Experiment	1.085	1.012	106.7	0.957	104.5	0.917

[a]Data extracted from [44].

yield reasonable geometries for most molecules; errors are typically of the order of 0.02 Å for bond lengths, and about 2° for bond angles.

NORMAL COORDINATES AND HARMONIC FREQUENCY ANALYSIS

In mass-weighted coordinates, the Hessian matrix becomes the harmonic force constant matrix, from which a normal coordinate analysis may be carried out to yield harmonic frequencies and normal modes, essentially a prediction of the fundamental IR transition frequencies. For frequency analysis, the Hessian matrix is derived analytically. The split valence basis sets, 4-31G, 6-31G*, and 6-31G** yield very similar values for the predicted harmonic frequencies, systematically too high by about 10%. It is possible to scale the elements of the force constant matrix individually to improve agreement with experiment but uniform scaling by 0.81 (corresponding to a decrease of 10% in the calculated frequencies) usually gives quite satisfactory results. The fundamental normal vibrational modes and calculated frequencies for ammonia are shown in Figure 2.5.

SUCCESSES AND FAILURES OF HARTREE–FOCK THEORY

Hartree–Fock theory is a rigorous *ab initio* theory of electronic structure and has a vast array of successes to its credit. Equilibrium structures of molecules are calculated almost to experimental accuracy, and reasonably accurate properties (dipole moments, IR and Raman intensities, etc.) can be calculated from HF wavefunctions. Relative energies of conformers and barriers to conformational changes are reproduced to experimental accuracy. Relative energies

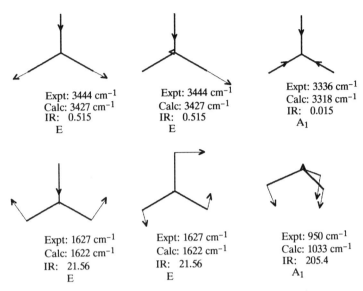

Figure 2.5. Experimental frequencies [45] and calculated (HF/6-31G**) normal modes, frequencies (× 0.9) and IR intensities (km/mol) of ammonia.

of structural isomers are also very well reproduced. However, the approximation involved in the derivation of HF theory (single determinantal many-electron wavefunction constructed of one-electron wavefunctions, i.e., as antisymmetrized sum of products) has a serious consequence which has been mentioned several times, namely the treatment of electron–electron interactions in an averaged way through the $J(1)$ and $K(1)$ integral operators with consequent loss of electron correlation. The *correlation error* is most severe in the region near the minimum of the potential energy hypersurface where the lowest energy solution is of the RHF type and the major part of the error involves each pair of electrons in the same orbital (pair correlation). The magnitude of the error, though large, is very constant and tends to cancel almost quantitatively if the number and nature of occupied MOs is the same in two structures being compared. HF theory will be expected to fail in cases where the number and nature of bonded electron pairs change. Thus, HF theory is unsatisfactory for any single electron process, such as reduction or oxidation (including ionization), or electronic excitation (photochemistry). It also fails to describe homolytic bond dissociation, a major component of most chemical reactions in the gas phase. A secondary consequence of the averaged treatment of electron correlation error is that the RHF solution is too stable. Bond dissociation in RHF theory is always heterolytic (to ions). As a result, the local curvature of the potential energy hypersurface near minima is too steep and force constants (and harmonic frequencies) are too large. In the next section, a very brief introduction to methods of improving HF theory is presented.

POST-HARTREE–FOCK METHODS

Although HF theory is useful in its own right for many kinds of investigations, there are some applications for which the neglect of electron correlation, or the assumption that the error is constant (and so will cancel) are not warranted. Post-Hartree–Fock methods seek to improve the description of the electron-electron interactions, using HF theory as a reference point. Improvements to HF theory can be made in a variety of ways, including the method of **Configuration Interaction (CI)** and by use of **Many-Body Perturbation Theory (MBPT)**. It is beyond the scope of this text to treat CI and MBPT methods in any but the most cursory manner. However, both methods can be introduced from aspects of the theory already discussed.

Configuration Interaction (CI) Theory

Earlier it was argued that the many-electron wavefunction (the true solution to the electronic Schrödinger equation) could be expanded in terms of an infinite series of single determinantal wavefunctions [equation (2.13)]:

$$\Psi(1, 2, \cdots, N_e) = \sum_{a=1}^{\infty} d_a \Phi_a \qquad (2.13)$$

where each of the determinants is of the form [equation (2.90)],

$$\Phi(1, 2, \cdots, N_e) = (N_e!)^{-1/2} \begin{vmatrix} \phi_1(1) & \phi_2(1) & \phi_3(1) & \cdots & \phi_{N_e}(1) \\ \phi_1(2) & \phi_2(2) & \phi_3(2) & \cdots & \phi_{N_e}(2) \\ \phi_1(3) & \phi_2(3) & \phi_3(3) & \cdots & \phi_{N_e}(3) \\ \vdots & \vdots & \vdots & & \vdots \\ \phi_1(N_e) & \phi_2(N_e) & \phi_3(N_e) & \cdots & \phi_{N_e}(N_e) \end{vmatrix}$$

(2.90)

differing only in their composition in terms of the MOs. If the MOs form an infinite complete orthonormal set, then so do the determinants constructed from them. The HF equations were solved in a finite basis of dimension, n, and so yielded n MOs which form an orthonormal set. Since $n > N_e$, ($> \frac{1}{2} N_e$, in the case of RHF), the "extra" MOs can be used to generate new determinants from the HF determinant (which is constructed from the N_e MOs of lowest energy) by replacement of the occupied (in Φ_{HF}) MOs by empty ("virtual") MOs. The determinants are called electron *configurations* because they describe the distribution of all of the electrons. A configuration constructed from Φ_{HF} by replacement of a single occupied MO by a virtual MO is called a *singly*

50 MOLECULAR ORBITAL THEORY

excited configuration because one can imagine it arising from the excitation of an electron from an occupied MO to an empty MO. If determinants are constructed from all possible single excitations, the number of singly excited determinants would be $N_e \times (n - N_e)$. For example, a calculation on the water molecule with the 6-31G* basis set would generate 19 MOs and 90 singly excited configurations (70 in an RHF calculation). If one generated the list of determinants from all possible replacements among the set of MOs, the set of configurations so obtained is said to be *complete*, and forms a finite complete orthonormal set with the dimension determined by the number of electrons and the number of MOS. The number, n_{CI}, is the same as the number of permutations of N_e objects among n bins with no more than one object per bin, namely $n_{CI} = n!/N_e!(n - N_e)!$, which for the above-mentioned water calculation would be 461,890. The many-electron wavefunction may be expanded in this finite set in the manner of equation (2.13) to yield a CI wavefunction,

$$\Psi^{CI}(1, 2, \cdots, N_e) = \sum_{a=1}^{n_{CI}} d_a \Phi_a \tag{2.91}$$

The variational method is used to find the optimum expansion in terms of the configurations, i.e., the energy is expressed as an expectation value as was done in equation (2.9),

$$E^{CI} = \frac{\int \Psi^{CI} H \Psi^{CI} \, d\tau}{\int |\Psi^{CI}|^2 \, d\tau}$$

$$= \frac{\sum_{a=1}^{n_{CI}} \sum_{b=1}^{n_{CI}} d_a d_b \int \Phi_a H \Phi_b \, d\tau}{\sum_{a=1}^{n_{CI}} d_a^2} \tag{2.92}$$

and differentiated with respect to each of the coefficients, d_a, and setting the result equal to zero,

$$\frac{\partial E^{CI}}{\partial d_a} = 2 \sum_{b=1}^{n_{CI}} d_b \left(\int \Phi_a H \Phi_b \, d\tau - E^{CI} \delta_{ab} \right) = 0 \tag{2.93}$$

The set of n_{CI} linear equations must then be solved for the energies and coefficients. This is accomplished by diagonalization of the *Hamiltonian matrix*,

H, whose elements are defined by,

$$\mathbf{H}_{ab} = \int \Phi_a H \Phi_b \, d\tau \qquad (2.94)$$

The elements of the Hamiltonian matrix can be expressed in terms of the MO energies and core, coulomb and exchange integrals between the MOs involved in the "excitations" which generated each configuration. The eigenvalues of **H** are the energies of different electronic states, the lowest energy being the energy of the ground state. Matrix diagonalization is a straight-forward procedure for small matrices, but is a formidable task for large matrices. Techniques exist to extract the lowest eigenvalues of large matrices, but in practice, complete CI calculations cannot be carried out except for the smallest molecules and some systematic selection procedure to reduce the size of n_{CI} must be used. It can be shown that most of the correlation error in HF theory, namely that associated with pairs of electrons in the same orbital, may be corrected if one includes in the CI calculation all singly and doubly excited configurations (SDCI). GAUSSIAN 86, 88, 90, etc., will perform SDCI if asked. The truncation of the CI expansion introduces an anomaly called a *size consistency* error. In other words, the sum of the (for example) SDCI energies for A and B calculated separately are not *exactly* the same as the SDCI energy of A and B handled as a single system. The size consistency error in SDCI is usually small and largely corrected by addition of the effects of some quadruple excitations by the *Davidson correction* [46]. The correlation errors of most ground state calculations are largely corrected by SDCI calculation with the Davidson correction. The computational time involved is approximately proportional to n^6.

Excited States from CI Calculations

Excited state energies and wavefunctions are automatically obtained from CI calculations. However, the quality of the wavefunctions is more difficult to achieve. The equivalent of the HF description for the ground state requires an all-singles CI (SCI). Singly excited configurations do not mix with the HF determinant, i.e.,

$$\mathbf{H}_{HF,b} = \int \Phi_{HF} H \Phi_b^{SE} \, d\tau = 0 \qquad \text{(Brillouin's theorem)} \qquad (2.95)$$

SCI may provide a very reasonable description for the electronic excitation process and of the excited state potential energy surface from which to study photochemical processes. GAUSSIAN 90 [33] is the first widely available quantum chemistry program which allows geometry optimization on SCI excited state potential energy surfaces. A description for an excited state which is equivalent to the SDCI description of the ground state requires all single, double, and many of the triple excitations. Some of these may be added by perturbation theory in a manner which is beyond the scope of the present approach. Quite accurate electronic transition energies and transition dipole and optical rotary strengths may be calculated at this level of theory.

MANY BODY PERTURBATION THEORY

There are many variations of many body perturbation theories. In this course we will only touch on one of these, the Møller–Plesset (MP) variation of Rayleigh–Schrödinger (RS) perturbation theory. Simply stated, perturbation theories attempt to describe differences between systems, rather than to describe the systems separately and then take the difference. The image is of a reference system which is suddenly subjected to a perturbation. The object is to describe the system in the presence of the perturbation in relation to the unperturbed system. The perturbation may be a real perturbation, such as an electric or magnetic field, electromagnetic radiation, the presence of another molecule or medium, a change in the geometry, etc., or it may be a conceptual device, such as a system in which the electrons did not interact, the perturbation being the turning on of the electron–electron interaction.

Rayleigh–Schrödinger Perturbation Theory

If the solutions (energies, $E_n^{(0)}$, and wavefunctions, $\Psi_n^{(0)}$) of the Schrödinger equation for the unperturbed system, $H^{(0)}\Psi_n^{(0)} = E_n^{(0)}\Psi_n^{(0)}$ are known, and the operator form of the perturbation, H^P, can be specified, the Rayleigh–Schrödinger perturbation theory will provide a description of the perturbed system in terms of the unperturbed system. Thus, for the perturbed system, the SE is

$$H\Psi_n = (H^{(0)} + \lambda H^P)\Psi_n = E_n\Psi_n \qquad (2.96)$$

The parameter, λ, is introduced in order to keep track of the order of the perturbation series, as will become clear. Indeed, one can perform a Taylor series expansion of the perturbed wavefunctions and perturbed energies, using λ to keep track of the order of the expansions. Since the set of eigenfunctions of the unperturbed SE form a complete and orthonormal set, the perturbed wavefunctions can be expanded in terms of them. Thus,

$$\Psi_n = \Psi_n^{(0)} + \lambda\Psi_n^{(1)} + \lambda^2\Psi_n^{(2)} + \cdots \qquad (2.97)$$

$$E_n = E_n^{(0)} + \lambda E_n^{(1)} + \lambda^2 E_n^{(2)} + \cdots \qquad (2.98)$$

The superscripts in parentheses indicate successive levels of correction. If the perturbation is small, this series will converge. Substitution of equations (2.97) and (2.98) into equation (2.96) and collecting powers of λ yields,

$$(H^{(0)} + \lambda H^p)(\Psi_n^{(0)} + \lambda \Psi_n^{(1)} + \lambda^2 \Psi_n^{(2)} + \cdots)$$
$$= (E_n^{(0)} + \lambda E_n^{(1)} + \lambda^2 E_n^{(2)} + \cdots)(\Psi_n^{(0)} + \lambda \Psi_n^{(1)} + \lambda^2 \Psi_n^{(2)} + \cdots)$$
(2.99)

$$(H^{(0)}\Psi_n^{(0)} - E_n^{(0)}\Psi_n^{(0)})\lambda^0 + (H^{(0)}\Psi_n^{(1)} + H^p\Psi_n^{(0)} - E_n^{(0)}\Psi_n^{(1)} - E_n^{(1)}\Psi_n^{(0)})\lambda^1$$
$$+ (H^{(0)}\Psi_n^{(2)} + H^p\Psi_n^{(1)} - E_n^{(0)}\Psi_n^{(2)} - E_n^{(1)}\Psi_n^{(1)} - E_n^{(2)}\Psi_n^{(0)})\lambda^2$$
$$+ \cdots = 0 \qquad (2.100)$$

Equation (2.100) is a power series in λ which can only be true if the coefficients in front of each term are individually zero. Thus,

$$H^{(0)}\Psi_n^{(0)} - E_n^{(0)}\Psi_n^{(0)} = 0 \qquad (2.101)$$

$$H^{(0)}\Psi_n^{(1)} + H^p\Psi_n^{(0)} - E_n^{(0)}\Psi_n^{(1)} - E_n^{(1)}\Psi_n^{(0)} = 0 \qquad (2.102)$$

$$H^{(0)}\Psi_n^{(2)} + H^p\Psi_n^{(1)} - E_n^{(0)}\Psi_n^{(2)} - E_n^{(1)}\Psi_n^{(1)} - E_n^{(2)}\Psi_n^{(0)} = 0 \qquad (2.103)$$

Equation (2.101) is just the Schrödinger equation for the unperturbed system. Equation (2.102) is the first-order equation. Multiplying each term of equation (2.102) on the left by $\Psi_n^{(0)}$ and integrating yields

$$\int \Psi_n^{(0)} H^{(0)} \Psi_n^{(1)} \, d\tau + \int \Psi_n^{(0)} H^p \Psi_n^{(0)} \, d\tau$$
$$- \int \Psi_n^{(0)} E_n^{(0)} \Psi_n^{(1)} \, d\tau - E_n^{(1)} \int |\Psi_n^{(0)}|^2 \, d\tau = 0 \qquad (2.104)$$

Since $H^{(0)}$ is a Hermitian operator and $\Psi_n^{(0)}$ is an eigenfunction of it, the first and third integrals are equal and cancel, leaving an expression for the first-order correction to energy,

$$E_n^{(1)} = \int \Psi_n^{(0)} H^p \Psi_n^{(0)} \, d\tau \qquad (2.105)$$

Multiplication of equation (2.102) by $\Psi_m^{(0)}$ ($m \neq n$) and integrating gives,

$$\int \Psi_m^{(0)} H^{(0)} \Psi_n^{(1)} \, d\tau + \int \Psi_m^{(0)} H^p \Psi_n^{(0)} \, d\tau$$
$$- \int \Psi_m^{(0)} E_n^{(0)} \Psi_n^{(1)} \, d\tau - E_n^{(1)} \int \Psi_m^{(0)} \Psi_n^{(0)} \, d\tau = 0 \qquad (2.106)$$

54 MOLECULAR ORBITAL THEORY

The last integral is zero because of the orthogonality of the unperturbed wavefunctions. Equation (2.106) simplifies to

$$\int \Psi_m^{(0)} \Psi_n^{(1)} \, d\tau = -\frac{\int \Psi_m^{(0)} H^p \Psi_n^{(0)} \, d\tau}{E_m^{(0)} - E_n^{(0)}} \quad (2.107)$$

Let the first-order correction to the perturbed wavefunction be expanded as a linear combination of unperturbed wavefunctions, i.e.,

$$\Psi_n^{(1)} = \sum_{l=0}^{\infty} \Psi_l^{(0)} a_{ln} \quad (2.108)$$

Substitution of equation (2.108) into equation (2.107) yields an expression for the expansion coefficient, namely,

$$\sum_{l=0}^{\infty} a_{ln} \int \Psi_m^{(0)} \Psi_l^{(0)} \, d\tau = a_{mn} = -\frac{\int \Psi_m^{(0)} H^p \Psi_n^{(0)} \, d\tau}{E_m^{(0)} - E_n^{(0)}} \quad (2.109)$$

Thus, the first-order correction to the zero-order (unperturbed) wavefunction is obtained by substituting equation (2.109) into equation (2.108) and changing the summation index

$$\Psi_n^{(1)} = -\sum_{\substack{m=n \\ m \neq n}}^{\infty} \frac{\int \Psi_m^{(0)} H^p \Psi_n^{(0)} \, d\tau}{E_m^{(0)} - E_n^{(0)}} \Psi_m^{(0)} \quad (2.110)$$

The diagonal term $m = n$ is excluded from the summation in equation (2.110) since that wavefunction is the zero-order term. The summation should converge at some finite value of m as the energy difference in the denominator becomes large. It is generally believed that a correction to the energy which is comparable to the first-order correction to the wavefunction would involve the second-order term, $E_n^{(2)}$, which may be extracted from the second-order equation (2.103). Multiply every term on the left by Ψ_n^0 and integrate

$$\int \Psi_n^{(0)} H^{(0)} \Psi_n^{(2)} \, d\tau + \int \Psi_n^{(0)} H^p \Psi_n^{(1)} \, d\tau$$
$$- E_n^{(0)} \int \Psi_n^{(0)} \Psi_n^{(2)} \, d\tau - E_n^{(1)} \int \Psi_n^{(0)} \Psi_n^{(1)} \, d\tau - E_n^{(2)} \int |\Psi_n^{(0)}|^2 \, d\tau = 0$$

$$(2.111)$$

As before, the first term and the third term are equal and cancel. The fourth term also is zero as can be verified by substitution of equation (2.110) into it. Thus,

$$E_n^{(2)} = \int \Psi_n^{(0)} H^p \Psi_n^{(1)} \, d\tau \qquad (2.112)$$

Substitution of equation (2.110) into equation (2.112) yields the usual expression for the second-order correction to the energy

$$E_n^{(2)} = - \sum_{m \neq n}^{\infty} \frac{\left| \int \Psi_m^{(0)} H^p \Psi_n^{(0)} \, d\tau \right|^2}{E_m^{(0)} - E_n^{(0)}} \qquad (2.113)$$

In summary, the wavefunction correct to first order, and the energy correct to second order, are

$$\Psi_n = \Psi_n^{(0)} - \sum_{m \neq n} \frac{\int \Psi_m^{(0)} H^p \Psi_n^{(0)} \, d\tau}{E_m^{(0)} - E_n^{(0)}} \Psi_m^{(0)}$$

(correct to first order) \qquad (2.114)

$$E_n = E_n^{(0)} + \int \Psi_n^{(0)} H^p \Psi_n^{(0)} \, d\tau - \sum_{m \neq n}^{\infty} \frac{\left| \int \Psi_m^{(0)} H^p \Psi_n^{(0)} \, d\tau \right|^2}{E_m^{(0)} - E_n^{(0)}}$$

(correct to second order) \qquad (2.115)

The parameter λ has been embedded in the definition of H^p. The wavefunction from perturbation theory [equation (2.114)] is not normalized and must be renormalized. The energy of a truncated perturbation expansion [equation (2.115)] is not variational, and it may be possible to calculate energies lower than "experimental."

Møller–Plesset Perturbation Theory (MPPT)

MPPT aims to recover the correlation error incurred in Hartree–Fock theory for the ground state whose zero-order description is Φ_{HF}. The Møller–Plesset zero-order Hamiltonian is the sum of Fock operators, and the zero-order wavefunctions are determinantal wavefunctions constructed from HF MOs. Thus the zero-order energies are simply the appropriate sums of MO energies. The "perturbation" is defined as the difference between the sum of Fock operators

MOLECULAR ORBITAL THEORY

and the exact Hamiltonian

$$H^{(0)} = \sum_{i=1}^{N_e} F(i)$$

$$= \sum_{i=1}^{N_e} \left\{ h(i) + \sum_{b=1}^{N_e} [J_b(i) - K_b(i)] \right\} \quad (2.116)$$

$$H^p = \sum_{i=1}^{N_e-1} \sum_{j=I+1}^{N_e} \frac{1}{r_{ij}} - \sum_{i=1}^{N_e} \sum_{b=1}^{N_e} [J_b(i) - K_b(i)] \quad (2.117)$$

We state without further derivation that the electronic energy corrected to second order in Møller–Plesset perturbation theory, E_{MP2}, is

$$E_{MP2} = \sum_{a=1}^{N_e} \epsilon_a - \frac{1}{2} \sum_{a=1}^{N_e} \sum_{b=1}^{N_e} (J_{ab} - K_{ab})$$

$$+ \frac{1}{4} \sum_{a=1}^{N_e} \sum_{b=1}^{N_e} \sum_{u=N_e+1}^{n} \sum_{v=N_e+1}^{n} \frac{|\langle ab| |uv \rangle|^2}{\epsilon_a + \epsilon_b - \epsilon_u - \epsilon_v} \quad (2.118)$$

where the notation $\langle ab| |uv \rangle$ means

$$\langle ab| |uv \rangle = \iint \phi_a(1)\phi_b(1) \frac{1}{r_{12}} \phi_u(1)\phi_v(2) \, d\tau_1 \, d\tau_2$$

$$- \iint \phi_a(1)\phi_b(2) \frac{1}{r_{12}} \phi_v(1)\phi_u(2) \, d\tau_1 \, d\tau_2 \quad (2.119)$$

If carried out with a good basis set (6-31G* or better), the benefits of MPPT, carried out to second order (MP2), include moderate improvements in structures and relative energies, and often significant improvement in the values of secondary properties such as dipole moments, vibrational frequencies and infrared and Raman absorption intensities. Modern quantum chemistry codes such as GAUSSIAN 92 incorporate analytical calculation of MP2 forces and force constants. Although it adds substantially to the time required to carry out the calculations, the results of MPPT usually make the extra effort worthwhile.

CHAPTER 3

ORBITAL INTERACTION THEORY

RELATIONSHIP TO THE HARTREE–FOCK EQUATIONS

Orbital interaction theory forms a comprehensive model for examining the structures and kinetic and thermodynamic stabilities of molecules. It is not intended to be, nor can it be, a quantitative model. However, it can function effectively in aiding understanding the fundamental processes in chemistry, and it can be applied in most instances without the use of a computer. Also known as Perturbative Molecular Orbital (PMO) theory, it was originally developed from the point of view of weak interactions [4], [5]. However, the interaction of orbitals is more transparently developed, and the relationship to quantitative MO theories is more easily seen, by straight-forward solution of the Hückel (independent electron) equations. From this point of view, the theoretical foundations lie in Hartree–Fock theory developed in Chapter 2 [47].

THE HÜCKEL APPROXIMATION

Orbital Energies and the Total Electronic Energy

Recall that the minimum requirement for a many-electron wavefunction is that it be written as a suitably antisymmetrized sum of products of one-electron wavefunctions, i.e., as a Slater determinant of MOs [equation (2.68)]

$$\Phi_{RHF}(1, 2, 3, \cdots, N_e)$$
$$= (N_e!)^{1/2} |\phi'_1(1)\alpha(1) \quad \phi'_1(2)\beta(2) \quad \phi'_2(3)\alpha(3) \cdots \phi'_M(N_e)\beta(N_e)|$$

(3.1)

In Chapter 2, we found that the condition that this be the best possible wavefunction of this form is that the MOs be eigenfunctions of a one-electron operator, the Fock operator [equation (2.42)], i.e., (without the primes)

$$F(1)\phi_a(1) = \epsilon_a\phi_a(1) \qquad (3.2)$$

where the Fock operator in *Restricted* form [RHF, the UHF form was given in equation (2.41)] is given by

$$F(1) = h(1) + \sum_{b=1}^{M} [2J_b(1) - K_b(1)] \qquad (3.3)$$

The closed shell ground state wavefunction [equation (3.1)] is constructed from the MOs with the lowest energies, enough for all of the electrons. Then the total RHF electronic energy [equation (2.69)] is given by

$$E_{RHF} = 2\sum_{a=1}^{M} \epsilon_a - \sum_{a=1}^{M}\sum_{b=1}^{M} (2\mathrm{J}_{ab} - \mathrm{K}_{ab}) \qquad (3.4)$$

In order to develop a "back-of-the-envelope" kind of theory, we make a series of approximations. The first is the Hückel or *Independent Electron Approximation* (IEA), in which all electron–electron interactions are ignored. Thus, the Fock operator reduces to just the *core Hamiltonian*, $h(1)$, and the total electronic energy is the sum of the MO energies times the appropriate occupation number, 2 (or 1 for radical or excited state). It is too gross an approximation just to drop the electron–electron interactions and keep $h(1)$ as is. In order to compensate for the loss of part of the repulsive potential, $h(1)$ must be modified to some extent, resulting in an *effective core Hamiltonian*, $h^{\mathrm{eff}}(1)$. Thus,

$$F(1) \approx h^{\mathrm{eff}}(1), \quad h^{\mathrm{eff}}(1)\phi_a(1) = \epsilon_a\phi_a(1), \quad E_{IEA} = 2\sum_{a=1}^{M}\epsilon_a \qquad (3.5)$$

Equation (3.5) defines the essence of the Hückel Molecular Orbital (HMO) theory. Notice that the total energy is the sum of the energies of the individual electrons. *Simple* Hückel Molecular Orbital (SHMO) theory requires further approximations that we will discuss in due course.

CASE STUDY OF A TWO-ORBITAL INTERACTION

Let us consider the simplest possible case of a system which consists of two orbitals, $\chi_A(1)$ and $\chi_B(1)$, with energies, ϵ_A and ϵ_B, which can interact. We wish to investigate the results of the interaction between them, i.e., what new wavefunctions are created and what their energies are. Let us be clear about what the subscripts A and B represent. The subscripts denote orbitals belonging to two *physically distinct* systems; the systems, and therefore the orbitals, are in separate positions in space. The two systems may in fact be identical, e.g., two water molecules or two sp^3 hybrid orbitals on the same atom or on different but identical atoms (say, both C atoms). In this case, $\epsilon_A = \epsilon_B$, or the two systems may be different in all respects. Even if the systems and the associated orbitals are identical, they will in general differ in the way they are oriented in space relative to each other, or their separation may vary. All of these factors will affect the way the orbitals of the two interact.

In the language of perturbation theory, the two orbitals will constitute the *unperturbed* system, the "perturbation" is the interaction between them, and the result of the interaction is what we wish to determine. The situation is seen in Figure 3.1(a). The diagram shown in Figure 3.1(b) conveys the same information in the standard representations of PMO or Orbital Interaction theory. The two interacting but unperturbed systems are shown on the left and the right, and the system after the interaction is turned on is shown between them. Our task is to find out what the system looks like after the interaction. Let us start with the two unperturbed orbitals and seek the best MOs that can be constructed from them. Thus,

$$\phi(1) = c_A \chi_A(1) + c_B \chi_B(1) \tag{3.6}$$

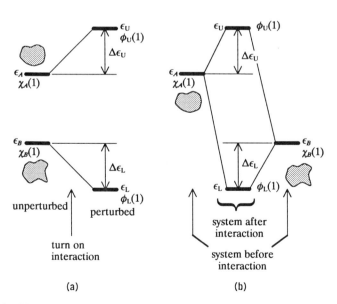

Figure 3.1. (a) "Perturbation" view of interaction of two orbitals; (b) standard interaction diagram.

60 ORBITAL INTERACTION THEORY

The energy of this orbital is given by the expectation value

$$\epsilon = \frac{\int \phi(1) h^{\text{eff}}(1) \phi(1) \, d\tau_1}{\int |\phi(1)|^2 \, d\tau_1}$$

$$= \frac{\int (c_A \chi_A(1) + c_B \chi_B(1)) h^{\text{eff}}(1)(c_A \chi_A(1) + c_B \chi_B(1)) \, d\tau_1}{\int (c_A \chi_A(1) + c_B \chi_B(1))^2 \, d\tau_1}$$

$$= \frac{c_A^2 \int \chi_A(1) h^{\text{eff}}(1) \chi_A(1) \, d\tau_1 + c_B^2 \int \chi_B(1) h^{\text{eff}}(1) \chi_B(1) \, d\tau_1}{c_A^2 \int |\chi_A(1)|^2 \, d\tau_1 + c_B^2 \int |\chi_B(1)|^2 \, d\tau_1} \cdots$$

$$\cdots \frac{+ 2 c_A c_B \int \chi_A(1) h^{\text{eff}}(1) \chi_B(1) \, d\tau_1}{+ 2 c_A c_B \int \chi_A(1) \chi_B(1) \, d\tau_1} \tag{3.7}$$

$$= \frac{c_A^2 h_{AA} + c_B^2 h_{BB} + 2 c_A c_B h_{AB}}{c_A^2 + c_B^2 + 2 c_A c_B S_{AB}} \tag{3.8}$$

In proceeding from equations (3.7) to (3.8) we notice that the integrals

$$h_{AA} = \int \chi_A(1) h^{\text{eff}}(1) \chi_A(1) \, d\tau_1 = \epsilon_A,$$

$$h_{BB} = \int \chi_B(1) h^{\text{eff}}(1) \chi_B(1) \, d\tau_1 = \epsilon_B \tag{3.9}$$

are just the energies of the electron *localized* to sites A and B, respectively. If sites A and B have identical orbitals, then $h_{AA} = h_{BB}$. In other words, the orbital energies are the same. The *interaction* integral, h_{AB},

$$h_{AB} = \int \chi_A(1) h^{\text{eff}}(1) \chi_B(1) \, d\tau_1 \tag{3.10}$$

provides a measure of the interference of the electron waves in energy units—in other words, how much energy may be gained if the wavefunctions at sites A and B overlap in a constructive manner (constructive interference or *bonding*) or how much energy must be added to the system to accommodate overlap in a destructive manner (destructive interference or *antibonding*). The wavefunctions at A and B are assumed to be normalized. In terms of overlap integrals,

$$S_{AA} = \int |\chi_A(1)|^2 \, d\tau_1 = 1, \qquad S_{BB} = \int |\chi_B(1)|^2 \, d\tau_1 = 1,$$

$$S_{AB} = \int \chi_A(1)\chi_B(1) \, d\tau_1 \qquad (3.11)$$

Notice that the overlap integral, S_{AB}, will depend on the position and orientation of the orbitals at sites A and B, as does the interaction integral, h_{AB}. The minimum energy solution is found by the variational method which we have used twice in the previous chapter. Equation (3.8) is differentiated with respect to c_A and with respect to c_B, resulting in two linear equations which can be solved. Thus,

$$\frac{\partial \epsilon}{\partial c_A} = \frac{\partial}{\partial c_A} \left[\frac{c_A^2 \epsilon_A + c_B^2 \epsilon_B + 2c_A c_B h_{AB}}{c_A^2 + c_B^2 + 2c_A c_B S_{AB}} \right] = 0 \qquad (3.12)$$

$$2 \frac{(\epsilon_A - \epsilon)c_A + (h_{AB} - S_{AB}\epsilon)c_B}{c_A^2 + c_B^2 + 2c_A c_B S_{AB}} = 0$$

$$(\epsilon_A - \epsilon)c_A + (h_{AB} - S_{AB}\epsilon)c_B = 0 \qquad (3.13)$$

Similarly,

$$(h_{AB} - S_{AB}\epsilon)c_A + (\epsilon_B - \epsilon)c_B = 0 \qquad (3.14)$$

A solution of equations (3.13) and (3.14) may be found by setting the determinant of the coefficients of the c's equal to zero. Thus

$$\begin{vmatrix} (\epsilon_A - \epsilon) & (h_{AB} - S_{AB}\epsilon) \\ (h_{AB} - S_{AB}\epsilon) & (\epsilon_B - \epsilon) \end{vmatrix} = 0 \qquad (3.15)$$

Or

$$(\epsilon_A - \epsilon)(\epsilon_B - \epsilon) - (h_{AB} - S_{AB}\epsilon)^2 = 0 \qquad (3.16)$$

Expanding equation (3.16) yields

$$(1 - S_{AB}^2)\epsilon^2 + [2S_{AB}h_{AB} - (\epsilon_A + \epsilon_B)]\epsilon + \epsilon_A\epsilon_B - h_{AB}^2 = 0 \qquad (3.17)$$

The two roots of equation (3.17) may be abstracted by routine application of the quadratic formula. Before we do that, let us examine the results from three simplified special cases.

Case 1: $\varepsilon_A = \varepsilon_B$, $S_{AB} = 0$

The simplest case arises when the orbitals at the two sites are identical or accidentally have the same energy. The overlap integral will not in general be equal to zero but often is very small and may be *approximated* as zero for mathematical convenience. The solution to the entitled case is easily found from equation (3.16) which becomes

$$(\epsilon_A - \epsilon)^2 - h_{AB}^2 = 0 \tag{3.18}$$

for which the two solutions are

$$\epsilon = \epsilon_A \pm h_{AB} \tag{3.19}$$

In order to determine which of the two roots is lower in energy, one needs to know whether the wavefunctions of A and B suffer constructive interference (positive overlap) or destructive interference (negative overlap). Let us assume that the wavefunctions are in phase (positive overlap). Then h_{AB} will be negative since $h^{\text{eff}}(1)$ is negative, and the upper and lower roots, ϵ_U and ϵ_L, respectively, are

$$\epsilon_U = \epsilon_A - h_{AB} \tag{3.20}$$

$$\epsilon_L = \epsilon_A + h_{AB} \tag{3.21}$$

Substitution of the lower energy solution, equation (3.21) into equation (3.13), and setting S_{AB} to zero, yields the relationship between the coefficients for the lower energy wavefunction (MO), $\phi_L(1)$, namely $c_A = c_B$. Requiring that the wavefunction be normalized,

$$c_A^2 + c_B^2 + 2c_A c_B S_{AB} = 1 \tag{3.22}$$

yields the result (with zero overlap)

$$\phi_L(1) = (2)^{-1/2}[\chi_A(1) + \chi_B(1)] \tag{3.23}$$

Using the upper root, equation (3.20), in a similar manner with equations (3.14) and (3.22), yields

$$\phi_U(1) = (2)^{-1/2}[\chi_A(1) - \chi_B(1)] \tag{3.24}$$

Notice that with Case 1 assumptions, the energies after the interaction are raised and lowered by equal amounts, h_{AB}, relative to the "unperturbed" energy, $\epsilon_A\ (=\epsilon_B)$, and the resulting wavefunctions are an *equal* admixture of the "unperturbed" wavefunctions.

Case 2: $\varepsilon_A = \varepsilon_B$, $S_{AB} > 0$, $S_{AB} \ll 1$

This is a more realistic case since, if the two orbitals can interact, the overlap will not be zero. However, the overlap may be very small if the two orbitals are far apart, and in this case, a mathematical simplification ensues, as seen below.

In the case that the overlap is not zero (assumed positive), the solutions may again conveniently be determined from equation (3.16). Thus

$$(\epsilon_A - \epsilon)^2 - (h_{AB} - S_{AB}\epsilon)^2 = 0 \tag{3.25}$$

for which the two solutions are

$$\epsilon_U = \frac{\epsilon_A - h_{AB}}{1 - S_{AB}} \approx \epsilon_A - h_{AB} + (\epsilon_A - h_{AB})S_{AB} \tag{3.26}$$

$$\epsilon_L = \frac{\epsilon_A + h_{AB}}{1 + S_{AB}} \approx \epsilon_A + h_{AB} - (\epsilon_A + h_{AB})S_{AB} \tag{3.27}$$

The second part of equations (3.26) and (3.27) arises from the expansion of $1/(1 \pm S_{AB})$ which converges rapidly if $S_{AB} \ll 1$. Thus it is apparent that inclusion of the overlap term reduces the amount of stabilization of the bonding combination of orbitals and also reduces the amount of destabilization of the antibonding combination. However, the reduction of the amount of destabilization, $|(\epsilon_A - h_{AB})S_{AB}|$, is *less* than the reduction in the amount of stabilization, $|(\epsilon_A + h_{AB})S_{AB}|$, leaving the antibonding combination somewhat higher in energy than the bonding combination relative to the reference unperturbed energy. The situation is shown in Figure 3.2. The last result, which is a consequence of overlap, is fundamental to the application of PMO theory. The precise amount of destabilization for the degenerate case may be calculated by

$$\Delta\epsilon = \epsilon_u + \epsilon_L - 2\epsilon_A$$

$$= \frac{\epsilon_A - h_{AB}}{1 - S_{AB}} + \frac{\epsilon_A + h_{AB}}{1 + S_{AB}} - 2\epsilon_A$$

$$= \frac{2}{1 - S_{AB}^2}(-h_{AB}S_{AB} + \epsilon_A S_{AB}^2) \tag{3.28}$$

The energy change after interaction depends on two terms according to equation (3.28). The first term, involving $-h_{AB}S_{AB}$, is positive and contributes to the destabilization. The second term which is proportional to the energy of the interacting orbitals, is negative since ϵ_A ($= h_{AA}$) is negative, and leads to a stabilization. The first term is expected to dominate since $S_{AB} < 1$ and for valence orbitals, the intrinsic interaction energy, h_{AB}, is of the same order of

64 ORBITAL INTERACTION THEORY

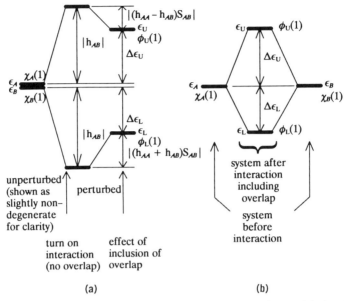

Figure 3.2. Case 2: (a) "Perturbation" view of interaction of two orbitals with inclusion of overlap; (b) standard interaction diagram showing relative destabilization of the upper orbital.

magnitude (though smaller) as the orbital energy, ϵ_A. We will return to this point in our Case 4 study.

It can be easily verified by insertion of the precise value of ϵ_L and ϵ_U, equations (3.27) and (3.26), respectively, that the orbitals, $\phi_L(1)$ and $\phi_U(1)$, again involve an equal admixture of the $\chi_A(1)$ and $\chi_B(1)$, in-phase and out-of-phase, respectively, as in Case 1, but the coefficients are slightly different, namely

$$\phi_L(1) = [2(1 + S_{AB})]^{-1/2}[\chi_A(1) + \chi_B(1)] \quad (3.29)$$

$$\phi_U(1) = [2(1 - S_{AB})]^{-1/2}[\chi_A(1) - \chi_B(1)] \quad (3.30)$$

Case 3: $\epsilon_A > \epsilon_B$, $S_{AB} = 0$

For this case study, we again adopt the "zero overlap" approximation, but consider the nondegenerate case, assuming that $\epsilon_A > \epsilon_B$. The secular equation, from equation (3.17) becomes

$$\epsilon^2 - (\epsilon_A + \epsilon_B)\epsilon + \epsilon_A\epsilon_B - h_{AB}^2 = 0 \quad (3.31)$$

Application of the quadratic formula and some algebra leads to

$$\epsilon = \frac{1}{2}(\epsilon_A + \epsilon_B) \pm \frac{1}{2}\sqrt{(\epsilon_A + \epsilon_B)^2 - 4\epsilon_A\epsilon_B + 4h_{AB}^2}$$

$$= \frac{1}{2}(\epsilon_A + \epsilon_B) \pm \frac{1}{2}\sqrt{(\epsilon_A - \epsilon_B)^2 + 4h_{AB}^2}$$

$$= \frac{1}{2}(\epsilon_A + \epsilon_B) \pm \frac{1}{2}(\epsilon_A - \epsilon_B)\sqrt{1 + \frac{4h_{AB}^2}{(\epsilon_A - \epsilon_B)^2}} \quad (3.32)$$

If $4h_{AB}^2 \ll (\epsilon_A - \epsilon_B)^2$, i.e., the interaction energy is much less than the energy difference between the interacting orbitals, one can simplify equation (3.32) by expansion of the square root.

$$\epsilon \approx \frac{1}{2}(\epsilon_A + \epsilon_B) \pm \frac{1}{2}(\epsilon_A - \epsilon_B)\left(1 + \frac{1}{2} \cdot \frac{4h_{AB}^2}{(\epsilon_A - \epsilon_B)^2}\right) \quad (3.33)$$

Notice that this situation is really only valid for separated systems. If the two orbitals are spatially close, say in the same molecule, it is unlikely that the last approximation would be appropriate. This situation is discussed further below.

In equation (3.33), the lower root is obtained by taking the $(-)$ sign. Thus

$$\epsilon_L \approx \frac{1}{2}(\epsilon_A + \epsilon_B) - \frac{1}{2}(\epsilon_A - \epsilon_B)\left(1 + \frac{1}{2} \cdot \frac{4h_{AB}^2}{(\epsilon_A - \epsilon_B)^2}\right) \quad (3.34)$$

or

$$\epsilon_L \approx \epsilon_B - \frac{h_{AB}^2}{(\epsilon_A - \epsilon_B)} \quad (3.35)$$

Similarly,

$$\epsilon_U \approx \epsilon_A + \frac{h_{AB}^2}{(\epsilon_A - \epsilon_B)} \quad (3.36)$$

Evidently, if the unperturbed orbitals are not of the same energy, and if the energy difference is large compared to the interaction energy, the effect of the interaction is to lower the lower orbital further by an amount proportional to the square of the interaction energy and inversely proportional to the energy difference. The upper orbital is raised by the same amount. One should suspect, correctly, that if overlap were to be taken into account, the amount of raising would be greater than the amount of lowering. Substitution of equations (3.35) and (3.36) separately into equations (3.13) and (3.14), yields the molecular orbitals. Specifically, insertion of equation (3.35) into equation (3.14), with $S_{AB} = 0$, yields

$$h_{AB}c_{AL} + (\epsilon_B - \epsilon_L)c_{BL} = 0, \quad (3.37)$$

or

$$h_{AB}c_{AL} + \left(\epsilon_B - \epsilon_B + \frac{h_{AB}^2}{\epsilon_A - \epsilon_B}\right)c_{BL} = 0 \quad (3.38)$$

Therefore,

$$c_{AL} = -\frac{h_{AB}}{\epsilon_A - \epsilon_B}c_{BL} \quad (3.39)$$

from which

$$\phi_L(1) = c_{BL}\left[-\frac{h_{AB}}{\epsilon_A - \epsilon_B}\chi_A(1) + \chi_B(1)\right] \quad (3.40)$$

Similarly, insertion of equation (3.36) into equation (3.13), with $S_{AB} = 0$, yields

$$\phi_U(1) = c_{AU}\left[\chi_A(1) + \frac{h_{AB}}{\epsilon_A - \epsilon_B}\chi_B(1)\right] \quad (3.41)$$

The factors c_{BL} and c_{AU} are determined by normalization, but this is not important for our purposes. Notice that the factor $h_{AB}/(\epsilon_A - \epsilon_B)$ is *negative* and less than 1 in absolute magnitude. Thus the lower MO is derived from the interacting orbital which has lower energy by the admixture of a small amount of the interacting orbital of higher energy. The signs of the coefficients of the two interacting orbitals are the same, i.e., $-h_{AB}/(\epsilon_A - \epsilon_B) > 0$. The upper MO is derived from the interacting orbital which has higher energy by the admixture of a small amount of the interacting orbital of lower energy. The signs of the coefficients of the two interacting orbitals in the antibonding MO are different. (Note: The assertion about the relative signs of the coefficients is contingent on the fact that the overlap of the two orbitals is positive.) The results from the Case 3 study are depicted in Figure 3.3. Some of the results obtained above have been reviewed and applied by Hoffmann in an investigation of orbital interactions through space and through bonds [48] and are used extensively in the excellent book by Albright, Burdett, and Whangbo [49].

Case 4: $\varepsilon_A > \varepsilon_B$, $S_{AB} > 0$

As stated earlier, the exact algebraic solution of equation (3.17) is readily obtained but does not add to the concepts already derived. The orbital energies and wavefunctions will resemble equations (3.35), (3.36), and (3.40), (3.41), respectively. The consequences of nonzero overlap will again prove to be that $|\Delta\epsilon_U| > |\Delta\epsilon_L|$. The dependence of the net destabilization, $|\Delta\epsilon_U| - |\Delta\epsilon_L|$, was discussed in the Case 2 study [equation (3.28)] and was found to depend on

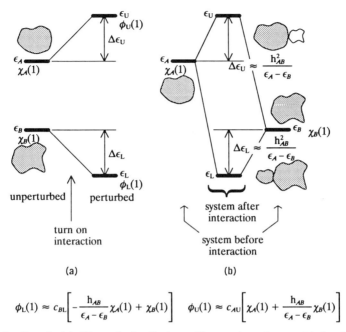

Figure 3.3. Case 3: **(a)** "Perturbation" view of interaction of two orbitals of unequal energy; **(b)** standard interaction diagram.

two terms, a dominant destabilizing term and a stabilizing term which is proportional to the energy of the degenerate pair. The more general expression is easily derived. The roots of equation (3.17) by the quadratic formula are:

$$\epsilon_U = \frac{-b}{2a} + \frac{1}{2a}\sqrt{b^2 - 4ac} \quad \text{and} \quad \epsilon_L = \frac{-b}{2a} - \frac{1}{2a}\sqrt{b^2 - 4ac} \quad (3.42)$$

where

$$a = (1 - S_{AB}^2), \quad b = 2S_{AB}h_{AB} - (\epsilon_A\epsilon_B), \quad \text{and} \quad c = \epsilon_A\epsilon_B - h_{AB}^2 \quad (3.43)$$

Thus the net destabilization is

$$|\Delta\epsilon_U| - |\Delta\epsilon_L| = \epsilon_U + \epsilon_L - (\epsilon_A + \epsilon_B)$$

$$= \frac{-b}{a} - (\epsilon_A + \epsilon_B)$$

$$= \frac{S_{AB}}{(1 - S_{AB}^2)}[-2h_{AB} + (\epsilon_A + \epsilon_B)S_{AB}] \quad (3.44)$$

As in the Case 2 study, two terms of opposite sign are involved. The sum of orbital energies before and after interaction is the same when the quantity in brackets [equation (3.44)] is zero, i.e.,

$$h_{AB} = \frac{(\epsilon_A + \epsilon_B)}{2} S_{AB} \qquad (3.45)$$

Equation (3.45) resembles the formula [equation (3.46)] adopted by Wolfsberg and Helmholtz [50] for the interaction matrix elements. This form, with the empirical factor, $k = 1.75$, was adopted by Hoffmann in his derivation of extended Hückel theory [51], namely

$$h_{AB} = k \frac{(h_{AA} + h_{BB})}{2} S_{AB}, \quad k = 1.75 \quad (\text{recall } h_{AA} = \epsilon_A \text{ and } h_{BB} = \epsilon_B)$$

$$(3.46)$$

That k should be between 1 and 2, and consequently that the sum of orbital energies after interaction should be greater than before the interaction, was offered theoretical justification by Woolley [52]. Substitution of equation (3.46) into equation (3.44), yields the result

$$|\Delta\epsilon_U| - |\Delta\epsilon_L| = \frac{-|k'|}{(1 - S_{AB}^2)} (\epsilon_A + \epsilon_B) S_{AB}^2 \qquad (k' \approx 0.75) \qquad (3.47)$$

In summary, at least for the frontier orbitals where equation (3.47) is expected to be valid, a net destabilization depending in a complex manner on the square of the overlap integral and proportional to the sum of unperturbed orbital energies ensues from the interaction.

THE EFFECT OF OVERLAP

The magnitude of the overlap of the interacting orbitals has an influence on the appearance of the interaction diagram because it affects the magnitude of the interaction integral [equation (3.46)], the difference between stabilization and destabilization [equation (3.47)], and the polarization of the resulting molecular orbitals, ϕ_L and ϕ_U. It is convenient to categorize the overlap dependence into two broad regimes, the regime of large overlap, which is usually the case when the interactions of interest are between adjacent regions of the same molecule, i.e., intramolecular, and the limit of small overlap, which is almost always the case when one is considering intermolecular interactions.

The magnitudes of $\Delta\epsilon_U$ and $\Delta\epsilon_L$ are directly related to the magnitude of h_{AB}; the larger the overlap, the larger is h_{AB}, and the larger are $\Delta\epsilon_U$ and $\Delta\epsilon_L$. Thus the energy changes which ensue from intramolecular interactions are usually much larger than those which originate intermolecularly. Larger overlap also

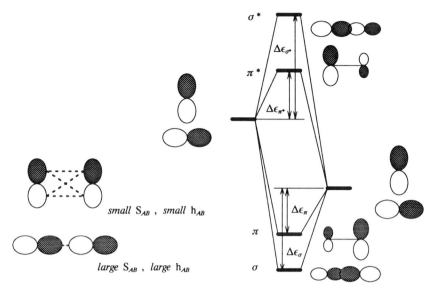

Figure 3.4. A σ- and π-type interaction between p orbitals: The σ orbitals are further apart and *less* polarized than the π orbitals. Also, $|\Delta\epsilon_{\sigma*} - \Delta\epsilon_\sigma| > |\Delta\epsilon_{\pi*} - \Delta\epsilon_\pi|$.

means that it is more difficult to meet the criterion for the results displayed in equations (3.35) and (3.36), namely that the ratio, $|h_{AB}/(\epsilon_A - \epsilon_B)|$, is much less than unity. In other words, $\Delta\epsilon_U$ and $\Delta\epsilon_L$ are more closely proportional to h_{AB} rather than its square, and will not exhibit the inverse dependence on the difference of the initial orbital energies. For *intra*molecular considerations, the orbital energy difference becomes less important than the magnitude of h_{AB} itself whereas for *inter*molecular interactions, orbital energy differences are more important, and the inverse dependence on the energy difference is expected. The polarization of the resulting orbitals is also affected by the overlap, although this is more difficult to demonstrate algebraically and we do not do so here. We simply state the results: for a given energy difference (not equal to zero), the larger the overlap, the less the polarization of ϕ_L and ϕ_U. Less polarization means greater mixing of the initial orbitals. A distinction between σ and π orbitals may be made on this basis. Since the overlap of two p orbitals in a π fashion is smaller than in σ fashion, one expects more polarization in the π orbitals than in the σ orbitals. The π and π^* orbitals are closer together than the σ and σ^* orbitals. The σ^* orbital is more antibonding relative to the σ orbital than the π^* orbital is relative to the π orbital. All three consequences are depicted in Figure 3.4.

THE RELATIONSHIP TO PERTURBATION THEORY

It should be apparent that the expressions for the wavefunctions after interaction [equations (3.40) and (3.41)] are equivalent to the Rayleigh–Schrödinger Perturbation Theory (RSPT) result for the perturbed wavefunction correct to first order [equation (2.114)]. Similarly, the parallel between the MO orbital energies [equations (3.35) and (3.36)], and the RSPT energy correct to second order [equation (2.115)] is obvious. The "missing" first-order correction emphasizes the correspondence of the first-order corrected wavefunction and the second-order corrected energy. Note that equations (3.35), (3.36), (3.40), and (3.41) are valid under the same conditions required for the application of perturbation theory, namely that the perturbation be weak compared to energy differences.

GENERALIZATIONS FOR INTERMOLECULAR INTERACTIONS

Equations (3.35), (3.36), (3.40), and (3.41) are valid under the condition that the perturbation be weak compared to energy differences, i.e., $|h_{AB}/(\epsilon_A - \epsilon_B)| \ll 1$. Case 1 and Case 2 studies revealed the situation in the degenerate case. We now interpolate the intermediate situation, remembering that overlap will not be zero except when dictated to be so by local symmetry. However, overlap will always be small when dealing with interactions between molecules. Refer to Figures 3.1–3.5 for the defined quantities.

Generalization 1: $\Delta\varepsilon_L \approx \Delta\varepsilon_U$

The destabilization of the upper MO relative to a higher energy interacting orbital is approximately the same as the stabilization of the lower MO relative to a lower energy interacting orbital.

Generalization 2: $|\Delta\varepsilon_L| < |\Delta\varepsilon_U|$

The destabilization of the upper MO relative to a higher energy interacting orbital is always greater than the stabilization of the lower MO relative to a lower energy interacting orbital. This is a direct consequence of nonzero overlap [equation (3.47)].

Generalization 3: $(|\Delta\varepsilon_U| - |\Delta\varepsilon_L|) < |\Delta\varepsilon_L|, |\Delta\varepsilon_U|$

The magnitude of the *difference* between destabilization and stabilization is small compared to the actual magnitudes of the stabilization and destabilization.

Generalization 4: $h_{AB} \approx k(\varepsilon_A + \varepsilon_B)S_{AB}$, (k a Positive Constant, S_{AB} Assumed Positive)

The interaction matrix element is not precisely proportional to the overlap integral but the behavior with respect to distance and symmetry is essentially the same. In other words, h_{AB} will be zero by symmetry when S_{AB} is zero by symmetry, but not otherwise. Also, h_{AB} decreases in magnitude as a function of increasing separation in much the same way as S_{AB}. Thus, **two orbitals will not interact if they behave differently toward local elements of symmetry.**

The proportionality to the sum of unperturbed orbital energies [equation (3.46)] may be misleading. Remember that the orbital energies are negative quantities. Therefore, the *higher* the orbital energy, the *smaller* the absolute magnitude of the intrinsic interaction.

Generalization 5: $\Delta\varepsilon_L \approx \Delta\varepsilon_U \approx (h_{AB}^2/\varepsilon_A - \varepsilon_B)$

The energy raising and lowering depend directly on the square of the interaction matrix element (the square of the overlap matrix element) and inversely on the energy separation of the interacting orbitals. This relationship is precisely true only in the limit of very small interaction. The energy raising and lowering are *maximum* when the energy difference is zero, in which case they have the value approximately h_{AB}.

Generalization 6: $\phi_L \approx \chi_b + d\chi_A$; $\quad \phi^U \approx \chi_a - d\chi_B$
$d = -(h_{AB}/\varepsilon_A - \varepsilon_B) \quad d < 1$

The lower MO is polarized toward the lower interacting orbital and is the in-phase (bonding combination) of the two. The upper MO is polarized toward the higher interacting orbital and is the out-of-phase (antibonding combination) of the two. In the limit of weak interaction, the amount of mixing of the minor component is approximately proportional to the interaction matrix element and inversely proportional to the energy difference between the two interacting orbitals. The orbitals mix equally ($d = 1$) when the orbital energy difference is zero. The above generalizations are summarized in Figure 3.5.

ENERGY AND CHARGE DISTRIBUTION CHANGES FROM ORBITAL INTERACTION

The energy change which takes place upon the mixing of two orbitals depends on the number and distribution of the electrons. The energy change after interaction is

$$\Delta E = n_U \epsilon_U + n_L \epsilon_L - n_A \epsilon_A - n_B \epsilon_B \qquad (3.48)$$

where n_I gives the number of electrons in orbital I, and $(n_U + n_L) = (n_A + n_B)$. If neither orbital is occupied, there is no energy change. There is no need to calculate energy changes quantitatively. Inspection of the orbital interaction diagrams will provide the qualitative assessment of the energy change, i.e., large or small, stabilizing or destabilizing. We consider separately the situations which may arise as a function of the number of electrons.

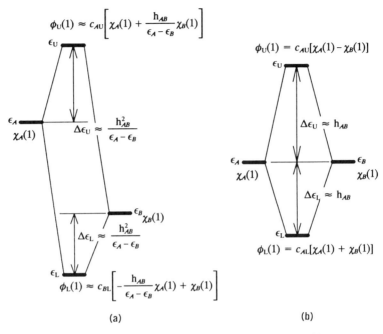

Figure 3.5. Orbital interaction diagrams: **(a)** Nondegenerate case; **(b)** degenerate case.

The Four-Electron–Two Orbital Interaction

With four electrons, the bonding and antibonding molecular orbitals are both filled (Figure 3.6). The interaction will be repulsive and the system will adjust in such a way as to minimize the interaction. Closed shell molecules will tend to repel each other. The common term is *Steric Interaction*. The net interactions between filled orbitals of two molecules separated by more than the van der Waals contact distance will be weak; the bonding and antibonding interactions which may individually be large, almost cancel.

Filled orbitals which cannot physically separate because they are part of the same molecule, may have substantially larger overlap than the intermolecular case, and the destabilizing interaction, $\Delta\epsilon_U$, may be much larger than the stabilizing interaction, $\Delta\epsilon_L$. The net repulsion will lead to conformational changes so as to minimize the repulsive interactions. If the interaction cannot be avoided by conformational change, then as the result of the interaction, a pair of electrons is raised in energy. The system has a lower ionization potential, is more easily oxidized, and is more basic in the Lewis sense. All of these features are observed in the experimental properties of norbornadiene (Figure 3.7). The in- and out-of-phase interactions of the two occupied π-bonding orbitals may be described by an interaction diagram shown in Figure 3.5(b). Norbornadiene has two ionization potentials, 9.55 eV and 8.69 eV [53], which are above and below the ionization potential of (Z)-2-butene (9.12 ev [54]), cyclohexene (9.11 eV [55]), or norbornene (8.95 eV [55]) as expected on the basis of the orbital

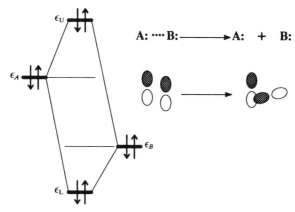

Figure 3.6. The repulsive four-electron–two-orbital interaction: Closed shell species will separate; filled orbitals will tend to align in each other's nodes.

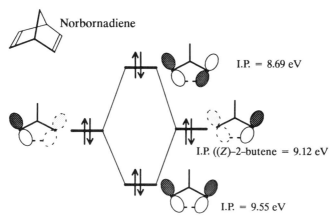

Figure 3.7. Four-electron–two-orbital interaction diagram for norbornadiene and its ionization potentials.

interaction diagram in Figure 3.7. The consequences of intramolecular interaction of σ-bonding orbitals is further discussed in Chapter 4.

The Three-Electron–Two-Orbital Interaction

With three electrons one is dealing with a free radical. The bonding MO is doubly occupied and the antibonding molecular orbital has a single electron (Figure 3.8). The orbital interaction will be attractive and the system will adjust in such a way as to maximize the interaction. The situation depicted in Figure 3.8(a) may arise from the interaction of a free radical species with an inert solvent or matrix, or intramolecularly when a radical site is adjacent to an electron donor substituent. The energy of the odd electron is raised and the

74 ORBITAL INTERACTION THEORY

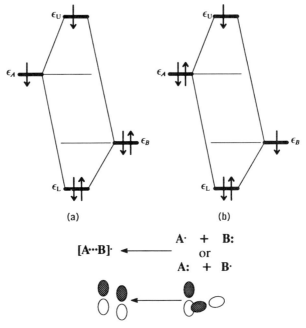

Figure 3.8. The three-electron–two-orbital interaction: **(a)** Odd electron in higher orbital—moderate stabilization; **(b)** the odd electron in the lower orbital—large stabilization and the possibility of electron transfer.

electron is delocalized to some extent. Such a radical would be more nucleophilic. The situation depicted in Figure 3.8(b) arises when a species with a relatively high ionization potential is ionized. A larger amount of energy is released upon complex formation and there is a net transfer of charge from A to B. If the complex is not stable and separates, the odd electron will stay with system A and an electron transfer from A to B will have taken place. Free radicals are discussed further in Chapter 7.

The Two-Electron–Two-Orbital Interaction

Maximum stabilization occurs in the two-electron–two-orbital interaction. A system will reorient itself to maximize such an interaction. In Figure 3.9 are depicted the two most common instances of this interaction. Figure 3.9(a) may depict the interaction of a Lewis base with a Lewis acid to form a dative bond, e.g., $NH_3 + BF_3 \rightarrow H_3N^+ - B^-F_3$, a hydrogen bond, or a tight complex, as between aryl systems and NO^+ [56]. The interaction is accompanied by charge transfer from B to A. Most chemical reactions are initiated by an interaction of this type between an occupied molecular orbital of one molecule and a virtual molecular orbital of another.

Figure 3.9(b) may also represent the interaction of a nonbonded ("lone

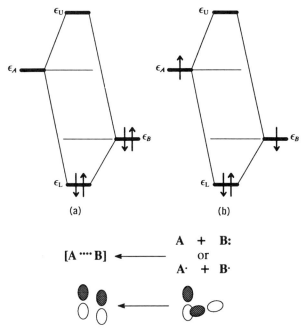

Figure 3.9. The two-electron–two-orbital interaction: **(a)** Both electrons in lower orbital—dative bond formation and electron transfer from B to A; **(b)** one electron in each orbital—large stabilization and covalent bond formation.

pair'') orbital with an adjacent polar π bond or σ bond [57]. If a polar π bond, one can explain stabilization of a carbanionic center by an "electron withdrawing" substituent (C=O), or the special properties of the amide group. If a polar σ bond, we have the origin of the anomeric effect. The interaction is accompanied by charge transfer from B to A, an increase in the ionization potential, and a decreased Lewis basicity and acidity. These consequences of the two-electron–two-orbital interaction are discussed in greater detail in subsequent chapters.

If Figure 3.9(a) represents bond formation between two molecules, the reverse process would correspond to heterolytic bond cleavage. In the gas phase, heterolytic bond cleavage is never observed. As we shall see, this interaction is the primary interaction between any pair of molecules whether it leads to bond formation or not. It is responsible for van der Waals attraction and hydrogen bonding.

Figure 3.9(b) portrays homolytic bond formation by the recombination of radicals and is accompanied by charge transfer from A to B. The radicals must be singlet-coupled. The interaction of triplet-coupled electron pairs is repulsive and does not lead to bond formation. The reverse process describes homolytic bond cleavage, and results in singlet-coupled free radicals.

76 ORBITAL INTERACTION THEORY

The One-Electron–Two-Orbital Interaction

The one-electron–two-orbital interaction, shown in Figure 3.10, again deals with a radical. The electron in the half-filled orbital is stabilized and delocalized. If it occurs intramolecularly, Figure 3.10 illustrates the stabilization of a radical site by an "electron-withdrawing" substituent. Such a radical would be more electrophilic. The intermolecular process may lead to complex formation. The energy of stabilization resulting from the interaction of the singly occupied MO of methyl radical with the empty σ^* antibonding orbital of HCl and HBr has been estimated theoretically as 2.53 kcal/mol and 0.67 kcal/mol, respectively [58]. Free radicals are discussed further in Chapter 7.

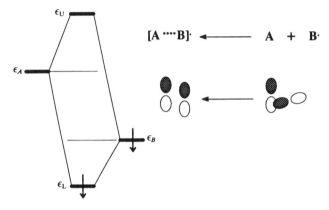

Figure 3.10. The one-electron–two-orbital interaction: The half-filled orbital is stabilized.

The Zero-Electron–Two-Orbital Interaction

The zero-electron–two-orbital interaction is shown in Figure 3.11. This interaction may seem trivial since no electrons are involved and therefore there is no change in the energy. This is far from the truth, however. Indeed, if the interaction of the empty orbitals occurs *inter*molecularly, Figure 3.11 illustrates the energy of the vacant MO which may determine the structure of a negatively charged complex, such as $(H_2O)_2^-$. More importantly from the perspective of organic chemistry, an *intra*molecular interaction between a site with a low-lying empty group orbital and a neighboring site that has a high empty orbital will result in a lower empty orbital, and therefore increased Lewis acidity overall. Furthermore, to the extent that ϕ_L is delocalized to the A site, acidification of the A site is also indicated. Such a diagram will describe the acidification of C—H bonds adjacent to a carbonyl group or the C—X bond of alkyl halides, as shown in some detail in Chapter 10.

INTERACTIONS BETWEEN MOLECULES: MANY ELECTRONS, MANY ORBITALS 77

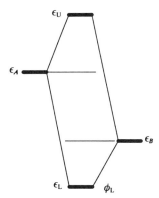

Figure 3.11. The zero-electron–two-orbital interaction: The system is more Lewis acidic, and some Lewis acidity is transferred to A.

INTERACTIONS BETWEEN MOLECULES: MANY ELECTRONS, MANY ORBITALS

Most molecules consist of many electrons and many orbitals, including "virtual" orbitals. When molecules interact, all MOs and all electrons are involved in principle. Before we discuss the possible relative importance of the various interactions, some general principles need to be stressed.

General Principles Governing the Magnitude of h_{AB} and S_{AB}

Recall that in the case of the hydrogen atom, the *radius* of an orbital is proportional to n^2, the *energy* is proportional to n^{-2}, and the *number of nodes* is equal to $n - 1$, where n is the principal quantum number. A similar effect occurs in molecules. The size (spatial extension) of MOs and the number of nodes generally increase as the energy increases. The lower the energy of the orbital, the more tightly bound is the electron, and the smaller is the average radius. The interaction matrix element, h_{AB}, depends approximately inversely on the average separation of the "centers of mass" of the MOs and directly on their spatial extension in the same way that the overlap integral S_{AB} does. Two orbitals will overlap more when the molecules bearing them are closer together. At a given separation, the extent of overlap depends on the size of the orbitals in a complex way. This is illustrated schematically in Figure 3.12. Very large orbitals overlap at large separations but the overlap does not reach a large absolute value, even at bonding distances. At the large separations, tight orbitals do not overlap very well. At shorter separations, the overlap of a small orbital with a large one may exceed the overlap of two large orbitals, and at shorter distances still, the largest overlap and therefore the largest interaction may be between two small orbitals.

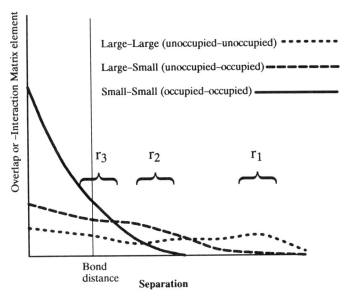

Figure 3.12. Schematic representation of overlap of MOs of different sizes as a function of separation.

Interactions of MOs

The interaction of two molecules is depicted in Figure 3.13. The possible orbital interactions are not shown in the form of an interaction diagram but are illustrated by dashed lines in Figure 3.13(a).

At large separations, r_1 (Figure 3.12), interactions between filled orbitals (**a, b, c**) will be negligible. The largest interactions will be between the virtual orbitals (**h**) but since there are no electrons, there are no energetic consequences. Notice that addition of an electron to the pair (or larger assembly) of neutral molecules would lead to maximum stabilization of the electron if interaction **h** is maximized. The solvated electron in liquid ammonia is an example of such a system. An electron solvated by two water molecules has been detected in the gas phase. The system has been studied theoretically and found to adopt a geometry expected on the basis of maximizing LUMO–LUMO interactions [59]; see also [60] for references to other studies of electrons attached to small clusters.

At intermediate separations, r_2 (Figure 3.12), interactions between occupied MOs (**a, b, c**) begin to become important but the four-electron–two-orbital interactions are only weakly repulsive and are outweighed by the attractive two-electron–two-orbital occupied-virtual interactions (**d, e, f, g**). All molecules encounter this attraction (van der Waals attraction [61]). In general, the larger the number of points of contact, the more favorable the interaction, hence the observed relationship between "surface area" and the strength of van der Waals forces. Nodal characteristics of the interacting occupied and

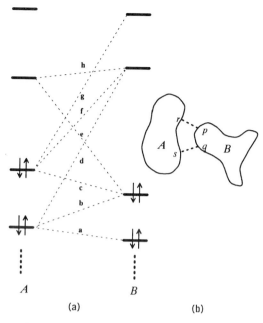

Figure 3.13. Interactions between two molecules: **(a)** Orbital interactions (see text); **(b)** schematic showing relatively few points of contact.

virtual orbitals determine the maximum magnitude of the interaction matrix element h_{AB}; the energy contribution is proportional to the square of h_{AB}. The energy contribution is also approximately proportional to the inverse of the energy separation. Both factors tend to make the highest occupied MO (HOMO)–lowest unoccupied MO (LUMO) interactions (**e, f**) dominant. The pair of MOs with the smaller HOMO–LUMO gap will tend to dominate the intermediate distance behavior of the molecule–molecule interaction, interaction **f** in Figure 3.13(a). The optimum distance of approach of two neutral molecules defines the van der Waals surface of the molecules. This occurs at the point where forces due to the attractive two-electron–two-orbital interactions are balanced by the repulsive forces of the four-electron–two-orbital interactions, approximately 1–1.5 Å from the outermost nuclei.

At bonding separations, r_3 (Figure 3.12), the four-electron–two-orbital interactions are strongly dominant, leading to the observed r^{-12} behavior at distances within the van der Waals separation. In order for two molecules in their ground states to undergo chemical reaction, there must be at least one exceptionally strong two-electron–two-orbital interaction that will permit close approach of the molecules. This interaction is necessarily accompanied by partial electron transfer.

Figure 3.13(b) attempts to emphasize that for large molecules, relatively few points of contact are possible, the other regions of the molecules being too far apart to interact. Two molecules which can interact favorably will tend to orient themselves to maximize the energy gain due to a two-electron–two-orbital interaction. This is the key to site specific reactions of electrophiles and nucleophiles and the "lock and key" mechanism of enzyme–substrate binding. Let us consider the interaction of MOs, ϕ_A and ϕ_B, on different molecules or in different parts of the same molecule, and explicitly take advantage of the approximate proportionality between h_{AB} and S_{AB}. Then, for a given energy separation, the interaction is proportional to h_{AB} (small energy separation) or $|h_{AB}|^2$ (large separation). Taking the former,

$$h_{AB} \approx kS_{AB}$$

$$= k \int \phi_A(1)\phi_B(1)\, d\tau_1$$

$$= k \sum_{a=1}^{n_A} \sum_{b=1}^{n_B} c_{aA} c_{bB} \int \chi_a(1)\chi_b(1)\, d\tau_1 \qquad (3.49)$$

$$\approx k c_{rA} c_{pB} \int \chi_r(1)\chi_p(1)\, d\tau_1 + k c_{sA} c_{qB} \int \chi_s(1)\chi_q(1)\, d\tau_1 \quad (3.50)$$

Equation (3.49) illustrates that ultimately the strength of the interaction between two *MOs* depends on the separation of the *AOs* of each MO (through the overlap integral) and on the magnitude of the product of the coefficients of those AOs. The last line [equation (3.50)] results if two atoms, r and s, of A can approach two atoms, p and q, respectively, of B, as shown by the dashed lines in Figure 3.13(b). This is the common situation for *pericyclic* reactions discussed in Chapter 12. In the vast majority of cases, the contacts between two molecules are even more restrictive. Usually, only two AOs, one from A and one from B, are close enough to overlap; then the strength of the interaction depends on the product of the coefficients of those orbitals and the degree of overlap that can be achieved. Conversely, two independent molecules which approach each other will orient themselves in such a way as to minimize the separation of the atomic orbitals which have the largest coefficients if the interaction is favorable, or to maximize it, if it is not.

ELECTROSTATIC EFFECTS

Whenever possible, we will attempt to identify the effects of interactions between relatively few frontier orbitals, which describe the distribution of relatively few of the electrons of a molecule. We effectively ignore the vast numbers of the rest of the electrons in the system, as well as the nuclear charges. In some cases this neglect is not justified and may lead to misleading results. When a molecule or parts of it are *charged*, coulombic interactions may dominate. The electrostatic energy of interaction of each pair of centers, i and j, with net charges, q_i and q_j, respectively, separated by a distance, R_{ij}, in a medium with dielectric constant, ϵ_m, has the form

$$E^{\text{electrostatic}} = \frac{q_i q_j}{R_{ij} \epsilon_m} \quad (3.51)$$

The energy required to separate an electron ($q_e = -1e$) from a proton ($q_p = +1e$) from a distance of 0.52917 Å (R_{ep} = radius of H atom) to infinity in vacuum ($\epsilon = 1$) is 13.6 eV or 319 kcal/mol (the ionization potential of H). This value is many times the strength of an average chemical bond. The energies involved are so large that one may generalize to say that in any medium of low dielectric constant, charged species in solution are always paired with oppositely charged species, surrounded by neutral dipolar species (if any are present), and in tight association with solvent molecules themselves.

Electrostatic effects cannot be ignored whenever a process takes place that changes the numbers of charged species. Heterolytic cleavage of σ bonds (see Chapter 4) would only be expected to take place in solvents of high dielectric constant, water being the best. Electron transfer between neutral species would not be observed in the gas phase except at contact separations where salt formation is possible. Reactions of charged species with neutral but dipolar species will, at least in part, be driven by electrostatic effects.

The energies of localized orbitals will be strongly affected by the presence of formal charges. The energies of all orbitals, filled or empty, will be lowered by the presence of a formal positive charge, and raised by a formal negative charge. The presence of highly polar bonds may significantly affect the energies of orbitals not directly involved in the bonding. For example, the energy of the nonbonded orbital on carbon would be expected to be lower in the CF_3 free radical than in the CH_3 free radical.

We will not attempt to quantify electrostatic effects but will need to be aware of possible influences as we consider our orbital interaction diagrams. Fortunately, the directions of electrostatic influences are easy to deduce from equation (3.51).

GROUP ORBITALS

The following is a summary of the local group molecular orbitals from which one may select the primary building blocks for interaction diagrams. The orbitals are classified according to the coordination number of the central atom. The local symmetry properties and relative energies are *independent* of the atomic number of the central atom itself. The group orbitals of a zero-coordinated atom are not just the set of four valence orbitals of the atom, namely s, p_x, p_y, and p_z, because we will assume that for the purpose of deducing orbital interactions, it is our intention to make σ and possibly π bonds to the uncoordinated atom. Because two orbitals of the atom, s and p_x, will each

0-Coordinated Atoms ····X

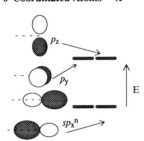

Notes:

(a) The higher energy orbitals are pure *p* orbitals with the same energy, i.e. they are degenerate.

(b) One of the *p* orbitals will *always* be perpendicular to the plane of an adjacent group, and be *antisymmetric* w.r.t reflection in the plane. The other (p_y) will lie in the plane and be *symmetric* w.r.t. reflection in the plane.

(c) the 'sp_x' hybrid orbitals are shown as degenerate for convenience. One will interact maximally with an adjacent group and form a σ bond. The other will not interact.

(d) in *neutral* atoms, X, the electron count for the heavy atoms is: B 3; C 4; N 5; O 6; F 7.

Figure 3.14. Group orbitals for zero-coordinated atoms.

interact in a σ fashion with a nearby atom, we mix these beforehand to form two new hybrid orbitals, one of which will interact maximally with the neighboring atom because it is pointed right at it, and another which will be polarized away from the second atom and therefore will interact minimally with it. The group orbitals of such a zero-coordinated atom are shown in Figure 3.14. Each element of coordination corresponds to the prior existence of a σ bond and the removal of one of the four orbitals and one of the valence electrons. In the figures that follow, the coordination is represented by an X—H σ bond. The group orbitals for a monocoordinated atom (excluding H) are shown in Figure 3.15. The existing bond is made from the equivalent of *one* of the hybridized valence orbitals of the atom (Figure 3.14), a hybrid orbital formed by mixing some proportion of s and p_x so as to maximize the energy associated with bond formation which is assumed to be in the x direction. This hybrid orbital is removed from the set of valence orbitals. The second combination of the s and p_x, mixed in inverse proportions, remains as one of the valence orbitals. It will be lower in energy than the pure p_y and p_z orbitals. In fact, the bond formation is accomplished using primarily the p_x orbital with minor admixture of the s orbital. The admixture (polarization) is maximum for B and C, much reduced in the case of N, and almost nonexistent for O, F, and any of the higher row elements. The remaining hybrid orbital, denoted sp_x^n in Figure 3.15, therefore is a true sp hybrid orbital for C, but has a much smaller amount of p character in the case of N ($n \leq 1$), and is almost a pure s orbital ($n \ll 1$) in the case of the other first and higher row atoms. The lack of hybridization is due to the increasing energy separation between the $2s$ and $2p$ orbitals as one proceeds across the first row, and due to the increased nodal characteristics of $3s$, $3p$, $4s$, $4p$, and higher principal quantum numbered orbitals of the higher row elements.

Dicoordinated atoms may have either of two distinct local geometries, linear or bent. In either case, only two valence orbitals remain and the number of valence electrons for the neutral atom is reduced by 2. The valence orbitals for

Monocoordinated Atoms —X

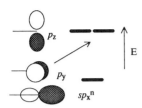

Notes:

(a) The higher energy orbitals are pure p orbitals with the same energy, i.e. they are degenerate.

(b) In a planar molecule, one p orbital (p_z) will *always* be perpendicular to the plane of the molecule, and be *antisymmetric* w.r.t reflection in the plane. The other (p_y) will lie in the plane and be *symmetric* w.r.t. reflection in the plane.

(c) the 'sp_x' hybrid orbital will be closest to the 'p' orbitals for B and C, and furthest for F. For O, and the halogens, there will be little 'p' character in this orbital.

(d) in *neutral* hydrides H–X, the electron count for the heavy atoms is: B 2; C 3; N 4; O 5; F 6.

Figure 3.15. Group orbitals for monocoordinated atoms.

dicoordinated atoms in linear and bent bonding arrangements are shown in Figures 3.16 and 3.17, respectively. Which coordination geometry an atom prefers is largely a function of constraints imposed by the coordinating atoms and the number of valence electrons. For the purpose of using the group orbitals in an interaction diagram, one should have prior knowledge of which geometry is appropriate in the particular case. As in the monocoordinated case, if the dicoordinated geometry is bent, then the same factors govern the extent of hybridization of the sp^n orbital (Figure 3.17): n will be close to 2 in the case of C, less than 2 for N, and much smaller for all other atoms. In other words, the orbital will be essentially an s orbital for all atoms higher in the periodic table than C and N.

Tricoordinated atoms may also have two distinct geometric configurations, planar or pyramidal. Only one valence orbital remains and it will be a pure p orbital in the planar case, hybridized in the pyramidal case. Both configurations for tricoordinated atoms are depicted in Figure 3.18.

In many instances, the interaction of a neighboring methylene group or methyl group influences the characteristics of a functional group. The appropriate group orbitals of $-CH_2-$ and $-CH_3$ are shown in Figures 3.19 and 3.20, respectively.

Dicoordinated Atoms —X—
Linear geometry

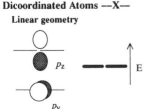

Notes:

(a) The two orbitals are pure p orbitals with the same energy, i.e. they are degenerate.

(b) In a planar molecule, one p orbital (p_z) will *always* be perpendicular to the plane of the molecule, and be *antisymmetric* w.r.t reflection in the plane. The other (p_y) will lie in the plane and be *symmetric* w.r.t. reflection in the plane.

(c) in *neutral* hydrides H_2X, the electron count for the heavy atoms is: B 1; C 2; N 3; O 4.

Figure 3.16. Group orbitals for a linear dicoordinated atom.

Bent Geometry

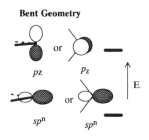

Notes:
(a) The two σ bonds define a local plane of symmetry. The p orbital (p_z) will *always* be perpendicular to the local plane of the molecule, and be *antisymmetric* w.r.t reflection in the plane.

(b) the 'sp' hybrid orbital will be closest to the 'p' orbitals for B and C, and furthest for F. For O, and the halogens, there will be little 'p' character in this orbital.

(c) The 'sp^n' orbital lies along the bisector of the bond angle, and is *symmetric* w.r.t. reflection in the local plane of symmetry defined by the bonds.

(d) in *neutral* hydrides H_2X, the electron count for the heavy atoms is: B 1; C 2; N 3; O 4.

Figure 3.17. Group orbitals for a dicoordinated atom with a nonlinear arrangement of σ bonds.

Tricoordinated Atoms

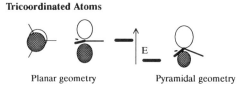

Planar geometry   Pyramidal geometry

Notes:
(a) In the planar geometry, the orbital is a pure p orbital and is *antisymmetric* w.r.t. reflection in the local plane of symmetry defined by the σ bonds.

(b) In the pyramidal geometry, the orbital is a sp^n hybrid which lies along the local three-fold axis of rotation.

(c) in *neutral* hydrides H_3X, the electron count for the heavy atoms is: B 0; C 1; N 2.

Figure 3.18. Group orbitals for a tricoordinated atom with planar (left) and pyramidal (right) arrangements of σ bonds.

The group orbitals of methylene as a substituent e.g. $-CH_2-$.

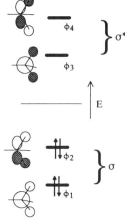

Notes:
(a) The bonding (σ) and antibonding (σ*) MOs form two group orbitals, one of which is symmetric with respect to the local plane of symmetry (ϕ_1, and ϕ_3,) and the other, a 'p-like' orbital which is antisymmetric to the local plane (ϕ_2 and ϕ_4)

(b) Of each group, the higher energy orbital is formed from pure p orbitals at the central atom.

(c) In a molecule of C_s symmetry, the 'p-like' orbitals e.g. ϕ_2 and ϕ_4, will *always* be perpendicular to the plane of the molecule, and be *antisymmetric* w.r.t reflection in the plane. The other, e.g. ϕ_1 and ϕ_3, will be *symmetric* w.r.t. reflection in the plane.

Figure 3.19. Group orbitals for a tetracoordinated atom as a substituent interacting through σ bonds. A methylene group is illustrated.

The methyl group as a substituent e.g. $-CH_3$

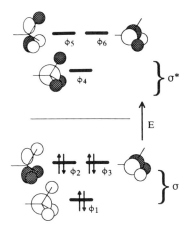

Notes:
(a) The bonding (σ) and antibonding (σ^*) MOs form two group orbitals, each with a unique axially symmetric orbital (ϕ_1, and ϕ_4,), and a pair of degenerate 'p-like' orbitals (ϕ_2,ϕ_3, and ϕ_5,ϕ_6)

(b) Of each group, the higher energy orbitals are formed from pure p orbitals at the central atom.

(c) In a molecule with C_s symmetry, one of the degenerate bonding and antibonding orbitals, e.g. ϕ_2 and ϕ_5, will *always* be perpendicular to the plane of the molecule, and be *antisymmetric* w.r.t reflection in the plane. The other, e.g. ϕ_3 and ϕ_6, will lie in the plane and be *symmetric* w.r.t. reflection in the plane.

Figure 3.20. Group orbitals for a tetracoordinated atom as a substituent with a tetrahedral arrangement of σ bonds. A methyl group is illustrated.

ASSUMPTIONS FOR APPLICATION OF QUALITATIVE MO THEORY

Gimarc has specified a set of rules for the application of qualitative MO theory [62], particularly for the determination of molecular structures and conformations, following the pioneering work of Walsh [63]. These are repeated here.

1. Electrons in molecules are completely delocalized and move in molecular orbitals which extend over the entire molecular framework.

2. For properties which can be explained by qualitative MO theory, only the valence electrons need be considered.

3. Satisfactory MOs can be formed from linear combinations of atomic orbitals. This is the well-known LCAO–MO approximation already discussed in Chapter 2.

4. The atoms which form the molecules of a particular series or class contribute the same kinds of valence AOs from which MOs can be constructed. Therefore the MOs for each series or type of molecular framework must be qualitatively similar and individually molecules differ primarily in the number of valence electrons occupying the common MO system.

5. The total energy of the molecule is the sum of the orbital energies of the valence electrons, or, more accurately, changes in the total energy parallel those in the orbital energy sum. Allen and co-workers [64] have shown that this assumption results from a fortuitous cancellation of energy terms.

6. No explicit consideration of electron–electron or nuclear–nuclear repulsions are included in this simple model.

86   ORBITAL INTERACTION THEORY

7. Molecular orbitals must be either symmetric or antisymmetric with respect to the symmetry operations of the molecule. These symmetry restrictions severely limit the number and kinds of AOs that combine in a particular MO. This makes the job of forming the MOs even easier, since most small molecules have high symmetry. To use qualitative MO theory, the only part of group theory necessary is a knowledge of symmetry classifications (see Chapter 1).
8. From properties of AOs available on component atoms it is possible to draw pictures of what the MOs must be like and to establish the approximate order of energies without calculations.
9. Changes in molecular shape which increase the in-phase overlap between two or more AOs in an MO tend to lower the energy of that MO. Conversely, changes in shape which decrease in-phase overlap or increase out-of-phase overlap among AOs in an MO tend to raise the energy of the MO. This can be called the overlap rule.
10. No *a priori* assumptions about orbital hybridization are needed. As we shall see, however, simple application of the idea of hybridization will come in handy from time to time.

AN EXAMPLE: THE CARBONYL GROUP

The carbonyl group is an important functional group in organic chemistry. It undergoes both nucleophilic and electrophilic additions and has a profound

influence on the properties of neighboring groups. In this book, a separate chapter (Chapter 8) is dedicated to it. At this point we will develop the complete bonding scheme for the carbonyl group using orbital interaction diagrams. These examples will serve to illustrate applications of the principles for construction of orbital interaction diagrams and also illustrate the wealth of information which may be deduced in the basis of the diagrams.

Construction of the Interaction Diagram

We begin the construction of the diagram by deciding on the groups which are interacting. If we were interested in only the π bonds, we note that the π bond is between a tricoordinated atom (the carbon atom) and a monocoordinated atom (the oxygen atom). However, we will not assume prior bonding between the carbon atom and the oxygen atom. Therefore, we will use a bent dicoordinated atom as one of our groups, and a zero-coordinated atom for the other. Refer to Figure 3.21 as the development of the interaction diagram pro-

AN EXAMPLE: THE CARBONYL GROUP

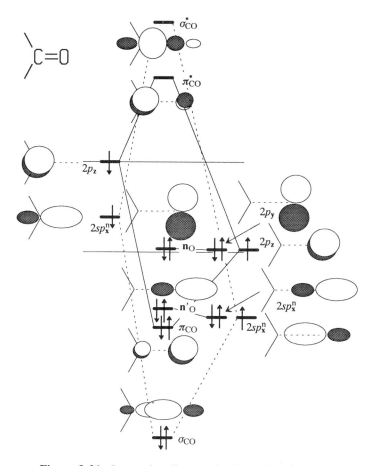

Figure 3.21. Interaction diagram for the carbonyl group.

gresses. We must next decide on the placement of the groups relative to each other. Since we are attempting to understand the bonding in the carbonyl group, we will position the dicoordinated C atom and the zero-coordinated O atom in the positions and orientations that they occupy in the finished carbonyl group, using our prior knowledge of the structure of the carbonyl group. We also chose to display the planar structure as lying in the plane of the paper (the xy plane), the most obvious choice. We know that the dicoordinated atom has two valence orbitals, a p orbital perpendicular to the plane of the bent σ bonds (a p_z orbital), and an sp^n hybrid orbital oriented along the bisector of the angle made by the σ bonds (the x direction) and polarized away from them (Figure 3.17). The zero-coordinated atom has 4 valence orbitals, two π orbitals oriented perpendicular to the direction of the incipient σ bond (p_y and p_z), and two sp^n hybrids oriented along the incipient σ bond but polarized in opposite directions (Figure 3.14).

We must then decide on the placement of these orbitals relative to each other

88 ORBITAL INTERACTION THEORY

on the same energy scale. We have no way of doing this precisely, but we note that oxygen is substantially more electronegative than C (Table 2.3 in Chapter 2) and so we place the oxygen group orbitals to *lower* energy than the carbon group orbitals. The placement on the energy scale (the vertical direction) is shown by short bold horizontal lines. The positioning in space, i.e., left or right side, should be consistent with the orientation of the two groups relative to each other in space. In the present case, we have chosen oxygen to be to the right.

Our choice of orientation of the σ framework (as determined by the bent coordination of the dicoordinated C atom and the intended C—O σ bond) in the xy plane (the plane of the page) determines the orientation of the valence group orbitals. These should be drawn on the diagram beside the appropriate energy levels, each superimposed on a separate sketch of the σ framework, and labeled fully, as shown in Figure 3.21. All of the orbitals should be drawn about the same size, since their "weights" are equal before the interaction (their coefficients are 1).

Next we must decide which orbitals of the right and left side interact, and approximately or relatively, how strong the interaction is. Consider first the sp^n hybrid orbital on the left-hand side. This will interact only with the sp^n hybrid orbital on the right-hand side which is pointed toward it (toward the left). The interaction is of σ type and because the two orbitals overlap strongly, a large amount of stabilization ($\Delta\epsilon_L$) and destabilization ($\Delta\epsilon_U$) ensue. Make $\Delta\epsilon_L$ at least as large as the energy separation of the orbitals on the left- and right-hand sides of your diagram (Figure 3.21), and draw a short bold horizontal line at this position midway between the left and right energy level lines. Make $\Delta\epsilon_U$ even larger (by about 20%) and draw it in place the same way. Connect these four levels by thin straight lines (shown as dashed lines in Figure 3.21). The molecular orbitals, ϕ_L and ϕ_U, which result from the interaction, must also be drawn. These should be positioned near the lines denoting their energies, and positioned on the σ framework in a manner consistent with the uninteracted orbitals. The nature of ϕ_L and ϕ_U are determined by the algebra of the first part of this chapter. The lower orbital, which we may label σ_{CO}, consists of the two orbitals interacting in phase, with a larger contribution from the oxygen side and a smaller one from the carbon side. The actual choice of phases does not matter, only that the phases where the orbitals overlap the most are the same. The phases are best shown by shading. The *sizes* of the orbitals drawn should be proportional to the coefficients expected for them. We do not know the exact values but we do know that the coefficients will be less than 1 as a result of the interaction, since the MO is now made up of two orbitals. Most importantly, we know that the MOs will be polarized. The lower orbital, σ_{CO}, is polarized toward O. The size of the contributing oxygen orbital should be drawn a little smaller than the uninteracted oxygen orbital, and the size of the in-phase carbon orbital smaller still. The upper orbital σ_{CO}^*, is polarized toward C. The size of the contributing carbon orbital should be drawn a little smaller than the uninteracted carbon orbital, and the size of the out-of-phase oxygen orbital smaller still.

Consider now the p_z orbital on the left. It can overlap and therefore interact, only with one orbital on the right, the p_z orbital of oxygen. The interaction is of π type. Because of the smaller overlap, the π-type interaction is intrinsically smaller than a σ-type interaction between similar orbitals. This must be borne in mind in positioning the π and π^* orbitals on the diagram relative to the already placed σ orbitals. As a rough guide, you may take $\Delta\epsilon_L$ for the π-type interaction to be about one half of $\Delta\epsilon_L$ for the σ-type interaction, similarly for the energies of destabilization. Having decided on the positions (Figure 3.21), draw the levels in, label them π_{CO} and π'_{CO}, and draw the connecting straight lines. Sketch in the MOs, bearing in mind the principles stated above in connection with the σ_{CO} and σ'_{CO} orbitals.

Two orbitals of the oxygen atom have not been involved in the interaction so far, the p_z orbital and the other sp^n hybrid. These are transferred to the middle of the diagram unchanged in energy or shape, although they should be redrawn in place, they are nonbonding orbitals of the carbonyl group, n'_O and n_O.

Lastly, one must occupy the MOs with the correct number of electrons. A neutral dicoordinated carbon atom has two valence electrons and a neutral uncoordinated oxygen atom has six, for a total of eight. Place electrons into the MOs two at a time. The HOMO is seen to be the higher of the nonbonding MOs, n_O, and the LUMO is π^*_{CO}.

Interpretation of the Interaction Diagram

Having carefully constructed the interaction diagram for the carbonyl group in Figure 3.21, we must now interpret it. We first make note of the frontier orbitals.

The Frontier Orbitals, HOMO and LUMO: The HOMO, n_O, is a nonbonding MO strongly localized to oxygen. It is predominantly a $2p$ orbital lying in the plane of the molecule. Within the (local) C_{2v} point group, it transforms as the b_2 irreducible representation. The LUMO, π^*_{CO}, is a π antibonding orbital strongly polarized toward the carbon atom, and relatively low in energy, at least compared to the position of the π^*_{CC} orbital of ethylene. We know this because the C—O $2p-2p$ π-type interaction matrix element, h_{AB}, should be similar in magnitude to the interaction between $2p$ of carbon atoms, but the oxygen and carbon $2p$ orbital energies are quite different, leading to a smaller $\Delta\epsilon_U$ for π^*_{CO}. Within the (local) C_{2v} point group, it transforms as the b_1 irreducible representation.

Bonding: Of the four occupied MOs, two are bonding and two are nonbonding, resulting in a net bond order of 2, i.e., a double bond. Both the σ and π bonds are polarized toward oxygen, the σ more than the π because of the smaller intrinsic overlap (π type overlap is smaller than σ type overlap).

Dipole Moment: A large bond dipole moment is expected, with the negative end at oxygen and the positive end at carbon.

Geometry: The combination of σ and π bonds forces coplanarity of the carbon atom, and the other two atoms attached to the carbon atom.

Basicity or Nucleophilicity: Since the HOMO is strongly localized to the oxygen atom (the coefficient of $2p_O$ is close to 1), and the oxygen atom is monocoordinated but uncharged, one should expect the Lowry–Bronsted basicity to be intermediate to that of alkoxides which are monocoordinated but charged, and neutral species with dicoordinated oxygen atom, such as water, alcohols or ethers. Attachment of a proton or other Lewis acid would occur at the oxygen atom, in the plane of the carbonyl group, and more or less perpendicular to the C—O bond. Since the donor orbital is nonbonding, addition of the Lewis acid to the oxygen is not accompanied by loss of either the σ or π bond (see Figure 8.2 in Chapter 8).

Acidity or Electrophilicity: The LUMO is a π-type orbital polarized toward the carbon atom (the coefficient of $2p_C$ is larger than the coefficient of $2p_O$). The LUMO is also relatively low in energy. The low energy and large coefficient at carbon both suggest that the carbonyl group should be reactive as a Lewis acid or electrophile, and that addition of nucleophiles (Lewis bases) would occur at the carbon atom, and from a direction approximately *perpendicular* to the plane of the carbonyl group. Attachment of a Lewis base means addition of 2 electrons to the π_{CO}^* orbital, and is accompanied by loss of the π bond (if both bonding and antibonding orbitals are equally occupied, there is no bond). See Figure 8.2 in Chapter 8.

Ionization: One cannot say anything about the magnitude of the ionization potential from just one interaction diagram. However, we can say with confidence that the lowest energy (strictly speaking, the ground electronic state) of the molecular ion, M^+, would correspond to a radical cation localized to the oxygen atom. One would expect to find mass spectral fragments which arise from the McLafferty rearrangement, and from rupture of the bond α to the carbonyl group.

UV Spectrum: The lowest energy electronic transition (HOMO–LUMO) is $n_O - \pi_{CO}^*$, leading to the nπ* state, which has A_2 symmetry in the local C_{2v} point group. The transition to this state is symmetry forbidden (it is quadrupole allowed but dipole forbidden) and will be expected to be weak.

Photochemistry: One might deduce that since the lowest electronic transition corresponds to transfer of an electron from oxygen atom to carbon atom, the nπ* state should have substantial diradical character, and should react also by a McLafferty-type rearrangement or α cleavage, as in the mass spectrometer. This is indeed the case. The photochemical α cleavage is called the Norrish Type I reaction, and the rearrangement is called the Norrish Type II reaction. Both are discussed in Chapter 14.

WHY DOES IT WORK, AND WHEN MIGHT IT NOT?

The application of orbital interaction theory to understand the electronic structures of molecules, as illustrated in the case of the carbonyl group above, is one of two conceptually distinct applications. Characteristics of the bonding and structure of the molecule were deduced from interactions of the group orbitals situated as they would be in the "finished" molecule. This application is in the regime of strong interactions and short distances, the regime in which the foundations of orbital interaction theory, the one-electron theory as exemplified by Hartree–Fock theory, have a firm footing and a demonstrated record of successes. Since the electrons are confined to a small region of space, lack of specific account of electron correlation is not serious in that its inclusion would not substantially affect the one-electron description, or the error incurred may be assumed to be reasonably constant when comparing different conformations.

The second broad area of application of orbital interaction theory is in the area of intermolecular interactions, from which many aspects of chemical reactivity may be inferred. This application is in the realm of weaker interactions and larger distances, conditions suitable to application of perturbation theories.

92 ORBITAL INTERACTION THEORY

The dominant orbital interactions in this regime are of the two-electron–two-orbital type, and usually a single orbital interaction, between the HOMO of one molecule and the LUMO of the other is sufficient for the purpose. Unfortunately, the one-electron theoretical foundation for this kind of long-range interaction involving electron transfer is much less sound. Accurate descriptions of weak interactions usually require theoretical procedures which go beyond Hartree–Fock theory for the reason that some account must be taken of the correlation of electron motions, particularly within the same orbital. The reason for this is easy to understand, and is illustrated in Figure 3.22. Prior to interaction (Figure 3.22(a)), the electrons are localized to the HOMO of the donor, B. At intermediate separations (Figure 3.22(b)), where the orbitals of A and B overlap, the space available to the electrons is greatly increased and the electrons are able to separate. After bond formation (Figure 3.22(c)), the space available to the electrons is again restricted. In the intermediate stage (Figure 3.22(b)), the lack of electron correlation inherent in one-electron theories is most strongly felt. In terms of valence bond structures, MO theory places comparable emphasis on the three structures, R_1, R_2, and R_3, whereas in reality, resonance structure, R_2 alone most accurately represents the true situation. Resonance structure R_2 corresponds to a configuration in which a single electron has been transferred from B to A. The reactivity of B as a donor (nucleophile) is expected to be correlated with its ionization potential [65]. Likewise, the reactivity of A as a Lewis acid should be correlated with the electron affinity of A. Post-Hartree–Fock theoretical analyses of heterolytic bond breaking [66], and the reverse, cation–anion combination reactions [67] have elucidated the role of the missing electron correlation. In the VBCM treatment of Shaik [8], [67], the valence bond configurations of reacting species are regarded as ''surfaces'' whose avoided crossing defines the barrier to the reaction. The Hartree–Fock model is expected to work well when the electronic description involves a single closed-shell configuration [Figure 3.22(a)

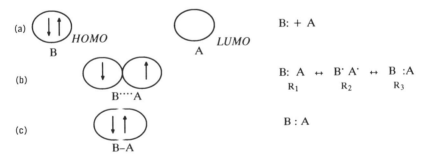

Figure 3.22. Stages in the interaction between two molecules, B and A, showing the effects of electron correlation and the equivalent resonance structures: **(a)** No interaction—the electrons are spin paired and confined to the orbital of B; **(b)** weak interaction (e.g., a transition structure)—the electrons can separate into a larger volume of space; **(c)** strong interaction—a bond is formed and the electron distribution is again confined.

and (c)] but overestimate the effect of electron–electron repulsion in the extended case [Figure 3.22(b)]. It is not surprising then that Hartree–Fock calculations tend to overestimate the energy of the transition state for intermolecular reactions. This failure should not be interpreted as a general failure of the one-electron model as an interpretive device, but rather as an inappropriate energy counting for the MO description due to the fact that electron–electron repulsions are accommodated in an average way. The MO description remains as a powerful and conceptually simple means for understanding the bonding even in the region of the transition state, and such MO descriptions are derivable by the simple rules of orbital interaction theory.

CHAPTER 4

SIGMA BONDS AND ORBITAL INTERACTION THEORY

THE C—X σ BONDS: X = C, N, O, F AND X = F, Cl, Br, I

Consider the interaction between two sp^n hybridized orbitals of unequal energy. The interaction diagram is illustrated in Figure 4.1(a). As discussed earlier, the lowest energy configuration has two electrons with energy ϵ_L. In terms of equation (3.48), $n_L = 2$ and $n_U = 0$. The magnitudes of $\Delta\epsilon_L$ and $\Delta\epsilon_U$ may be deduced as follows. For the purpose of deciphering of trends, the first series (C, N, O, F) involves atoms which are approximately the same size but differ greatly in electronegativity. The interaction matrix element, h_{cx}, will be essentially constant across the series, X = C, O, N, and F. However, relatively large changes will occur in the difference in orbital energies, the quantity ($\epsilon_C - \epsilon_X$) increasing in the series. The consequent energies of the resulting σ and σ* orbitals are shown in Fig. 4.2. The energies of both the bonding σ orbital and the antibonding σ orbital decrease, being the lowest for the C—F bond. The greatest polarization also occurs in the case of the C—F bond. In particular, the low-lying σ* orbital is highly polarized toward the C end of the bond.

The second series, involving bonding of C to different halogens (F, Cl, Br, I) involves atoms which are very different in size and also differ greatly in electronegativity. The halogen ends of the bonds are constructed from $2p$, $3p$, $4p$, and $5p$ orbitals, respectively. The overlap, and therefore the interaction matrix element, h_{cx}, will decrease very rapidly across the series, X = F, Cl, Br, and I. Relatively large changes will also occur in the difference in orbital energies, the quantity ($\epsilon_C - \epsilon_X$) decreasing in the series. The relative properties and reactivities of C—X bonds are consistent with dominance of the interaction matrix element (i.e., overlap dependence) in determining the resulting orbital energies. The consequent energies of the resulting σ and σ* orbitals

The C—X σ BONDS: X = C, N, O, F AND X = F, Cl, Br, I   95

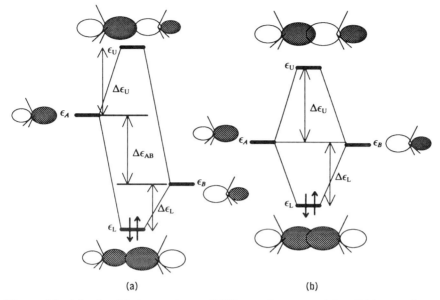

Figure 4.1. (a) σ bond between atoms of different electronegativity; (b) homopolar σ bond.

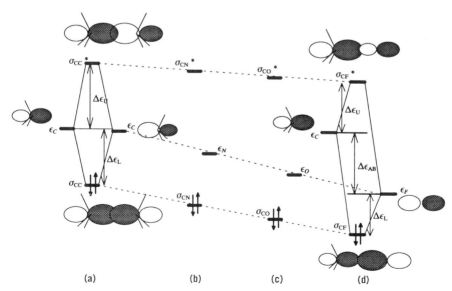

Figure 4.2. σ_{CX} bond and σ^*_{CX} antibond between atoms of the first row: (a) X = C; (b) X = N; (c) X = O; (d) X = F.

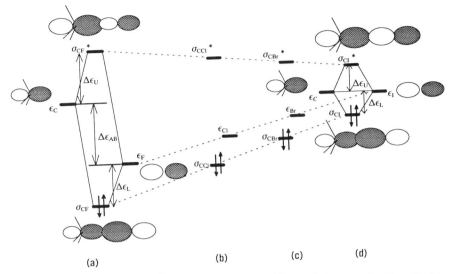

Figure 4.3. σ_{CX} bond and σ_{CX}^* antibond between different halogens. **(a)** X = F; **(b)** X = Cl; **(c)** X = Br; **(d)** X = I. No attempt has been made to show the increased size or nodal character of the *p* orbitals of Cl, Br, or I.

are shown in Figure 4.3. The predicted trend in σ_{CX} bond energies is observed in the bond ionization potentials [68]. The lowest-lying σ^* orbital is expected to be that of the C—I bond.

CASE STUDY 1: σ BONDS: HOMOLYTIC VERSUS HETEROLYTIC CLEAVAGE

Heterolytic Cleavage of σ Bonds

The energy change associated with heterolytic bond dissociation ($n_A = 0$ and $n_B = 2$) is $\Delta E^{het} = 2(\epsilon_L - \epsilon_B) = 2\Delta\epsilon_L \approx 2h_{AB}^2/\Delta\epsilon_{AB}$ (Figure 4.1). One may ask how ΔE^{het} would be expected to change in the series C—C, C—N, C—O, and C—F, or the series H—C, H—N, H—O, H—F. In the crudest approximation, one may say that the orbitals of C, N, O, and F are all approximately the same size and therefore the interaction matrix element, h_{AB}, will be approximately the same size for any A—B pair. The dominant factor determining the heterolysis energy therefore is the difference in orbital energies in the denominator, and one has directly the prediction (Figure 4.2) that ease of heterolytic cleavage for C—X is in the order, C < N < O < F, as observed.

Similarly, one may ask how ΔE^{het} would be expected to change in the series C—F, C—Cl, C—Br, and C—I. The situation is more complicated for this series, since both the electronegativity and orbital size vary. Variation in orbital size is approximately proportional to n^2, where n is the principal quantum number of the valence shell. The larger the discrepancy in orbital sizes, the smaller the effective overlap and the smaller the interaction matrix element, h_{AB}. Evidently for halides, the change in ΔE^{het} is dominated by the change in h_{AB}, and heterolytic cleavage of the C—I bond is easier than of the C—Cl bond. The C—Br bond is intermediate. The most difficult to cleave heterolytically is the C—F bond. One must bear in mind that heterolysis occurs only in polar media where additional bond forming (to solvent) takes place. Heterolysis of the C—F bond is rarely observed. Likewise, the observed Lowry-Bronsted acidity of the H—X acids is in the order, HI > HBr > HCl > HF, as is consistent with the relative energies of the σ^* orbitals or dominance of overlap consideration (Figure 4.3).

Heterolytic cleavage of neutral molecules in the gas phase is never observed. The energy required to separate charges, of the order of the ionization potential of H atom (13.6 eV, 1300 kJ/mol), is prohibitively high. However it is readily compensated by solvation in polar solvents, especially water. Since solvation energies are very large, the simple theory proposed to this point could only rationalize gross trends. Heterolytic cleavage of *charged* species in mass spectrometric and negative ion cyclotron resonance experiments is commonly observed.

Homolytic Cleavage of σ Bonds

The energy change associated with homolytic bond dissociation ($n_A = 1$ and $n_B = 1$) is $\Delta E^{\text{hom}} = 2\epsilon_L - \epsilon_A - \epsilon_B = 2\Delta\epsilon_L + \Delta\epsilon_{AB} \approx (2h_{AB}^2/\Delta\epsilon_{AB}) + \Delta\epsilon_{AB}$ (Figure 4.1). Thus, the energy difference of the final singly occupied orbitals must also be considered. How would ΔE^{hom} be expected to change in the series C—C, C—N, C—O, and C—F or the series H—C, H—N, H—O, H—F? Recognizing as before, that the orbitals of C, N, O, and F are all approximately the same size and therefore the interaction matrix element, h_{AB}, will be approximately the same size for any A—B pair, the dominant factor determining the homolytic bond dissociation energy therefore is the difference in orbital energies, which appears in the denominator of the first term and as the second term. The two terms have the opposite effect on the bond dissociation energy. To a first approximation, the bond energies are expected to be similar if $\Delta\epsilon_{AB}$ is small compared to $|h_{AB}|$, e.g., C—C and C—N, and dominated by the second term if $\Delta\epsilon_{AB} > |h_{AB}|$, e.g., C—O and C—F.

Similarly, one may ask how ΔE^{hom} would be expected to change in the series C—F, C—Cl, C—Br, and C—I, or the series HF, HCl, HBr, HI. The situation is simpler for these series, since both the electronegativity and orbital size vary but have the same effect on the bond dissociation energy. Variation in orbital size is approximately proportional to n^2, where n is the principal quantum number of the valence shell. The larger the discrepancy in orbital sizes, the smaller the effective overlap and the smaller the interaction matrix element, h_{AB}. It was argued above that for halides, the first term is dominated by the change in h_{AB}. Since the electronegativity, and therefore $\Delta\epsilon_{AB}$, also decreases, one expects homolytic bond dissociation energies of the C—X bonds

98 SIGMA BONDS AND OBITAL INTERACTION THEORY

TABLE 4.1. Average Homolytic Bond Dissociation Energies (kJ/mol) of R—X

	X						
R	CH_3	NH_2	OH	F	Cl	Br	I
H	435	460	498	569	432	366	299
CH_3	368	364	381	456	351	293	234

Source: *CRC Handbook.*

to be in the order, C—I < C—Br < C—Cl < C—F [69]. Likewise, the observed strengths of the bonds of the H—X acids are in the order, HI < HBr < HCl < HF, as is consistent with the dominance of overlap considerations. Actual bond strengths at 298 K are shown in Table 4.1.

BONDING IN CYCLOPROPANE

The above discussion applies to "normal" σ bonds in saturated compounds, as occur in molecules where the internuclear angles can reasonably approach the tetrahedral interorbital angle expected of sp^3 hybridization. Many molecules exist which encompass small rings, the prototypical molecule being cyclopropane, where the internuclear angle, 60°, is far from "ideal." For the purpose of understanding the properties of cyclopropanes using orbital interaction diagrams, we adopt the Walsh picture [70], shown in Figure 4.4. Each carbon atom is considered to be sp^2 hybridized [Figure 4.4(a)]. Two of the sp^2 hybrid orbitals are used for the external σ bonds, as expected for the wider

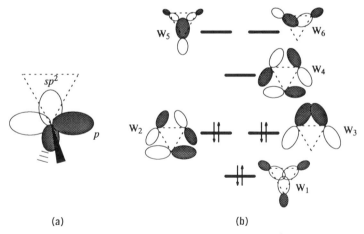

Figure 4.4. The bonding in cyclopropane: (**a**) The $2p$ and sp^2 hybrid orbitals; (**b**) the Walsh MOs. One of the degenerate HOMOs, W_2, and the LUMO, W_4, will interact in a π fashion with a substituent on the ring.

bond angle (HCH angle = 114° in cyclopropane). The set of three sp^2 hybrid orbitals, one from each center, are directed toward the center of the ring and interact strongly to form the unique bonding combination, W_1, and the pair of antibonding MOs, W_5, and W_6. The set of three (unhybridized) $2p$ orbitals are oriented tangentially to the ring and interact weakly to form the degenerate pair of bonding MOs, W_2, and W_3, which constitute the HOMOs, and the unique antibonding MO, W_4, which becomes the LUMO. In the absence of substituents, the orientation of the nodal surfaces of the pairs of degenerate MOs is arbitrary, but the orientation will be as shown in Figure 4.4(b) for a substituent at the lower vertex. Notice that one of the HOMOs, W_2, and the LUMO, W_4, will interact with a substituent in this position in a π fashion. Because of the poor overlap of the tangentially oriented $2p$ orbitals, the HOMO energy will be quite high, and the LUMO energy will be low. The cyclopropyl ring will therefore be expected to act both as a good π donor and a good π acceptor.

INTERACTIONS OF σ BONDS

The reactivity of σ bonds will be discussed in greater detail in other chapters but general principles can be expounded here. For most organic compounds (excluding those with three-membered rings), the σ bonding orbital is too low in energy for the bond to function effectively as an electron donor toward unoccupied orbitals of other molecules or other parts of the same molecule. Likewise, the σ^* antibonding orbital is too high in energy to function as an electron acceptor from occupied orbitals inter- or intramolecularly. However, there is ample evidence that σ bonds do interact with other *occupied* orbitals, and that σ^* orbitals interact with other *empty* orbitals. Evidence for the latter case is primarily in the form of chemical activation of C—H bonds by adjacent groups with low-lying empty or half-filled orbitals. The increased acidity of C—H bonds, i.e., reactivity with Lewis bases, under these circumstances is discussed in Chapter 10. Several examples will serve to exemplify the interaction of σ bonds with adjacent filled orbitals.

The energies of orbitals may be inferred by measurements of ionization potentials by Koopmans' theorem. As discussed at the end of Chapter 3, one group orbital of a methyl or methylene group will always have the correct nodal characteristics to interact with an adjacent π orbital, or with an adjacent sp^n orbital in a "π fashion." Thus, the methyl groups adjacent to the π bond in (Z)-2-butene (I.P. = 9.12 eV [54]) raise the energy of the π orbital by 1.39 eV relative to that of ethylene (I.P. = 10.51 eV [71]). A similar effect is observed in cyclohexene [55].

The through space interaction of the two π bonds of norbornadiene was presented in Chapter 3 as exemplifying a four-electron–two-orbital interaction. The interaction of the nonconjugated π bonds of 1,4-cyclohexadiene cannot be treated in the same way. The planar molecular skeleton reduces the direct interaction between the two π bonds while increasing the interaction of the π bonds with the intervening CH$_2$ groups. As a consequence, the *in-phase* combination of the two π bonds suffers a repulsive interaction with the *in-phase*

1,4–Cyclohexadiene's HOMO   DABCO's HOMO

combination of the CH$_2$ group orbitals and is raised above the *out-of-phase* combination to become the HOMO of the molecule. The interaction diagram is shown in Figure 4.5. The consequences of such through-space and through-bonds interactions on the ionization potentials of a number of nonconjugated dienes as a function of skeletal structure and substitution has been investigated [72].

The spectroscopic and chemical properties of 1,4-diazabicyclo[2.2.2]octane (DABCO) are consistent with a strong interaction of the *in-phase* combination of the nonbonded electron pairs of the nitrogen atoms with the symmetric combination of the C—C σ bonds, one consequence of which is that the *in-phase* combination is the HOMO. The two lowest ionization potentials are 7.6 eV and 9.7 eV [73]. Compare these to the I.P.'s of trimethyl amine (8.44 eV)

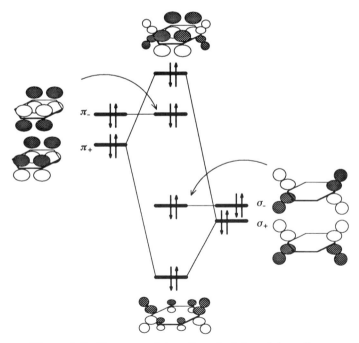

Figure 4.5. The $\sigma - \pi$ interactions in 1,4-cyclohexadiene.

and ammonia (10.5 eV) [74]. The relative importance of intramolecular orbital interactions through space and through bonds has been reviewed by Hoffmann [48].

σ BONDS AS ELECTRON DONORS OR ACCEPTORS

With respect to the ability of σ bonds to accept or donate electrons, several circumstances may act to change the norm. Substituents adjacent to the σ bond may act to raise the energy of the bonding orbital, or to lower the energy of the σ* antibonding orbital, and thereby increase the involvement of the bond in two-electron–two-orbital interactions. Whether the bond acts as a donor in an interaction with a low-lying virtual orbital, or as an acceptor in an interaction with a high-lying occupied orbital, the consequences for the bond are the same, a reduction of the bond order and a consequent weakening of the bond. The extreme consequence is a rupture of the bond, as occurs in a hydride transfer, or a nucleophilic substitution (S_N2). When the interaction does not "go all the way," the bond weakening is often apparent in the infrared spectrum as a shift to longer wavelength of the bond stretching frequency, and a lengthening of the bond itself. The fact that both consequences are usually observed has suggested an inverse relationship between bond strength as measured by the force constant, and the bond length [75]. This relationship which is widely accepted has no direct theoretical derivation and exceptions, particularly in the case of N—F bonds, have been noted and attributed to "polar" effects [76].

σ Bonds as Electron Acceptors

The acceptor ability may be improved in two ways, by lowering the energy of the σ* orbital, and by polarizing the orbital toward one end. The first improves the interaction between it and a potential electron donor orbital by reducing the energy difference, $\epsilon_A - \epsilon_B$, the second by increasing the possibility of overlap and therefore increasing the value of the intrinsic matrix element, h_{AB}.

As a σ Acceptor

The general features (orbital distribution and energy) for a σ bond between C and a more electronegative element or group, X, are shown in Figure 4.1(a), and Figures 4.2 and 4.3, where center A is carbon, and center B is X. The energy of the σ* orbital will be relatively low and the orbital is polarized toward carbon. Optimum interaction between the σ* orbital and a localized nonbonding orbital of a Lewis base will occur along the axis of the orbital and from the carbon end, as shown in Figure 4.6(a). The two-orbital–two-electron interaction is accompanied by charge transfer into the σ* orbital and consequent reduction of the bond order, as well as partial σ bond formation between

102 SIGMA BONDS AND OBITAL INTERACTION THEORY

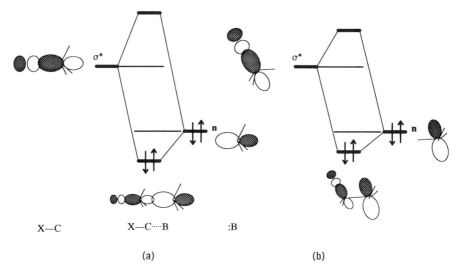

Figure 4.6. (a) Sigma-type acceptor interaction between the HOMO, **n**, of a base and the LUMO, σ^*, of a polarized σ bond; (b) Pi-type acceptor interaction between the HOMO, **n**, of an X: type group and the LUMO, σ^*, of an adjacent polar σ bond.

C and the incoming base. Carried to the extreme, a substitution reaction with inversion of configuration at C (S_N2) ensues, as discussed in Chapter 9. If the center A is H rather than C, the corresponding reaction is a proton abstraction, i.e., a Lowry–Bronsted acid–base reaction. If the interaction falls short of proton abstraction, the attractive interaction is called a *hydrogen bond*. Both aspects are discussed further in Chapter 10.

Interhalogen bonds are accompanied by very low LUMOs (σ_{XX}^*) and thus can function as good σ electron acceptors. The donor–acceptor complexes between ammonia and F_2, Cl_2, and ClF have been investigated theoretically [77] and found to have linear structures as expected on the basis of the above discussion.

As a π Acceptor

The σ^* orbital associated with a σ bond between C and a more electronegative element or group, X, is polarized toward carbon. Optimum interaction between the σ^* orbital and an adjacent localized nonbonding orbital of an X-type group will occur if the σ^* orbital and the p (or sp^n) nonbonded orbital of the neighboring group are coplanar, as shown in Figure 4.6(b). The two-orbital–two-electron interaction is accompanied by charge transfer into the σ^* orbital and consequent reduction of the bond order, as well as partial π bond formation between C and the adjacent group. Carried to the extreme, an elimination reaction (E1cb) ensues, as discussed in Chapter 10. The tendency of the neighboring group to assist the departure of the leaving group is called *anchimeric* *assistance* or the *neighboring group effect*. If the interaction falls short of elimination, the attractive interaction is called *negative hyperconjugation* [57]. The tendency to alter geometry or change conformation so as to maximize the interaction is called the *anomeric effect* [78].

σ BONDS AS ELECTRON DONORS

The donor ability of a σ bond may be improved in two ways, by raising the energy of the σ orbital, and by polarizing the orbital toward one end. The first improves the interaction between it and a potential electron acceptor orbital by reducing the energy difference, $\epsilon_A - \epsilon_B$, the second by increasing the possibility of overlap and therefore increasing the value of the intrinsic interaction matrix element, h_{AB}. Specifically, a bond between C and an element (e.g., a metal, Li, Na, Mg, etc.) or group (e.g., $-BR_3^-$) less electronegative than C will have the required features. Referring to Figure 4.1(a), the role of C will be played by group *B*. Since H is also somewhat less electronegative than C, C—H bonds are also potential electron donors when the electron demand is high. So are bonds to H from metals or metal-centered groups (e.g., NaH, LiAlH$_4$, NaBH$_4$, etc.).

As a σ Donor

The general features (orbital distribution and energy) for a σ bond between C and a less electronegative element or group, M, are shown in Figure 4.1(a), where center *A* is M, and center *B* is C. The energy of the σ orbital will be relatively high and the orbital is polarized toward carbon. Optimum interaction between the σ orbital and a localized empty orbital of a Lewis acid will occur between the C and M, closer to the carbon end, as shown in Figure 4.7(a), or from the backside if steric hindrance by M is severe. The two-orbital–two-electron interaction is accompanied by charge transfer from the σ orbital and consequent reduction of the σ bond order, as well as partial σ bond formation between C and the incoming acid. Carried to the extreme, a substitution reaction ensues in which the configuration at C may be *retained*. If the center *B* is H rather than C, the corresponding reaction is a hydride transfer (abstraction), as in the Cannizzaro or Meerwein–Pondorf–Verley reactions. If the interaction falls short of abstraction, a hydride bridge may be formed. Both aspects are discussed further in Chapter 10.

As a π Donor

A σ bond between C and a less electronegative element or group, M, is polarized toward carbon. Optimum interaction between the σ orbital and an adjacent localized empty orbital of a Z-type group will occur if the σ orbital and the empty *p* orbital of the neighboring group are coplanar, as shown in Figure

104 SIGMA BONDS AND OBITAL INTERACTION THEORY

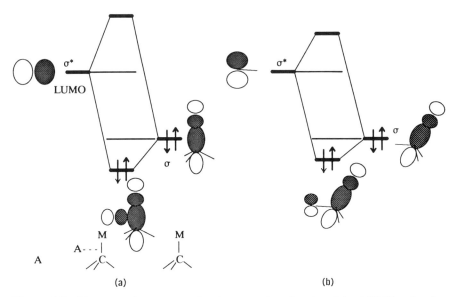

Figure 4.7. (a) σ-type donor interaction between the σ bond and the LUMO, σ^*, of a polarized σ bond (shown as a p orbital); (b) π-type donor interaction between the σ bond and the LUMO (shown as a p orbital), which may be the π^* orbital of an adjacent Z-type substituent, or σ^* of a polarized σ bond.

4.7(b). The two-orbital–two-electron interaction is accompanied by charge transfer from the σ orbital and consequent reduction of the bond order, as well as partial π bond formation between C and the adjacent group. The competitive π donation involving C—H and C—C σ bonds in cyclohexanones has been the subject of much discussion in the recent literature [79]. The carbonyl group distorts from planarity so as to achieve better alignment of the carbon $2p$ component of its π^* orbital with the σ bond, a secondary consequence of which may be substantial preference for attack by nucleophiles at the face opposite to the σ bond involved. Where the adjacent group is a carbocationic center, distortion of the donor σ bond toward the acceptor site, or migration of the bond (i.e., the group, M, attached to the other end) may occur. In the latter case, the reaction is a 1,2 hydride shift (M=H) or a Wagner–Meerwein rearrangement (M=alkyl). The weakening of the bond may also result in elimination of M$^+$. If the interaction falls short of elimination or migration, the attractive interaction is called *hyperconjugation*. The tendency to alter the geometry or change the conformation so as to maximize the attractive interaction is well documented [80]. In an extreme case, the group M may adopt a position midway between the C and the adjacent group, forming a two-electron–three-center bond as in nonclassical structures of carbocations (ethyl [81] (M=H), norbornyl [82]) M=alkyl)). The bonding in (car)boranes is another example [83].

If the LUMO is the *sp* hybrid orbital at the C end of a polarized σ bond, such as to a halide, geometric distortion also occurs, particularly a lengthening of the receiving σ bond. Carried to the extreme, an elimination reaction occurs (E1cb, as discussed in Chapter 10). Migration of M to the adjacent group does not occur.

CHAPTER 5

SIMPLE HÜCKEL MOLECULAR ORBITAL THEORY

In this chapter, Simple Hückel Molecular Orbital (SHMO) theory is developed. The reference energy, α, and the energy scale in units of β are introduced.

THE SIMPLE HÜCKEL ASSUMPTIONS

SHMO theory was originally developed to describe planar hydrocarbons with conjugated π bonds. Each center is sp^2 hybridized and has one unhybridized p orbital perpendicular to the trigonal sp^2 hybrid orbitals. The sp^2 hybrid orbitals form a rigid unpolarizable framework of equal C—C bonds. Hydrogen atoms are part of the framework and are not counted. The Hückel equations (3.5) described in the first part of Chapter 3 apply, namely

$$F(1) \approx h^{\text{eff}}(1); \quad h^{\text{eff}}(1)\phi_a(1) = \epsilon_a \phi_a(1), \quad E_{IEA} = 2 \sum_{a=1}^{M} \epsilon_a \quad (5.1)$$

Each MO is expanded in terms of the unhybridized p orbitals, one per center. The overlap integral between two parallel p orbitals is small and is approximated to be exactly zero. Thus,

$$\phi(1) = \sum_{A=1}^{N_N} c_A \chi_A(1), \quad \int \chi_A(1)\chi_B(1)\, d\tau_1 = \delta_{AB} \quad (5.2)$$

where N_N is the number of carbon atoms which is the same as the number of orbitals. Equation (5.2) is just a generalization of equation (3.6). The subsequent steps are precisely those which were followed in Chapter 3. The energy is expressed as an expectation value of the MO [equation (5.2)] with the effective Hamiltonian

$$\epsilon = \frac{\int \phi(1) h^{\text{eff}}(1) \phi(1)\, d\tau_1}{\int |\phi(1)|^2\, d\tau_1}$$

$$= \frac{\sum_{A=1}^{N_N} \sum_{B=1}^{N_N} c_A c_B \int \chi_A(1) h^{\text{eff}} \chi_B(1)\, d\tau_1}{\sum_{A=1}^{N_N} \sum_{B=1}^{N_N} c_A c_B \int \chi_A(1) \chi_B(1)\, d\tau_1}$$

$$= \frac{\sum_{A=1}^{N_N} c_A^2 h_{AA} + \sum_{A=1}^{N_N} \sum_{B \neq A}^{N_N} c_A c_B h_{AB}}{\sum_{A=1}^{N_N} c_A^2 + \sum_{A=1}^{N_N} \sum_{B \neq A}^{N_N} c_A c_B S_{AB}} \quad (5.3)$$

$$= \frac{\sum_{A=1}^{N_N} \left(c_A^2 h_{AA} + \sum_{B \neq A}^{N_N} c_A c_B h_{AB} \right)}{\sum_{A=1}^{N_N} c_A^2} = \frac{N}{D} \quad (5.4)$$

and the variational method applied. Differentiating equation (5.4) with respect to each of the coefficients, c_A,

$$\frac{\partial}{\partial c_A}(ND^{-1}) = \frac{\partial N}{\partial c_A} D^{-1} - ND^{-2} \frac{\partial D}{\partial c_A} = 0$$

$$\frac{\partial N}{\partial c_A} - \epsilon \frac{\partial D}{\partial c_A} = 0$$

$$(h_{AA} - \epsilon) c_A + \sum_{B \neq A}^{N_N} h_{BA} c_B = 0 \quad \text{for each } A = \{1, \cdots, N_N\} \quad (5.5)$$

The condition that the N_N linear equations have a solution is that the determinant of coefficients (of the c's) be equal to zero.

$$\begin{vmatrix} (h_{11} - \epsilon) & h_{12} & h_{13} & \cdots & h_{1N_N} \\ h_{21} & (h_{22} - \epsilon) & h_{23} & \cdots & h_{2N_N} \\ h_{31} & h_{32} & (h_{33} - \epsilon) & \cdots & h_{3N_N} \\ \vdots & \vdots & \vdots & & \vdots \\ h_{N_N 1} & h_{N_N 2} & h_{N_N 3} & \cdots & (h_{N_N N_N} - \epsilon) \end{vmatrix} = 0 \quad (5.6)$$

Within the SHMO approximations, all of the diagonal Hamiltonian matrix elements are equal and are designated "α." The Hückel "α" is the energy of an electron in a $2p$ orbital of a trigonally (sp^2) hybridized carbon atom. The off-diagonal matrix elements are all equal if the atoms involved are bonded together (since all bond distances are assumed to be equal) and these are designated "β." The Hückel "β" is the energy of interaction of two $2p$ orbitals of a trigonally (sp^2) hybridized carbon atom which are attached to each other by a σ bond. If the two atoms are not nearest neighbors, then h_{AB} is set equal to zero. In summary,

$$h_{AA} = a$$
$$h_{AB} = \beta \quad \text{if centers } A \text{ and } B \text{ are bonded}$$
$$h_{AB} = 0 \quad \text{if centers } A \text{ and } B \text{ are not bonded}$$

Thus, all of the diagonal elements are "$\alpha - \epsilon$." The off-diagonal elements are "β" if the two atoms involved are bonded, and "0" if they are not. It is usual to divide each row of the determinant by "β." This corresponds to a change of energy units and leaves either 1's or 0's in the off-diagonal positions which encode the connectivity of the molecule. The diagonal elements become $(\alpha - \epsilon)/\beta$, which is usually represented by x. While the determinant was expanded in Chapter 3 to yield the secular equation, it is more convenient in general to diagonalize the determinant using a computer. An interactive computer program, SHMO, is provided with this book. A brief description of the program is found in Appendix B.

The SHMO calculation on ethylene yields the results shown in Figure 5.1(a). The π and π^* orbitals are precisely 1 "β" unit above and below α. The SHMO results are presented in Figure 5.1(b) in the form of an interaction diagram. In this case, $\Delta\epsilon_L$ and $\Delta\epsilon_U$ are assigned the same value, namely $1|\beta|$, in the spirit of SHMO theory, but we know that the effect of proper inclusion of overlap would yield $\Delta\epsilon_U > \Delta\epsilon_L$.

The SHMO results for the series of "linear" π systems, allyl, butadiene, and pentadienyl are shown in Figure 5.2. The molecular species are portrayed with realistic angles (120°), and in the case of the last two, in a specific conformation. SHMO theory does not incorporate any specific geometric information since all non-nearest-neighbor interactions are set equal to zero. As a result, the SHMO results (MO energies and coefficients) are independent of whether the conformation of butadiene is *s-trans* as shown in Figure 5.2, or *s-cis*, as may be required for the Diels–Alder reaction. Likewise the results for the pentadienyl system shown in the "u" conformation in Figure 5.2, are identical to the results for the "w" or "sickle" conformations. The MOs are displayed as linear combinations of $2p$ atomic orbitals seen from the top on each center, with changes of phase designated by shading. The relative contribution of each atomic $2p$ orbital to the π MO is given by the magnitude of coefficient of the eigenvector from the solution of the SHMO equations. In the

THE SIMPLE HÜCKEL ASSUMPTIONS 109

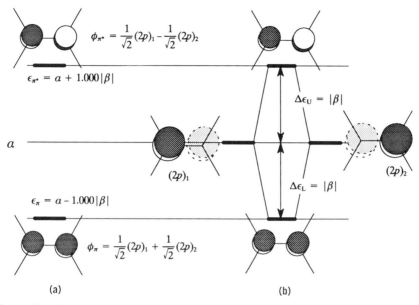

Figure 5.1. (a) SHMO results for ethylene; (b) the interaction diagram for ethylene ($\Delta \epsilon_L = \Delta \epsilon_U$ because overlap is assumed to be zero).

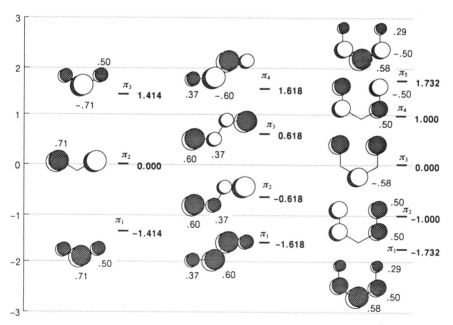

Figure 5.2. The SHMO orbitals of allyl, butadiene, and pentadienyl. The vertical scale is energy in units of $|\beta|$, relative to α. Coefficients not specified may be obtained by symmetry.

display, the size of the 2p orbital is proportional to the *magnitude* of the coefficient. In some MOs, such as π_2 of allyl, the node passes through a nucleus and a 2p orbital is not shown because its coefficient is identically zero. The orbitals which are near α will be of most interest in various applications. In the case of allyl, this orbital, π_2, is LUMO in allyl cation, SOMO in allyl radical, and HOMO in allyl carbanion. In the case of the pentadienyl system, π_3 plays the same role. In the case of butadiene, the HOMO is π_2. Since the energy of the HOMO is higher than the energy of the HOMO of ethylene, one might conclude that butadiene is more basic than ethylene, and more reactive toward electrophilic addition. Caution should be exercised in jumping to this conclusion, however, since the largest coefficient of butadiene's HOMO, 0.60, is smaller than the coefficient of the 2p orbital of the HOMO of ethylene, 0.71. The smaller coefficient would imply a weaker intrinsic interaction (h_{AB}) with Lewis acids and therefore reduced reactivity. Clearly, the energy factor and the intrinsic interaction (as judged from the coefficients) are in opposition, the first predicting higher reactivity and the second, lower. As is often the case in orbital interaction theory, one must resort to experimental observations in order to evaluate the relative importance of opposed factors. Since experimentally, dienes are more susceptible to electrophilic attack than unconjugated alkenes, we can conclude that the energy factor is more important than the relatively small difference in coefficients.

SHMO results for the series of cyclic π systems, cyclopropenyl, cyclobutadiene, cyclopentadienyl, and benzene, are shown in Figure 5.3. Several points may be noted. The lowest MO in each case has energy identical to $\alpha - 2|\beta|$, a result which can be proved to be general for any regular polygon. Each of the rings have degenerate pairs of MOs as a consequence of the three-, or higher-fold axis of symmetry. The orbitals of each degenerate pair may appear very different. The orientation of the nodal surfaces of the degenerate MOs is entirely arbitrary. Two equivalent orientations are shown for cyclobutadiene. The orientations shown are those which should be adopted for the purposes of interaction diagrams involving a single substituent on the ring. The pairs of degenerate MOs which form the HOMO of benzene is at exactly the same energy as the HOMO of ethylene. In orbital interaction terms, we would predict that, even though the HOMOs are of the same energy, benzene would be less susceptible to electrophilic attack then ethylene for the reason that the largest available coefficient, 0.58 in π_3, is smaller than 0.71, the coefficient in the HOMO of ethylene.

Conjugated π systems which do not contain any odd membered rings are called *alternant*, and, provided all the atoms are the same, alternant systems have a symmetrical distribution of orbital energies about the mean (α for C). The coefficients also repeat in magnitude in MOs which are equidistant from the mean. These features are readily apparent in the orbitals portrayed in Figures 5.1, 5.2, and 5.3.

Of the four cyclic conjugated π systems shown in Figure 5.3, only benzene, with six electrons in the π orbitals, is stable kinetically and thermodynami-

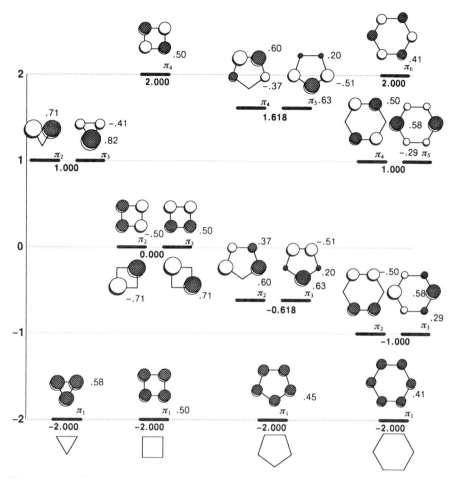

Figure 5.3. SHMO orbitals for cyclopropenyl, butadiene, cyclopentadienyl, and benzene. The energies are in units of $|\beta|$ relative to α. Two alternative but equivalent representations are shown for the degenerate π orbitals of cyclobutadiene. Sizes of the $2p$ orbitals are shown proportional to the magnitudes of the coefficients whose numerical values are given. Coefficients not specified may be obtained by symmetry.

cally. Neutral cyclopropenyl and cyclopentadienyl, with three and five π electrons, respectively, are free radicals. The cyclopropenyl cation, with two π electrons, and the cyclopentadienyl anion, with six, have filled shells and constitute aromatic systems in that they exhibit unusual stability, compared to other carbocations and carbanions, respectively. Cyclobutadiene is a special case. With two electrons in the degenerate HOMOs, one would expect that the electrons would separate and that the ground state would be a triplet. However, a distortion of the geometry from square to rectangular would eliminate the degeneracy and permit a singlet ground state. The ground state of cyclobutadiene has been shown experimentally to be singlet with a barrier for the rectangular to square deformation in the range $7 \leq \Delta H^{\ddagger} \leq 42$ kJ mol^{-1} [84]. Theoretical computations suggest that the higher value may be correct [85].

THE CHARGE AND BOND ORDER IN SIMPLE HÜCKEL MO THEORY: (S_{AB} = 0, ONE ORBITAL PER ATOM)

It is of interest to inquire how the electrons are redistributed during an interaction, and how a bond is affected. Using a simplified Mulliken Population Analysis (simplified by dropping all terms involving the overlap of atomic orbitals and assuming that in any given MO, there is only *one* atomic orbital on any given center), we may define a measure of the *electron population* on each center, and the *bond order* between two centers. Thus, we may assume that the following relations hold

$$\phi_a = \sum_{A=1}^{n} \chi_A c_{Aa}, \quad S_{AB} = \delta_{AB}, \quad \sum_{A=1}^{n} c_{Aa}^2 = 1 \quad (5.7)$$

where n is the number of atomic orbitals (= the number of nuclear centers, since there is one orbital per center).

P_A, the Electron Population of Center A; C_A, the Net Charge of Center A

The electron population of center A is defined as

$$P_A = \sum_{a=1}^{n} n_a c_{Aa}^2 \quad (5.8)$$

where n_a (= 0, 1, or 2) is the number of electrons in the ath MO, and the sum runs over the MOs (there are as many MOs as there are AOs, and atomic centers). The net charge, C_A, depends on the number of electrons, n_A, which are required for a neutral atomic center A. Thus

$$C_A = n_A - P_A \quad (5.9)$$

Notice that C_A is negative if the population of A, P_A, exceeds the number of electrons required to give a neutral center. It is easily verified that

$$\sum_{A=1}^{n} P_A = N_e, \quad \sum_{A=1}^{n} C_A = \text{net charge on the molecule} \quad (5.10)$$

FACTORS GOVERNING THE ENERGIES OF MOS: SIMPLE HÜCKEL MO THEORY

Exercise 5.1. Verify that the net charge at each carbon atom of each of the *neutral* ring systems shown in Figure 5.3 is zero (to 2 significant figures).

B_{AB}, the Bond Order Between Centers A and B

The *bond order*, B_{AB}, between centers A and B is defined as

$$B_{AB} = \sum_{a=1}^{n} n_a c_{Aa} c_{Ba} \tag{5.11}$$

where the quantities are defined as above. A positive value indicates bonding. Small negative values of B_{AB} may result. These are indications of *antibonding* or *repulsive* interaction between the centers concerned.

Exercise 5.2. Show that the maximum value for the bond order due to a single bond is 1 and that it occurs when

$$n_a = 2 \quad \text{and} \quad c_A = c_B = \frac{1}{\sqrt{2}}.$$

Exercise 5.3. Show that the bond order for benzene is 0.67 from the data in Figure 5.3.

FACTORS GOVERNING THE ENERGIES OF MOS: SIMPLE HÜCKEL MO THEORY

The Reference Energy, α, and Energy Scale, $|\beta|$

Most organic molecules are made up of the elements C, H, N, and O, with lesser amounts of S, P, and X (Cl, Br, I). Molecular orbitals are built up by the interaction of the atomic orbitals of these elements held together at bonding separations. It is convenient at this point to adopt an energy scale derived from Simple Hückel MO theory, in which the "coulomb integral," α ($= \alpha_C$) is the reference point on the energy scale,

$$a = a_C = \int [2p_c(1)] h^{\text{eff}}(1) [2p_c(1)] \, d\tau_1 \tag{5.12}$$

and the absolute value of the "resonance integral," $|\beta|$ ($= |\beta_{CC}|$), is the unit of energy,

$$\beta = \beta_{CC} = \int [2p_c(1)]_A h^{\text{eff}}(1) [2p_c(1)]_B \, d\tau_1 \tag{5.13}$$

Thus, α (= α_C) is the core energy of an electron localized to the 2p atomic orbital of a carbon atom, and β (= β_{CC}) is the energy associated with the interaction of two carbon 2p orbitals overlapping in a π (parallel) fashion at the C—C separation of benzene (or ethylene).

Heteroatoms in SHMO Theory

In SHMO, the core energies of heteroatoms, X, are specified in terms of α and β, and the interaction matrix elements for p orbitals overlapping in a π fashion on any pair of atoms, X and Y, are specified in terms of β. Thus,

$$a_X = a + h_X|\beta| \qquad (5.14)$$

$$\beta_{XY} = k_{XY}|\beta| \qquad (5.15)$$

In SHMO theory the energy of π bond formation in ethylene is 2β (since the strength of a C—C π bond is about 280 kJ/mol, one may consider $|\beta|$ to be about 140 kJ/mol). The energies of electrons in 2p orbitals of N and O, normally bound in π bonding environments (i.e., as —N—, —O—, and F·) are given by $h_{N2} = -0.51$ and $h_{O1} = -0.98$. The π-type interaction matrix elements of most pairs, except those involving F, are approximately given by $k_{XY} = -1$. The energies of electrons in 2p orbitals of N and O in normal (saturated) bonding environments (i.e., as >N— and —O—) are given by $h_{N3} = -1.37$ and $h_{O2} = -2.05$. The π-type interaction matrix elements of these kinds of orbitals with the normal C 2p orbitals are approximately given by $k_{CN} = -0.8$ and $k_{CO} = -0.67$, reflecting the smaller size of the orbitals and greater bond lengths. The value for the C—F interaction, $k_{CF} = -0.5$, is small for the same reasons. The change in effective electronegativity of the 2p orbitals is a consequence of the increase in the number of atoms to which they are coordinated (see below), although in usual SHMO usage, the distinction is made in terms of the number of electrons which the atom contributes to the π system. A more complete list of h_X and k_{XY} parameters, derived on the basis of PPP calculations by Van Catledge [86], is given in Table 5.1.

The Effect of Coordination Number of α and β

A decrease in the coordination number at N from three to two reduces the effective electronegativity of the remaining nonbonded p orbitals at the center. The change in h_N is $0.86|\beta|$. A decrease in the coordination number at O from two to one similarly reduces the effective electronegativity of the remaining nonbonded p orbitals at the O center. The change in h_O is larger, $1.12|\beta|$. Within the same molecule, the coordination number of N or O may readily be changed by the process of protonation or deprotonation, as a consequence of a change of pH, for instance. The results of SHMO calculations on the enolate anion and on enol are shown in Figure 5.4. In the enolate case, $h_{O1} = -0.97$,

TABLE 5.1. SHMO Values for Heteroatoms: $\alpha_X = \alpha + h_X|\beta|$; $\beta_{XY} = k_{XY}|\beta|$[a]

X	h_X #electr	Y	C[b]	B[b]	N2[c]	N3[b]	O1[d]	O2[c]	k_{XY} F[d]	Si[b]	P2[c]	P3[b]	S1[d]	S2[c]	Cl[d]
C[b]	1	0.00	−1.00												
B[b]	0	0.45	−0.73	−0.87											
N2[c]	1	−0.51	−1.02	−0.66	−1.09										
N3[b]	2	−1.37	−0.89	−0.53	−0.99	−0.98									
O1[d]	1	−0.97	−1.06	−0.60	−1.14	−1.13	−1.26								
O2[c]	2	−2.09	−0.66	−0.35	−0.80	−0.89	−1.02	−0.95							
F[d]	2	−2.71	−0.52	−0.26	−0.65	−0.77	−0.92	−0.94	−1.04						
Si[b]	1	0.00	−0.75	−0.57	−0.72	−0.43	−0.65	−0.24	−0.17	−0.64					
P2[c]	1	−0.19	−0.77	−0.53	−0.78	−0.55	−0.75	−0.31	−0.21	−0.62	−0.63				
P3[b]	2	−0.75	−0.76	−0.54	−0.81	−0.64	−0.82	−0.39	−0.22	−0.52	−0.58	−0.63			
S1[d]	1	−0.46	−0.81	−0.51	−0.83	−0.68	−0.84	−0.43	−0.28	−0.61	−0.65	−0.65	−0.68		
S2[c]	2	−1.11	−0.69	−0.44	−0.78	−0.73	−0.85	−0.54	−0.32	−0.40	−0.48	−0.60	−0.58	−0.63	
Cl[d]	2	−1.48	0.62	−0.41	−0.77	−0.80	−0.88	−0.70	−0.51	−0.34	−0.35	−0.55	−0.52	−0.59	−0.68

[a]Ref 86.
[b]Tricoordinated, planar geometry.
[c]Dicoordinated.
[d]Monocoordinated.

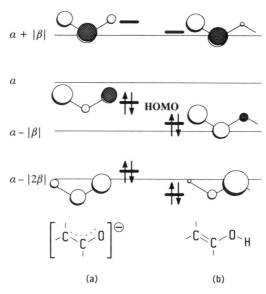

Figure 5.4. SHMO results for enolate (a) using O1 parameters and enol (b) with O2 parameters.

$k_{CO} = -1.06$, while for enol, the values, $h_{O2} = -2.09$, $k_{CO} = -0.66$, were used according to Table 5.1. The contributions of the individual $2p$ orbitals of C and O are displayed with sizes proportional to the magnitudes of the coefficients. The effect of the protonation can be seen as a lowering of the energy and changed polarization of all of the MOs, including the HOMO.

Normally no distinction is made between the kind of atom or group which is coordinated to the center of interest, but this may be a gross oversimplification in extreme cases. It is not reasonable to expect that the $2p$ orbital of a methyl group will have approximately the same energy as the $2p$ orbital of a trifluoromethyl group (assuming it were planar). Because of the strong inductive effect of the electronegative fluorine atoms acting in the σ framework, the carbon atom of trifluoromethyl would be significantly denuded of electrons. The $2p$ orbital is in effect more electronegative and falls below α.

Can one deduce reasonable values for the effective electronegativity of the p orbitals of C upon reduction of the coordination number from three to two (i.e., C2), as in alkynes, allenes, nitriles (R—CN), or carbenes, or even to one, as in CO, isonitriles (R—NC) or acetylides? A linear extrapolation from dicoordinated O ($\alpha_{O2} = \alpha - 2.09|\beta|$) and dicoordinated N ($\alpha_{N2} = \alpha - 0.51|\beta|$) to dicoordinated C yields an estimate of the energy of the $2p$ orbital as $\alpha + 0.86|\beta|$. This value is probably too high. It places the energy of the $2p$ orbital of a dicoordinated carbon above that of the $2p$ of a tricoordinated boron but the same is not true in the case of a dicoordinated N and a tricoordinated C, and the electronegativity differences in the series B, N, C (Table

2.3) are very similar. A reasonable compromise is to place α_{C2} midway between α_B and α_C,

$$a_{C2} = a + 0.23|\beta| \tag{5.16}$$

Hybridization at C in Terms of α and β

The 2s to 2p promotion energy of atomic C is about 800 kJ/mol. In a molecular environment, this value is expected to be somewhat less where the presence of other nuclei may stabilize p orbitals relative to s. The coordination number of the carbon atom has a direct effect on the orbital energies just as it had on the energies of heteroatom orbitals discussed in the previous section. Mullal has estimated the group electronegativities of CH_3, $CHCH_2$, and CCH to be 2.32, 2.56, and 3.10, respectively [87]. The last value is similar to his estimate for NH2. Boyd has placed all three values near 2.6 [88]. Allen, using his bond polarity index, has assigned values of 0.000, 0.027, and 0.050, respectively (compared to H -0.032 and F 0.189) [38]. Without attempting to be too quantitative, convenient values of the core energies of "hybrid" atomic orbitals, in $|\beta|$ units, recognizing that changes in coordination number also occur, are approximately

$$a_{sp} = a - 0.50|\beta| \quad \text{(coordination number = 1)} \tag{5.17}$$

$$a_{sp^2} = a - 0.33|\beta| \quad \text{(coordination number = 2)} \tag{5.18}$$

$$a_{sp^3} = a - 0.25|\beta| \quad \text{(coordination number = 3)} \tag{5.19}$$

The interaction energies of the pairs of hybridized orbitals interacting in a σ fashion would be strongly distance dependent. At typical single bond distances, one may adopt $k_\sigma = 1.5k_\pi|\beta|$ for all of them, but the value rises steeply as the separation is decreased. When the separation is that of a double bond, a value of $k_\sigma = 2.0k_\pi|\beta|$ is more appropriate. These values are suggested only to help place σ bonds or σ^* orbitals more or less correctly relative to π bonds and π^* orbitals when both may have similar energies.

GROSS CLASSIFICATION OF MOLECULES ON THE BASIS OF MO ENERGIES

Frontier orbital energy is not the only criterion which governs the chemical characteristics of a compound. For example, the magnitudes of the atomic orbital coefficients on any given atom may be responsible for the reduced basicity of benzene relative to ethylene, both of which have the same HOMO energy. Nevertheless, the energy criterion may be applied to deduce gross features. Figure 5.5 are shows four extreme cases into which molecules can

118 SIMPLE HÜCKEL MOLECULAR ORBITAL THEORY

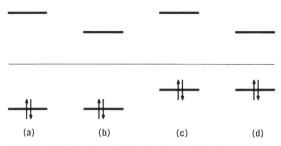

Figure 5.5. (a) High LUMO, low HOMO, large HOMO–LUMO gap: thermodynamically stable and chemically inert; (b) Low LUMO, low HOMO, large HOMO–LUMO gap: thermodynamically stable and chemically reactive as Lewis acid; (c) High LUMO, high HOMO, large HOMO–LUMO gap: thermodynamically stable and chemically reactive as Lewis base; (d) Low LUMO, high HOMO, small HOMO–LUMO gap: may be thermodynamically stable but chemically amphoteric and self-reactive.

be categorized on the basis of their frontier orbital energies. Compounds which have a large HOMO–LUMO gap ($>1.5|\beta|$) will be stable against self reaction, i.e., dimerization, polymerization, intramolecular rearrangements, etc. If the HOMO is low in an absolute sense ($<\alpha - 1|\beta|$), the compound will be chemically resistant to reaction with Lewis acids. If the LUMO is high in an absolute sense ($>\alpha + 1|\beta|$), the compound will be chemically resistant to reaction with Lewis bases.

- Compounds with a high LUMO and a low HOMO [Figure 5.5(a)] will be chemically inert. Saturated hydrocarbons, and fluorocarbons, and to some extent ethers, fall in this category.
- Compounds with a low HOMO and LUMO [Figure 5.5(b)] tend to be stable to self-reaction but are chemically reactive as Lewis acids and electrophiles. The lower the LUMO, the more reactive. Carbocations, with LUMO near α are the most powerful acids and electrophiles, followed by boranes and some metal cations. Where the LUMO is the σ^* of an H—X bond, the compound will be a Lowry–Bronsted acid (proton donor). A Lowry–Bronsted acid is a special case of a Lewis acid. Where the LUMO is the σ^* of a C—X bond, the compound will tend to be subject to nucleophilic substitution. Alkyl halides and other carbon compounds with "good leaving groups" are examples of this group. Where the LUMO is the π^* of an C=X bond, the compound will tend to be subject to nucleophilic addition. Carbonyls, imines, and nitriles exemplify this group.
- Compounds with a high HOMO and LUMO [Figure 5.5(c)] tend to be stable to self-reaction but are chemically reactive as Lewis bases and nucleophiles. The higher the HOMO, the more reactive. Carbanions, with HOMO near α are the most powerful bases and nucleophiles, followed by amides and alkoxides. The neutral nitrogen (amines, heteroaromatics)

and oxygen bases (water, alcohols, ethers, and carbonyls) will only react with relatively strong Lewis acids. Extensive tabulations of gas phase basicities, or *proton affinities* (i.e., $-\Delta G°$ of protonation) exist [89], [90]. These will be discussed in subsequent chapters.

- Compounds with a narrow HOMO–LUMO gap [Figure 5.5(d)] are kinetically reactive, subject to dimerization (e.g., cyclopentadiene), or reaction with Lewis acids or bases. Polyenes are the dominant examples of this group. The difficulty in isolation of cyclobutadiene lies not with any intrinsic instability of the molecule but with the self-reactivity which arises from an extremely narrow HOMO–LUMO gap.

CHAPTER 6

OLEFIN REACTIONS AND PROPERTIES

CASE STUDY 2: REACTIONS OF OLEFINS

Ethylene is the template for olefin reactions, but ethylene itself is rather unreactive, undergoing electrophilic attack by moderately strong Lewis acids. Nucleophilic attack on the π bond even by the strongest Lewis bases has not been reported. The following sequence involves intramolecular addition of a carbanion to an unactivated olefin [91], [92]. The reaction is undoubtedly facilitated by active participation of the lithium cation as a Lewis acid [93].

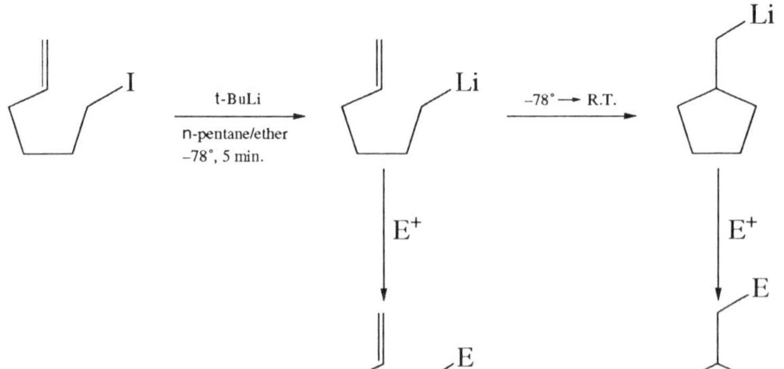

The normal course of reaction of alkenes involves addition of Lewis acids (electrophiles) yielding an intermediate carbocation which is trapped by a weak nucleophile [94]. The most common electrophilic addition reactions are sum-

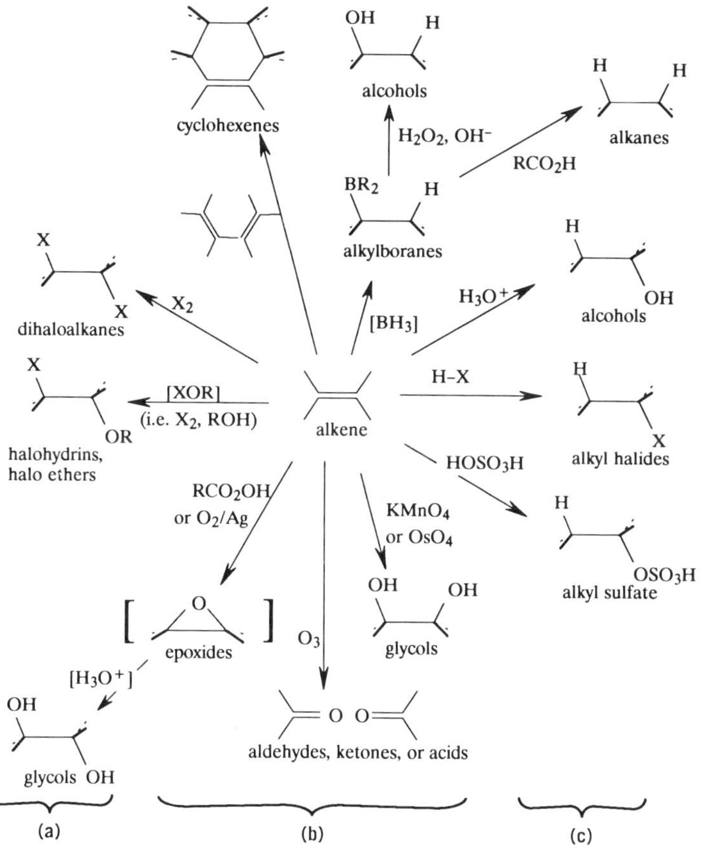

Figure 6.1. A summary of the most common electrophilic addition reactions of olefins. In each case the olefin reacts as a Lewis base. All reactions are regioselective. The overall stereochemistry is: **(a)** stereospecific *anti*; **(b)** stereospecific *syn*; **(c)** not stereospecific, in general.

marized in Figure 6.1. If the olefin is unsymmetrically substituted, the question of regioselectivity arises. We begin by examining the effects on the olefin π system of three classes of substituents as defined by Fleming, namely X: (π-electron donors), Z (π-electron acceptors), and C (conjugating) [7]. An interaction diagram showing the interaction of a C—C π bond with each type of substituent is shown in Figure 6.2. We note the effect of the substituent on the energy and polarization of the π bond *at its original site*.

Effect of X: Substituents

As shown in Figure 6.2(a), an X: substituent, which has a p orbital, or other suitable doubly occupied orbital which will interact with the π bond, raises the energy of the HOMO and LUMO, thus rendering the olefin more reactive as

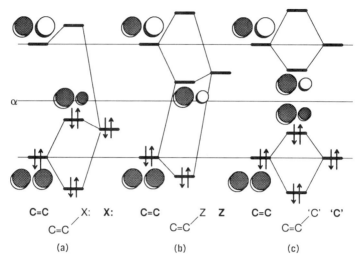

Figure 6.2. Interaction of C=C with: **(a)** X:; **(b)** Z; **(c)** C-type substituents.

a Lewis base. Of course, the electrons of the HOMO are also delocalized onto X. The probability of attack by an electrophile will be governed by the magnitude of the coefficient at the particular atomic position. Polarization of the HOMO away from the point of attachment of the X substituent directs electrophilic attack to that carbon. Attack may also be directed to X itself in certain cases, although this is usually reversible, and may have no net consequences. The X substituted olefin π system is isoelectronic to that of the allyl anion (Figure 5.2). The polarization of the π bond can be deduced by *averaging* the HOMOs of the olefin and allyl anion, i.e.,

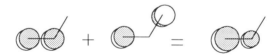

HOMO of alkene   HOMO of allyl anion

X: substituents are $-NR_1R_2$, $-OR$, $-SR$, $-F$, $-Cl$, $-Br$, $-I$, $-CH_3$ (or any alkyl). The R's may be H or alkyl, or aryl, or even acyl. Thus,

$$CH_3CH=CH_2 + H-OSO_2OH \rightarrow$$

$$CH_3C^+H-CH_3 + {}^-OSO_2OH \text{ (not } CH_3CH_2-C^+H_2)$$

As a special case, consider the aldol reaction

$$CH_2=CHO^- + CH_3CHO \rightarrow CH_3C(O^-)H-CH_2CHO$$

Effect of Z Substituents

As shown in Figure 6.2(b), a Z substituent, which has a p or π^* orbital, or other suitable empty low-lying orbital which will interact with the π bond, lowers the energy of the HOMO and LUMO, thus rendering the olefin more reactive as a Lewis acid. The electrons of the HOMO are also delocalized onto Z. The probability of attack by an electrophile will be governed by the magnitude of the coefficient at the particular atomic position. Polarization of the LUMO away from the point of attachment of the Z substituent directs nucleophilic attack to that carbon. Attack may also be directed to Z itself in certain cases, and this may be irreversible, providing an alternate pathway for the reaction. The Z-substituted olefin π system has some characteristics of allyl cation (Figure 5.2). The polarization of the π bond can be deduced by *averaging* the LUMOs of the olefin and allyl cation, i.e.,

LUMO of alkene + LUMO of allyl cation =

Z substituents are $-COR$, $-SOR$, $-SO_2R$, $-NO$, $-NO_2$. The R's may be H, alkyl, aryl, or even X: substituents. Thus, the Michael addition is the best example

$$CH_2=CH-CO_2CH_2CH_3 + CH^-(CO_2CH_2CH_3)_2 \rightarrow CH_2-C^-H-CO_2CH_2CH_3$$
$$|$$
$$CH(CO_2CH_2CH_3)_2$$

Consider also

Effect of "C" Substituents

As shown in Figure 6.2(c), a "C" substituent, which has evenly spaced π and π^* orbitals, raises the energy of the HOMO and lowers the LUMO, thus rendering the olefin more reactive both as a Lewis base and a Lewis acid. The electrons of the HOMO and LUMO are also delocalized onto "C." The probability of attack by an electrophile or nucleophile will be governed by the magnitude of the coefficient at the particular atomic position. Polarization of both the HOMO and LUMO away from the point of attachment of the "C" substituent directs attack to that carbon. Attack may also be directed to "C" itself in certain cases. The "C" substituted olefin π system is assumed to be that of butadiene (Figure 5.2), hence the polarization of the π and π^* orbitals, i.e.,

HOMO of 'C' substituted alkene   LUMO of 'C' substituted alkene

"C" substituents are conjugated alkenes, alkynes, or aryl groups.

Effect of Distortion of the Molecular Skeleton

Olefins may be synthesized in which the π bond must be superimposed on a σ framework which deviates significantly from the ideal geometry, namely coplanarity with a two sp^2 hybridized carbon atoms, or in the case of alkynes, collinearity. *Twisting* of the two ends of the double bond relative to each other has the consequence of reducing the π overlap and hence the resonance integral is less that $|\beta_{CC}|$. The higher HOMO means that the twisted olefin is more susceptible to electrophilic attack, the lower LUMO implies an increased susceptibility to nucleophilic attack, and the smaller HOMO–LUMO gap suggests a bathochromic shift for the $\pi\pi^*$ electronic transition [95]. The increased susceptibility of twisted (strained) alkenes toward electrophilic attack has been demonstrated experimentally for MCPBA (meta-chloroperbenzoic acid) epoxidation [96] of a variety of strained alkenes. The rate enhancement was attributed to relief of strain in the transition state, but a correlation was noted with ionization potential, and hence the energy of the HOMO. Conjugated alkenes which have high HOMO energies also were less reactive than expected on the basis of the correlation with twisted monoalkenes. In orbital interaction terms, the reduced reactivity of conjugated systems is attributed to the smaller orbital coefficients, and hence lower intrinsic interaction matrix elements [equation (3.50) of Chapter 3].

In fact, it is difficult to effect a major perturbation to the olefin π system by twisting. For instance, the σ framework of *trans*-cyclooctene is twisted by 40° out of planarity, but pyramidalization at each end of the double bond is such as to reduce the twist of the p orbitals to only about 11° [97]. The ionization potential, 8.69 eV, is lower than in the case of *cis*-cyclooctene (8.98 eV) or cyclohexene (9.12 eV), as expected. The highly strained anti-Bredt olefin, 11-bromo-*endo*-9-chloro-7-ethoxybicyclo[5.3.1]-undec-1(11)-ene has been synthesized and its structure determined by X-ray diffraction [98]. The twist in the σ framework is approximately 60°. As in the case of *trans*-cyclooctene, both ends of the π bond are pyramidally distorted in such a way as to reduce the twist angle of the p orbitals to 29°.

trans-Cyclooctene

11-Bromo-endo-9-chloro-7-ethoxybicyclo-[5.3.1]undec-1(11)-ene

Alkynes

At the level of simple orbital interaction theory, alkynes differ from alkenes in only two respects. First, the carbon atoms are dicoordinated, with the consequence that the *p* orbitals are at higher energy than α_C. Second, the shorter C—C separation implies that the intrinsic interaction will be larger than β_{CC}. The level of the degenerate π MOs is expected to be about the same as in alkenes but the π^* MOs will be higher in energy. Additional factors such as increased coulombic repulsion of the two electrons in each π MO (due to the shorter separation) may further destabilize the π MOs. In fact, the reactivity of alkynes toward electrophilic attack is rather similar to that of alkenes when the electrophile would be expected to form an acyclic intermediate (e.g., HX) but slower when a cyclic intermediate is formed (e.g., Br$_2$) [94]. Addition of nucleophilic free radicals to alkynes proceeds more slowly than to alkenes, an observation consistent with the expected higher π^* MO of the former. However, addition of nucleophiles is faster in general than to alkenes ([99], pp. 670–672). Since strong charged nucleophiles are always associated with a metallic counter ion, the Lewis acidity of the cation toward the second "nonreacting" π MO may be responsible for the increased reactivity.

CHAPTER 7

REACTIVE INTERMEDIATES

REACTIVE INTERMEDIATES, $[CH_3]^+$, $[CH_3]^-$, $[CH_3]\cdot$, AND $[:CH_2]$

The major carbon centered reaction intermediates in multistep reactions are carbocations (carbenium ions), carbanions, free radicals, and carbenes. Formation of most of these from common reactants is an endothermic process and is often rate-determining. By the Hammond principle, the transition state for such a process should resemble the reactive intermediate. Thus, although it is usually difficult to assess the bonding in transition states, factors which affect the structure and stability of reactive intermediates will also be operative to a parallel extent in transition states. We examine the effect of substituents of the three kinds discussed above on the four different reactive intermediates, taking as our reference, the parent systems, $[CH_3]^+$, $[CH_3]^-$, $[CH_3]\cdot$, and $[:CH_2]$.

Carbocations

In Figure 7.1, are shown the interaction diagrams relevant to a carbocation substituted by X:, Z, and "C" substituents.

- A carbocation is strongly stabilized by an X: substituent [Figure 7.1(a)] through a π-type interaction which also involves partial delocalization of the nonbonded electron pair of X to the formally electron-deficient center. At the same time, the LUMO is elevated, reducing the reactivity of the electron deficient center towards attack by nucleophiles. The effects of substitution are cumulative. Thus, the more X:–type substituents there are, the more thermodynamically stable the cation and the less reactive it is as a Lewis acid. As an extreme example, guanidinium ion, which may

REACTIVE INTERMEDIATES, $[CH_3]^+$, $[CH_3]^-$, $[CH_3]\cdot$, AND $[:CH_2]$

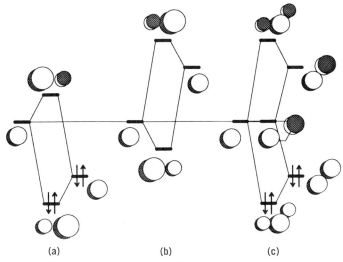

Figure 7.1. A carbocationic center interacting with: **(a)** an X: substituent; **(b)** a Z substituent; **(c)** a "C" substituent.

be written as $[C(NH_2)_3]^+$, is stable in water. Species of the type $[-C(OR)_2]^+$ are common intermediates in acyl hydrolysis reactions. Even cations stabilized by fluorine have been reported and recently studied theoretically [100].

The filled bond orbitals of adjacent alkyl groups may donate electrons by π-type overlap (Figures 3.19 and 3.20 in Chapter 3) to an adjacent carbocationic center. Thus, an alkyl group may be considered to be an X-type substituent. The highest combination of the C—H bonding orbitals of a methyl group has a π-donor capability intermediate to that of the nonbonded orbitals of O and F. The donor abilities of σ bonds were discussed in Chapter 4.

- A carbocation is only weakly stabilized by a Z substituent [Figure 7.1(b)] through a π-type interaction with the π bond of the Z group [not shown in Figure 7.1(b)]. The interaction is weak because the π bond of a Z substituent is very low in energy and polarized away from the cationic center. The dominant interaction is with the LUMO of Z, which does not add to thermodynamic stabilization, but greatly enhances the Lewis acidity of the cation, increasing the reactivity of the electron deficient center towards attack by nucleophiles. Z-substituted cations are being increasingly reported as intermediates in solvolysis reactions [101].
- A carbocation is strongly stabilized by a "C" substituent [Figure 7.1(c)] through π-type interactions which involve substantial delocalization into the substituent. The LUMO energy is relatively unchanged, but the reactivity of the electron deficient center towards attack by nucleophiles is

128 REACTIVE INTERMEDIATES

reduced because the orbital coefficients are smaller. The effects of substitution are cumulative. Thus, the more "C"-type substituents there are, the more thermodynamically stable the cation and the less reactive it is as a Lewis acid. A prime example is triphenyl carbocation.

Intermolecular Reactions of Carbocations

Carbocations are strong Lewis acids which occur as intermediates in reactions following the S_N1 (Chapter 9) or E1 (Chapter 10) mechanistic routes. The most obvious and common reaction is recombination with a nucleophile (a Lewis base) to form a σ bond. If the nucleophilic site (HOMO) involves a nonbonded pair of electrons (Path a), a stable covalently bonded complex will form. If the HOMO is a σ bond, direct reaction is unlikely unless the bond is high in energy and sterically exposed, as in a three-membered ring, but if the bond is to H, hydride abstraction may occur (Path b, steps 1 and 2), or a hydride bridge may form (Path b, step 1). The last two possibilities are discussed further in Chapter 10. If the HOMO is a π bond, a "π complex" may result (Path c, step 1), or, more commonly, donation of the π electrons results in the formation of a σ bond at the end where the π-electron density was higher, the other end becoming Lewis acidic in the process (Path c, steps 1 and 2). The effects of substituents on olefin reactivity were discussed in Chapter 6.

X = halide → alkyl halides
X = H_2O → alcohols
X = HOR or HOAr → ethers
X = RCO_2H or $ArCO_2H$ → esters
X = NR_3 → alkylammonium salts
X = RSH → thioethers (thiols if R = H)

Intramolecular Reactions of Carbocations

Intramolecular reactions of carbocations are shown in the scheme below. The combination of steps 1 and 2 corresponds to a 1,2 hydride shift (R = H) or a Wagner–Meerwein rearrangement (R = alkyl). The intermediate bridged nonclassical structure reached after step 1 may correspond to a transition state for the rearrangement, or the stable form of the carbocation. If R = H, the bridged form is known to be stable only in the case of ethyl cation, where there are no substituents on either carbon atom (other than H). In other cases, stable bridged forms are obtained if the R group has a relatively loosely bound pair of electrons in a *p*- or π-type orbital for back donation into the π*-like orbital of the C—C fragment, as in the case of R = halide (e.g., bromonium ion), R = aryl (e.g., ethylbenzenium ion [102]), R = vinyl (cyclopropylcarbinyl cation). More or less symmetrical bridging by R = alkyl is not likely but has been shown to be the case in the 2-norbornyl cation [82], [103]. Depending on substituents in R, elimination of R$^+$ may occur (step 3). Each of the cationic species may react intermolecularly as shown in the previous scheme at sites labeled ''A.'' If R = H, the hydrogen end of the σ bond in the two ''classical'' structures is also liable to attack by nucleophile (the E1 mechanism, Chapter 10). Of course, step 3 would not occur in this case.

Carbanions

Except for the most highly stabilized carbanions, carbanion chemistry in solution is always complicated by the presence of the counterion, usually a metal, which is a Lewis acid and almost invariably is involved in the course of the reaction. Relative stabilities of carbanions in solution are difficult to establish for the same reason. In recent years, much information has been gathered about carbanion stabilities, structures, and reactivities in the gas phase [104]. Figure 7.2 shows the interaction diagrams relevant to a carbanion substituted by X:, Z, and ''C'' substituents.

- A carbanion is destabilized by an X: substituent [Figure 7.2(a)] through a two-orbital–four-electron π-type interaction. The Lewis basicity and nucleophilicity are greatly increased. Because the HOMO is so high in energy, some involvement may be observed by the X: group's LUMO, particularly in the case of alkyl-substituted carbanions where gas-phase basicities suggest that in some cases, the alkyl group may act to stabilize a carbanion [104]. Carbanions are easily oxidized and may spontaneously autoionize in the gas phase. Methyl carbanion is stable in the gas phase, but ethyl, 2-propyl, and *tert*-butyl carbanions have not been observed in the gas phase [104]. Accordingly, radical intermediates should always be

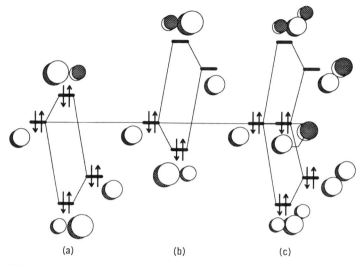

Figure 7.2. A carbanionic center interacting with: **(a)** an X: substituent; **(b)** a Z substituent; **(c)** a "C" substituent.

suspected in X: substituted carbanionic reactions in nonpolar media. The π effects of X: substitution are somewhat offset by the inductive effect of the electronegative X:. There is some evidence that the inductive effect accumulates more rapidly than the resonance effect. Carbanions next to two sulfur atoms are common and carbanions next to two oxygen atoms have been reported. Halogenation favors carbanion formation.

- A carbanion is strongly stabilized by a Z substituent [Figure 7.2(b)] through the π-type interaction with the LUMO of the Z group. Z-substituted carbanions are common intermediates in solvolysis reactions, and as nucleophiles in C—C bond forming reactions. The involvement of —CF$_3$ groups as Z-type substituents has been well documented [105]. Phosphonium (R$_3$P$^+$—) and sulfonium (R$_2$S$^+$—) groups stabilize carbanionic centers, forming ylides. The stabilization is due in part to electrostatic effects, and in part to orbital interactions via σ^* or 3d involvement, both of which may be lowered in energy by the formal (+)ve charge. In the last sense, these groups are acting as Z-type substituents. The synthetic potential of trigonal boron as a Z-type substituent in stabilizing carbanionic centers has been demonstrated [106].

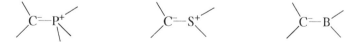

The very important, reactive intermediate, the enolate ion is an example of a Z-substituted carbanion. The charge distribution and HOMO obtained by SHMO calculation, are shown below. The oxygen atom bears

The enolate ion: $\diagup\!\!\diagup$ -0.76 O   Net charges -0.587 ◯ $\epsilon_{HOMO} = a - 0.412\,|\beta|$
$+0.09$
-0.33 0.309
                                          0.749

the majority of the negative charge but the HOMO has the highest contribution from the carbon atom. Accordingly, charged electrophiles (hard electrophiles) should preferentially add to the oxygen atom and neutral (soft) electrophiles would be expected to add to the carbon atom.

- A carbanion is also stabilized by a "C" substituent [Figure 7.2(c)] through π-type interactions which involve substantial delocalization into the substituent. The LUMO energy is relatively unchanged, but the reactivity of the electron rich center towards attack by electrophiles is reduced because the orbital coefficients are smaller. "C"-type substituents are predominantly hydrocarbons and cannot easily support a negative charge unless other factors are present.

Carbon Free Radicals

Free radicals are molecules with an odd number of electrons. In our simple theory, all electrons are considered to be paired up in molecular orbitals, leaving one orbital with a single electron. The molecular orbital which describes the distribution of the "odd" electron is designated the SOMO (singly occupied MO). In the ground state of the radical, the SOMO is also the highest occupied MO. Often the SOMO is strongly localized. If the localization is to a tricoordinated (trigonal) carbon atom, then the radical species is described as a carbon-free radical. If the SOMO is relatively low in energy, the principal interaction with other molecules will be with the occupied MOs (three-electron–two-orbital type, Figure 3.8). In this case the radical is described as *electrophilic*. If the SOMO is relatively high in energy, the principal interaction with other molecules may be with the unoccupied MOs (one-electron–two-orbital type, Figure 3.10). In this case the radical is described as *nucleophilic*. Substituents on the radical center will affect the electrophilicity or nucleophilicity of free radicals as shown below.

Figure 7.3 illustrates the interaction diagrams relevant to a carbon-free radical substituted by X:, Z, and "C" substituents. The figure also applies to free radicals centered on other atoms if one takes into account the orbital energies appropriate for the heteroatom.

- A free radical is stabilized by an X: substituent [Figure 7.3(a)] through a two-orbital–three-electron π-type interaction. The nucleophilicity of the radical is greatly increased. X:-substituted free radicals are more easily oxidized. The π effects of X: substitution are somewhat augmented by the inductive effect of the electronegative X: in stabilizing the radical.
- A carbon-free radical is stabilized by a Z substituent [Figure 7.3(b)]

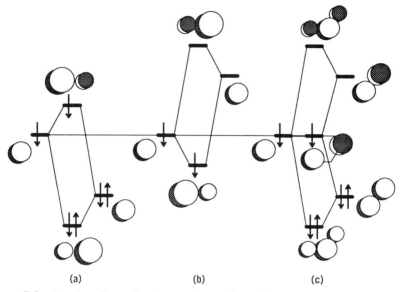

Figure 7.3. A carbon-free radical center interacting with: **(a)** an X: substituent; **(b)** a Z substituent; **(c)** a "C" substituent.

through the π-type interaction with the LUMO of the Z group. The SOMO is lowered in energy and the free radical is more electrophilic as a consequence.

- A carbon-free radical is also stabilized by a "C" substituent [Figure 7.3(c)] through π-type interactions which involve substantial delocalization into the substituent. The SOMO energy is relatively unchanged, but the reactivity of the odd-electron center is reduced because the orbital coefficients are smaller.

Free radical polymerization of a 1 : 1 mixture of dimethyl fumarate and vinyl acetate, resulting in a highly regular alternating co-polymer, illustrates the importance of substitution on the properties of both the free radical and the olefinic substrate [107]. Dimethyl fumarate is a "Z"-substituted olefin. It has a

low-lying LUMO which is not polarized due to the symmetrical substitution. Because of the low-lying LUMO, dimethyl fumarate is susceptible to nucleophilic attack. On the other hand, vinyl acetate is an olefin with an "X": substituent at one end. The HOMO is polarized away from the substituent acetoxy group, OAc. Since the HOMO is also raised in energy as a result of the interaction, vinyl acetate would be particularly susceptible to electrophilic attack, the preferred site being the C atom β to the substituent. Addition of a radical to dimethyl fumarate generates a "Z"-substituted carbon-free radical. Addition of a radical to vinyl acetate generates an "X:"-substituted carbon-free radical. Each of the two possible types of radicals has an equal probability of encountering either olefin. The situation is depicted in Figure 7.4. The low-lying SOMO of the electrophilic fumarate radical (left) interacts preferentially with the high HOMO of vinyl acetate leading to selective formation of the acetoxy radical (right). The high SOMO of the acetoxy radical interacts most strongly with the low LUMO of dimethyl fumarate with selective formation of the fumarate radical, etc.

Radical stabilities may be measured experimentally by the determination of homolytic bond dissociation energies (BDEs) in the gas phase [108], or in solution by relating them empirically to the pK_{HA} and the oxidation potentials, $E_{ox}(A^-)$ of weak acids, H—A [109].

$$BDE_{HA} = 1.37pK_{HA} + 23.1E_{ox}(A^-) + 73.3$$

A quantity called the radical stabilization energy (RSE) may be defined to

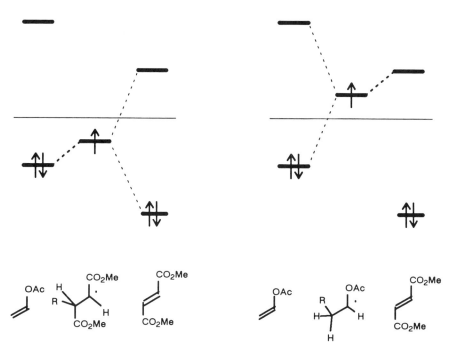

Figure 7.4. Interactions which determine the relative reactivities of carboxyalkyl (left) and acyloxy radicals (right).

related the stabilities of substituted carbon radicals to methyl radicals. The effects of adjacent X:, Z, and C substituents on the RSEs of carbon-centered radicals has been widely investigated [110], [111]. The expectations based on simple orbital interaction theory as espoused above are widely supported by the experimental findings, except that when the π-donor or- acceptor ability of the group is weak and the inductive electron withdrawing power is large, as in $F_3C\cdot$ and $(Me)_3N^+CH_2\cdot$, the net effect is to destabilize the radical relative to the methyl radical [111].

Carbenes

Carbenes are species which contain a dicoordinated carbon atom formally with two valence electrons. The possible electronic structures of the parent carbene, methylene $:CH_2$, are shown in Figure 7.5. The $2p$ orbital of the dicoordinated C atom is placed above α by about $0.25\,|\beta|$ in order to accommodate the lower electronegativity.

One would expect on the basis of Hückel MO theory that the lowest energy configuration, i.e., the electronic ground state of $:CH_2$, is $S_0(^1A_1)$. This is incorrect. In this case we are done in by the assumption that the electron–electron coulomb repulsion can be neglected. The coulomb repulsion is most severe when two electrons are constrained to the same small MO. Although one pays a price for separation of the electrons into different MOs, this is largely compensated by the relief of electron–electron repulsion which ensues by virtue of the orthogonality of the MOs. Thus, while it is true that S_0 is lower in energy than S_1, the difference is not as great as might have been expected. The *triplet* configuration, T_1 ($^3B_1$), is additionally stabilized by relief of "exchange repulsion," the self avoidance of two electrons with the same spin, i.e., the principle underlying Hund's rule. T_1 falls 37.7 kJ/mol *below*

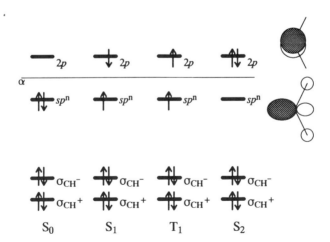

Figure 7.5. The electronic configurations and state designations for $:CH_2$.

S_0 in the case of $:CH_2$ [112]. The geometries of singlet and triplet carbene are different: singlet, HCH = 103° [113]; triplet, HCH = 136° [114]. The chemical consequences of whether the carbene is in its lowest singlet state (S_0) or in its lowest triplet state (T_1) are dramatic. With the electrons spin paired, the possibility exists for concerted reactions which preserve stereochemistry. This possibility is precluded if the carbene is in the triplet state. Chemical reactions of triplet carbenes are nonconcerted, and accompanied by undesirable side reactions, possible isomerizations and loss of stereochemical integrity. The olefin insertion reaction which yields cyclopropanes is discussed in more detail in Chapter 13.

The small difference between the energies of S_0 and T_1 may easily be overturned by the effects of substituents on the carbene center. Figure 7.6 shows the interaction diagrams which are relevant to the interaction of the carbene center with each of the three types of substituent. Because of more favorable overlap, the interaction of the Carbon $2p$ orbital with substituent p or π orbitals is expected to dominate. The sp^n orbital which lies in the nodal plane of the substituent p or π orbital with not interact, except more weakly with substituent σ orbitals. Its energy is shown in Figure 7.6 as unperturbed by the substituent.

- The $2p$ orbital of the carbene is raised in energy by an X: substituent [Figure 7.6(a)] thereby increasing the separation of the $2p$ and sp^n orbitals. The ground state of an X:-substituted carbene is S_0. In :CHF, S_0 is

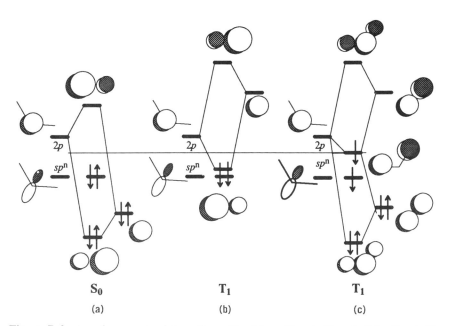

Figure 7.6. A carbene center interacting with: **(a)** an X: substituent; **(b)** a Z substituent; **(c)** a "C" substituent. Only the two highest MOs of the carbene are shown.

lower than T_1 by 61.5 kJ/mol [115]. In :CF_2, the separation is 237 kJ/mol [116]. Many carbenes in this class are known. The most familiar is dichloro carbene, :$CCl_2\cdot$, whose lower singlet electronic states have been investigated by *ab initio* calculations [117], [118]. The ground states of :CBr_2 and :CI_2 have also been predicted to be singlet [118]. Chloro methoxy carbene, Cl—C—OCH_3, has been studied spectroscopically [119]. See also [120] for an experimental and theoretical study of :$C(OCH_3)_2$ and F—C—OCH_3. A stable crystalline carbene with two N substituents has been prepared and characterized [121].

1,3-di-1-adamantylimidazol-2-ylidene

The carbene, 1,3-di-1-adamantylimidazol-2-ylidene, is additionally stabilized by aromaticity of the five-membered ring, and by steric protection by the neighboring adamantyl substituents. Similar carbenes with the adamantyl groups replaced by aryl groups are also stable [122].

- As shown in Figure 7.6(b) and (c), Z and "C" substituents either lower the $2p - sp^n$ gap or leave it about the same. In either case, the ground state for Z or "C"-substituted carbenes is expected to be T_1. For a recent investigation of phenyl carbene, C_6H_5CH, see [123]; ethynyl carbene, HC≡C—CH [124]; carboxylate substituted carbenes, R—C—CO_2R' [125]; formyl carbene, H—C—C(O)H [126].

Nitrenes, [NH], and Nitrenium Ions, $[NH_2]^+$

Nitrenes are the neutral nitrogen analogs of carbenes, while nitrenium ions are isoelectronic to carbenes. Many of the reactions which are observed for carbenes have parallels in nitrene and nitrenium ion chemistry. Like carbenes, nitrenes and nitrenium ions can exist in both singlet and triplet states. There are some interesting divergences in chemical properties and in the effects of substituents, however, which are readily understood on the basis of orbital interaction diagrams.

Nitrenes

Nitrenes are neutral species which contain a monocoordinated nitrogen atom formally with four valence electrons. The possible electronic structures of the parent nitrene, NH, are shown in Figure 7.7. On the basis of extrapolation from the energies of the $2p$ orbitals of tricoordinated ($\alpha - 1.37|\beta|$) and dicoordinated ($\alpha - 0.51|\beta|$) nitrogen atoms, one would expect the energy of the degenerate $2p$ orbitals of monocoordinated N atom to be close to α. The

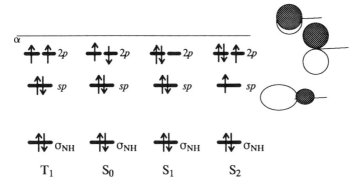

Figure 7.7. The electronic configurations and state designations for NH.

ground state is the triplet, T_1. T_1 falls 155 kJ/mol below S_0 which has the same orbital description [127]. The singlet state, S_1, which is analogous S_0 of :CH_2, has been found spectroscopically to lie 261 kJ/mol above T_1 [127]. The energy separations, which are much larger than in the corresponding carbene, represent relief of exchange and coulomb repulsion of the smaller orbitals of nitrogen. The gap between the nonbonding orbital, sp, and the degenerate $2p$ orbitals is sufficiently large that states such as S_2 (Figure 7.7) do not figure into the chemistry of nitrenes.

S_1 is the first state reached when a nitrene is generated by most methods, including photolysis of the corresponding azide, $R-N_3$. The reactions of singlet nitrenes are similar to analogous reactions of singlet carbenes, with the exception that nitrenes have a much greater propensity for dimerization, yielding the azo compound, $R-N=N-R$.

Figure 7.8 shows the interaction diagrams which are relevant to the interaction of the nitrene nitrogen with each of the three types of substituent. The situation is somewhat more complex than in the case of carbenes because some X: substituents, i.e., halogens, are axially symmetric and do not lift the degeneracy of the $2p$ orbitals. Alkyl substituents are also approximately threefold symmetric and probably would not lift the degeneracy enough to make the singlet state more stable. The amino group will raise the π system sufficiently to result in a singlet closed-shell ground state. Theoretical studies on aminonitrene, H_2N-N, find the singlet state to lie 63 kJ/mol below the triplet state [128]. However, the nitrene is very strongly basic and nucleophilic. The relatively small HOMO–LUMO gap suggests that such nitrenes may be colored and have a strong tendency to dimerize. Indeed, dialkylaminonitrenes, also called 1,1-diazines, are well known, and dimerize to tetrazenes [129].

$$\underset{\text{aminonitrene (1,1-diazine)}}{\overset{R}{\underset{R}{\diagdown}}N-N\diagup} \longleftrightarrow \overset{R}{\underset{R}{\diagdown}}N^+=N^- \xrightarrow{\text{dimerize}} \underset{\text{tetrazene}}{\overset{R}{\underset{R}{\diagdown}}N-N=N-N\overset{R}{\underset{R}{\diagup}}}$$

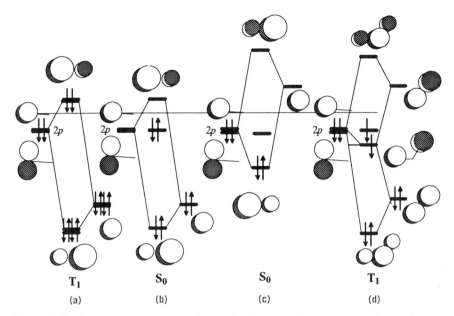

Figure 7.8. A nitrene center interacting with: **(a)** an axially symmetric X: substituent; **(b)** a planar X: substituent; **(c)** a Z substituent; **(d)** a "C" substituent. Only the two highest MOs of the nitrene are shown.

- As shown in Figure 7.8(c), Z substituents may also yield nitrenes with singlet ground states, in this case by lowering the energy of the π system. The remaining the $2p$ orbital is relatively low in energy. Thus acyl nitrenes will be strong Lewis acids at nitrogen in the σ plane. Such nitrenes have a strong tendency to rearrange by a 1,2 migration of the group attached to the acyl carbon to yield isocyanates. If the group is alkyl, the group migration is concurrent with nitrene formation, as in the Curtius, Hofmann, and Lossen rearrangements [130].

As with carbenes, "C" substituents will not alter the $2p_\pi - 2p_\sigma$ gap appreciably [Figure 7.8(d)]. Thus, the ground state for "C" substituted nitrenes is expected to be T_1. Theoretical [131], [132] and experimental [133] studies of phenylnitrene, C_6H_5N, are in agreement that the ground state is the triplet π radical, $^3A_2$ [T_1 of Figure 7.8(d)], which lies 75 kJ mol^{-1} below the open shell singlet, $^1A_2$. The closed shell state, S_0, i.e., $^1A_1$, is predicted to lie 162 kJ mol^{-1} above the triplet ground state [131].

Nitrenium Ions

Nitrenium ions are isoelectronic to carbenes [134]. They contain a dicoordinated nitrogen atom formally with two valence electrons. The possible electronic structures of the parent nitrenium ion $:NH_2^+$, are shown in Figure 7.9. One would expect on the basis of Hückel MO theory that the lowest energy configuration, i.e., the electronic ground state of $:NH_2^+$, is S_0 ($^1A_1$). As with $:CH_2$, this is incorrect. T_1 falls 126 kJ/mol *below* S_0 in the case of $:NH_2^+$ [135], [136].

Figure 7.10 illustrates the interaction diagrams which are relevant to the interaction of the nitrenium ion with each of the three types of substituent. The $2p$ and sp^n orbitals of the nitrenium ion are lower in energy than the analogous orbitals of a carbene. Several differences in the effects of substituents ensue.

- The $2p$ orbital of the nitrenium ion will interact strongly with the p orbital of an X: substituent [Figure 7.10(a)]. The nitrenium ion is considerably stabilized by the two-electron–two-orbital interaction. The $2p - sp^n$ gap is widened. The ground state of an X: substituted nitrenium ion is expected to be S_0. Apparently, according to *ab initio* studies, a methyl group is not a sufficiently good π donor. The ground state of CH_3NH^+ was found computationally to be T_1 [136], although the cyclic dialkyl nitrenium ion, $\overline{CH_2(CH_2)_3N^+}$, is predicted to have ground state S_0 [136]. The LUMO remains very low in energy and the X:-substituted nitrenium ion will be a strong Lewis acid. In fact, both CH_3NH^+ and $(CH_3)_2N^+$ are predicted to rearrange without activation by a 1,2 hydride migration from methyl to N on the single potential energy surface [136]. The X:-substituted phosphorus and arsenic analogs of nitrenium ions (phosphenium [137] and arsenium [138], respectively), are also known.

- As shown in Figure 7.10(b), Z substituents will interact relatively weakly because of the larger energy separation that is found in Z-substituted carbenes. The interaction results in some stabilization but lowers the $2p -$

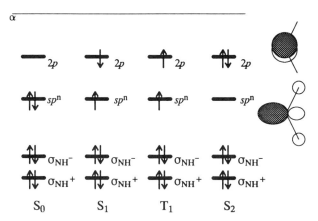

Figure 7.9. The electronic configurations and state designations for $:NH_2^+$.

140   REACTIVE INTERMEDIATES

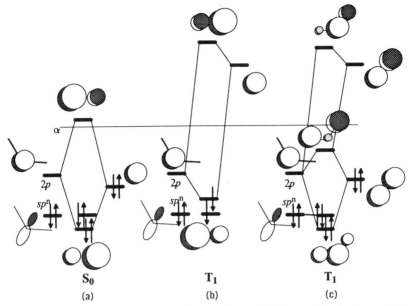

Figure 7.10. A nitrenium ion center interacting with: **(a)** an X: substituent; **(b)** a Z substituent; **(c)** a "C" substituent. Only the two highest MOs of the nitrenium ion are shown.

sp^n gap. Thus, the ground state for Z-substituted nitrenium ions is expected to be T_1. The triplet diradical will be strongly electrophilic. The S_0 state, with both electrons in the sp^n orbital would gain no stability from the interaction with the carbonyl *in the planar configuration*. In fact, on the S_0 potential surface, rotation through 90° about the acyl—N bond occurs, placing the sp^n orbital into conjugation with the Z substituent. Where the Z substituent is a carbonyl group, the perpendicular orientation

$$\underset{S_0 \text{ state}}{\overset{R_2}{\underset{R_1}{N^+\!-\!C}}\!\!\overset{O}{=\!\!\!=}} \longrightarrow R_2\!\!-\!\!\overset{}{\underset{R_1}{N^+\!-\!C}}\!\!\overset{O}{=\!\!\!=} \longrightarrow \underset{R_1}{\overset{R_2}{N\!-\!C^+\!\!=\!O}}$$

places the other acyl substituent parallel to the vacant $2p$ orbital, and a rearrangement ensues if the second substituent on N is not a good π donor [134], [139].

- As shown in Figure 7.10(c), because of the lower energy of the nitrenium ion orbitals, interaction of the $2p$ orbital with the π-bonding orbitals of "C" substituents will be stronger than the interaction with the π^* orbitals. Substantial stabilization ensues. The interaction raises the separation of the $2p$ and sp^n orbitals. Thus, the ground state for ''C''-substituted nitrenium ions is expected to be S_0, unlike the situation in ''C''-substituted carbenes. This expectation has been supported by *ab initio* calculations on phenylnitrenium ion [140]. Because the LUMO remains very low, this type of nitrenium ion will be very electrophilic.

CHAPTER 8

CARBONYL COMPOUNDS

CASE STUDY 3: REACTIONS OF CARBONYL COMPOUNDS

The frontier orbitals of a carbonyl group may be derived from the interaction of a tricoordinated carbon atom and a monocoordinated oxygen, assuming the σ_{CO} bond to be already in place. The interaction diagram is shown in Figure 8.1. The π and π^* orbitals are derived from a SHMO calculation although this is not necessary for a qualitative analysis. The nonbonded HOMO, n_O, is raised in energy slightly relative to an oxygen $2p$ orbital as a result of a four-electron–two-orbital interaction with the out-of-phase combination of the other two σ bonds to carbon (not shown). A broader analysis of the carbonyl group was presented in Chapter 3. Refer to Figure 3.21 and the discussion following it.

Electrophilic Attack on a Carbonyl Group

The π-bonding orbital is polarized toward the oxygen atom. The results of a SHMO calculation yield $\pi = 0.540(2p_C) + 0.841(2p_O)$. Thus the smaller coefficient of the carbon $2p$ orbital means that the carbonyl π bond will interact rather weakly with substituents attached to the carbon atom. The energy of the HOMO, $\epsilon_\pi = \alpha - 1.651|\beta|$, suggests that the π bond is significantly less basic than that of ethylene. Of course, the π bond is not the HOMO. Indeed, the HOMO, n_O, is slightly higher in energy than the π bond of ethylene and more polarized, suggesting significantly greater basicity [141]. The electrophile will approach the molecule from a direction which will maximize its interaction with the HOMO, namely in the plane of the carbonyl group, toward the oxygen atom, in a direction more or less perpendicular to the C=O bond, as shown in Figure 8.2(a). The approach vector is relatively unhindered. In-

CASE STUDY 3: REACTIONS OF CARBONYL COMPOUNDS   143

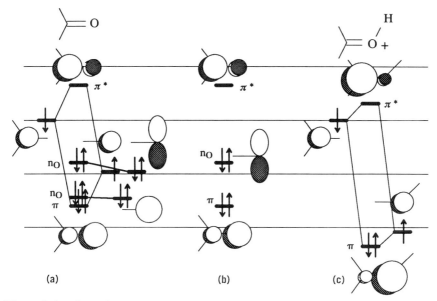

Figure 8.1. The carbonyl group: **(a)** interaction diagram; **(b)** the active orbitals (note: the lower n_O has been omitted); **(c)** interaction diagram for a protonated carbonyl.

volvement of the HOMO in bonding to the electrophile does not rupture the π bond. However, the coordination number of the oxygen atom increases from 1 to 2, and consequently its effective electronegativity will increase significantly as a consequence of attachment of the electrophile (e.g., a proton). Orbital interaction analysis predicts that the π^* (LUMO) is lowered in energy, and is more highly polarized. Both factors enhance the reactivity of the carbonyl group toward nucleophilic attack, and this fact is routinely employed in the design of synthesis. The π system of a protonated carbonyl is shown in Figure 8.1(c).

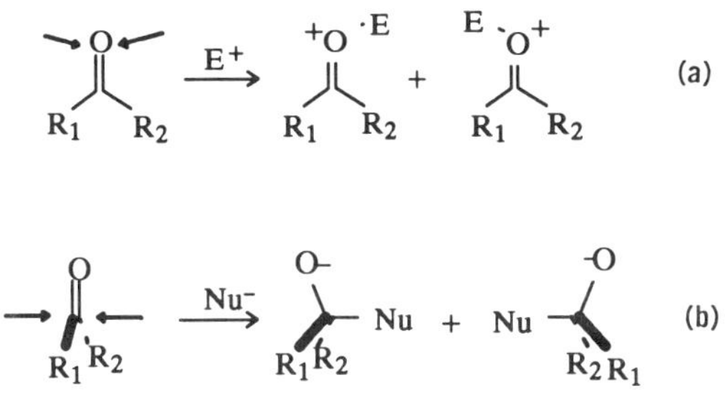

Figure 8.2. Reaction of a carbonyl compound with: **(a)** an electrophile (Lewis acid); and **(b)** a nucleophile (Lewis base). The π bond is formally broken in the reaction with the nucleophile **(b)**, but not in the reaction with the electrophile **(a)**. Stereoisomeric forms may be generated in either case.

Basicity and Nucleophilicity of the Oxygen Atom

Simple orbital interaction theory is only partially informative in distinguishing the relative basicities (nucleophilicities) of carbonyl groups in different bonding environments, since the n_O energy will be perturbed only by interaction with the in-plane σ-bonding orbitals of the attached sp^2 hybridized carbon atom. Comparison of ionization potentials of formaldehyde (10.9 eV), acetaldehyde (10.2 eV), acetone (9.7 eV), and methyl acetate (10.5 eV) reveals the effect of the σ interaction. The HOMO ($n_O = 2p$) of water, alcohols, and ethers is affected by a π-type interaction with neighboring σ bonds, as reflected in the I.P.s of water (12.6 eV), methanol (11.0 eV), and dimethyl ether (9.9 eV). The prediction with respect to the $2p$ orbital energies of the oxygen atoms of the two groups, according to the heteroatom parameters of SHMO theory (Chapter 5) which suggest that the $2p$ orbital of a monocoordinated oxygen is higher in energy than the $2p$ orbital of a dicoordinated oxygen, is thus only partially borne out.

Basicity in the gas phase is measured by the proton affinity (PA) of the electron donor, and in solution by the pK_b. Nucleophilicity could be measured in a similar manner, in the gas phase by the affinity for a particular Lewis acid (e.g., BF_3), and in solution by the equilibrium constant for the complexation reaction. Table 8.1 shows the available data for a number of oxygen systems. It is clear from the data in the table that the basicities of ethers and carbonyl compounds, as measured by PA and pK_b, are similar. However, the nucleo-

TABLE 8.1. Ionization Potentials (I.P.), Proton Affinities (PA), pK_b Values, and BF_3 Affinities

Molecule	I.P.[a] eV	PA[b] kJ mol^{-1}	pK_b^c	BF_3 affinity[b] kJ mol^{-1}
Water, H_2O	12.6	733	15.74	52
Methanol, CH_3OH	11.0	792	16	71
Dimethyl ether, CH_3OCH_3	9.9	825	17.5	77
Tetrahydrofuran, $(CH_2)_4O$	9.4	859		90
Oxetane, $(CH_2)_3O$	9.7	851		91
Formaldehyde, CH_2O	10.9	741	18	39
Acetaldehyde, $CH_3C(O)H$	10.2	797	24	56
Acetone, $CH_3C(O)CH_3$	9.7	839	21	59
Acrolein, $CH_2=CHC(O)H$		828		59
Butenone, $CH_2=CHC(O)CH_3$	10.1	861		61
(E)-Methylacrolein, $CH_3CH=CHC(O)H$				66
Dimethylacrolein, $(CH_3)_2C=CHC(O)H$				72
Methyl acetate, $CH_3C(O)OCH_3$	10.5	846	20.5	58
Methyl acrylate, $CH_2=CHC(O)OCH_3$		851	20	58

[a]See [55].
[b]Computed values, MP3/6-31G*//HF/6-31G*, [142].
[c]$pK_b = 14 - pK_a$ (of conjugate acid), [99].

philicity of ethers, as measured by the BF_3 affinity, is greater than that of carbonyl compounds, the latter values being depressed by steric interactions.

Nucleophilic Attack on a Carbonyl Group

The π^*-antibonding orbital is polarized toward the carbon atom. The results of a SHMO calculation yield $\pi^* = 0.841\,(2p_C) - 0.540\,(2p_O)$. Thus the larger coefficient of the carbon $2p$ orbital means that the carbonyl π^* orbital will interact strongly with substituents attached to the carbon atom. The carbonyl group as a substituent is a Z-type substituent, as discussed earlier. Since the polarized π^*-antibonding orbital is the LUMO, with energy, $\epsilon_\pi^* = \alpha - 0.681\,|\beta|$ (by SHMO), carbonyl compounds will be much better Lewis acids than ethylene (which, as we have seen, is not Lewis acidic at all), and the Lewis acidity will be sensitive to the presence and nature of the substituents, as shown by the interaction diagrams in Figure 8.3. Interaction with X: substituents will raise the energy of the LUMO and interaction with Z and C substituents will lower the LUMO energy. Approach of the nucleophile (Lewis base) will be toward the C atom of the carbonyl in a direction perpendicular to the plane of the carbonyl group. If the nucleophile is negatively charged, it will tend to be ion-paired with the counterion (usually a metal ion such as Li^+, Na^+). The counterion will be actively involved in the reaction by coordinating to the oxygen atom [143]. The product of the reaction, the "tetrahedral intermediate," has a σ bond between the carbon and the base. Occupation of the π^* orbital results in rupture of the π bond, the original bonding electrons being localized to the oxygen atom. The resulting alkoxide is a strong base and a

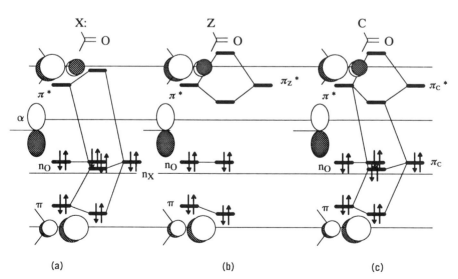

Figure 8.3. The substituted carbonyl group: **(a)** X: substituent; **(b)** Z substituent; **(c)** C substituent.

strong π donor. If the tetrahedral intermediate has a polarized bond, i.e., a bond with a low-lying σ^* orbital, that bond will probably rupture, regenerating the stable carbonyl π bond and displacing the electronegative element at the end of the polarized bond as a Lewis base. The possible scenarios following nucleophilic attack on a carbonyl group are summarized in Scheme 1. Case 1,

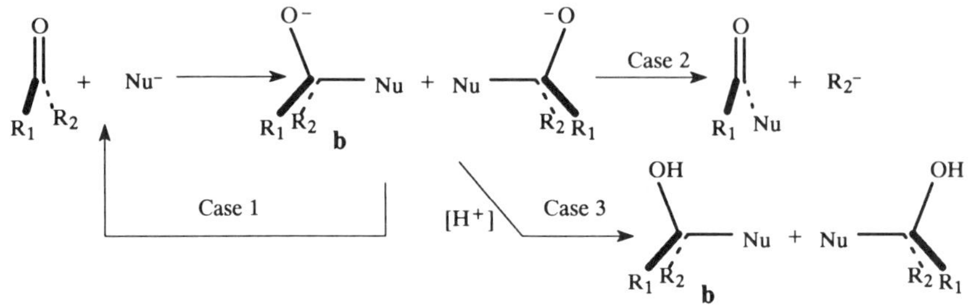

Scheme 1

the reversal, is the normal situation if R_1 and R_2 are H, alkyl, or aryl, and Nu^- is halide, although if the halide is delivered by a strong Lewis acid such as HX or SOX_2, the oxygen may be eliminated (as water or sulfite) in a second step, resulting on a dihalide. If R_1 is H, alkyl, or aryl, and R_2 is a "leaving group" such as halide, alkoxide, acyl, etc., then Case 2 (nucleophilic acyl substitution) is the probable course. Hydrolyses of esters, amides, acid halides, and acid anhydrides fit this case, as do partial reductions of acids or esters to aldehydes or ketones, or the addition of organometallics to esters. If R_1 and R_2, are H, alkyl, or aryl, and Nu^- is an H, alkyl, or aryl anion (a metal hydride or organometallic), then the intermediate alkoxide will be the stable product and this is usually protonated to form the alcohol (Case 3). Approach of the nucleophile to the carbon via a path perpendicular to the planar face of the carbonyl is relatively unhindered. However, steric hindrance may be sufficiently severe to prevent reaction if the carbonyl is embedded in a crowded region of the molecule, or if R_1 and/or R_2 are sufficiently bulky that increased crowding in the transition structure (which will resemble the tetrahedral intermediate) raises the activation barrier to too high a value. The carbonyl group is particularly sensitive to attack by nucleophiles which themselves contain an X: substituent. The rate enhancement due to the presence of the X: substituent is greater than observed in the case of nucleophilic attack at saturated carbon (S_N2, see Chapter 9). Several theories have been proposed for the "extra" reactivity [144], [145], [146]. The simplest proposes a four-center interaction between the LUMO (π^*_{CO}) and the HOMO of the nucleophile (n_{Nuc-X}) which is in fact an occupied π^* orbital and has the correct nodal characteristics for effective overlap. The four-electron repulsive component of the interaction is easily relieved if the geometry of the transition state is as shown in Figure 8.4.

Cyclohexanones in which the chair inversion is constrained by substitution undergo diastereoselective nucleophilic addition, the nature of which (i.e.,

CASE STUDY 3: REACTIONS OF CARBONYL COMPOUNDS   147

Figure 8.4. The optimal structure of the transition state for addition of an X:-substituted nucleophile to a carbonyl group according to orbital interaction considerations.

preferentially axial or preferentially equatorial) depends on the nature of the substituents. The explanation of this effect has been extensively explored [79], [143], [147], [148], [149], [150], [151]. The simplest explanation, shown in Figure 8.5, involves a distortion of the carbonyl group from planarity in such a way as to improve π-type donation from the ring C—C σ bond, or the axial σ bond (often a C—H bond) in the α position, whichever is the better donor. A secondary effect is the improved interaction between the distorted π^* orbital and the HOMO of the attacking nucleophile on the side opposite to the selected σ bond. In unactivated carbonyl compounds, the distortion is too small realistically to account for the observed diastereoselectivity. However, the pyramidalization may be significant at carbonyls to which a Lewis acid is coordinated as has been shown in a crystallographic structure determination [79]. The explanation of Cieplak [147] addresses stabilization of the transition state for the addition of the nucleophile (or the pathway approaching the transition state), using the interactions depicted in Figure 8.5, but substituting the incipient σ^* orbital of the carbonyl–nucleophile complex for the π^*_{CO} orbital. Substituents at the three or four positions can increase or reduce the π-donor ability of the C—C bond. The unsubstituted C—C bond is not as good a π donor as the C—H bond, but X:-type substitution, including additional alkyl groups,

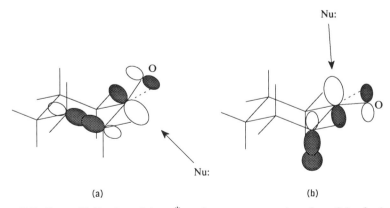

Figure 8.5. Pyramidalization of the π^*_{CO} to improve π-type donation of the β σ bonds: **(a)** C—C participation favors equatorial attack by nucleophile; **(b)** C—H (axial) bond participation favors axial attack by nucleophiles.

can reverse this through a repulsive four-electron–two-orbital π-type interactions with the σ bond.

The Amide Group

The amide group is a special case of a carbonyl group substituted by a very good electron-donating X: substituent, X: = R_1R_2N-. It forms the backbone of every protein and of the most important commercial polymer–nylon. The bonding in an amide group is shown in Figure 8.6. The molecular π orbitals and their placement are as obtained by SHMO theory, and are similar to those obtained by *ab initio* calculations [152], except for the relative energies of n_O and π_2. The *ab initio* calculations have the HOMO to be the π_2 MO which is slightly above the nonbonded MO, n_O. SHMO places π_2 at $\alpha - 1.18|\beta|$, slightly below the energy of the $2p$ orbital of a monocoordinated oxygen atom (Table 5.1, Chapter 5). The π_2 MO (HOMO) is marginally polarized toward N. With either result, the basicity of the oxygen atom would be expected to be *greater* than the basicity of the nitrogen atom because n_O is more localized and therefore will overlap better, and interact more strongly, with a typical electrophile. In other words, the coefficient of the orbital on oxygen in n_O is larger than the coefficient of the orbital on nitrogen in π_2 [equation (3.50) and the discussion after it]. The strong involvement of the nitrogen $2p$ orbital in conjugation with the π^* orbital of the carbonyl group causes the nitrogen atom to adopt a nearly trigonal planar geometry and forces all of the substituents on both N and C to be approximately coplanar [153]. The barrier hindering rotation about the N—C(O) bond is 75–84 kJ/mol (75 kJ/mol in a N,N-dimethylformamide [154]). Because of the high energy LUMO, nucleophilic

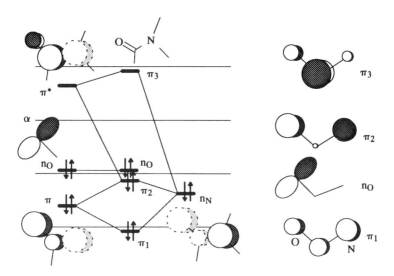

Figure 8.6. The amide group orbitals as an X:-substituted carbonyl group.

substitution at the carbonyl is greatly slowed down compared to other X: substituted derivatives [155]. If the geometric constraints of the molecular framework force the nitrogen to be distorted from planarity, the n_N-π^* interaction is reduced. As a consequence, reactivity of the carbonyl toward nucleophilic attack is increased [156] as is the nucleophilicity of the nitrogen atom [157].

THERMODYNAMIC STABILITY OF SUBSTITUTED CARBONYL GROUPS

Some aspects of the effects of substituents in the thermodynamic stability of carbonyl compounds may be deduced from the interaction diagrams of Figure 8.3. Substitution by X:-type substituents [Figure 8.3(a)] results in increased stabilization since there is a substantial lowering of energies of the occupied π orbitals, principally due to the $n_{X:}$-π^*_{CO} interaction. Stabilization will decrease in the series X: = $-NH_2$, $-OH$, $-F$, since the $n_{X:}$-π^*_{CO} separation increases, and if the corresponding second-row elements are substituted, X: = $-PH_2$ $-SH$, $-Cl$, because of the decreased ability to make π bonds between $2p$ orbitals (part of π^*_{CO}) and $3p$ (or higher np) orbitals, which results in lower intrinsic interaction integrals. The phosphine group is especially weakly stabilizing because of the high s character of n_P. Substitution of Z-type substituents (Figure 8.3(b)) is predicted to lead to modest stabilization, principally as a result of the π_{CO}-π^*_Z interaction which lowers the energy of π_{CO}-bonding orbital. Substitution by "C"-type substituents [Figure 8.3(c)] is predicted to lead to greater stabilization compared to Z substituents as a result of the π_{CC}-π^*_{CO} interaction which lowers the energy of π_{CC} bonding orbital. Additional factors such as electronegativity of the group, and consequences of possible rehybridization are beyond the capabilities of simple orbital interaction theory, and may affect the predictions, particularly in the case of Z and C substituents. Experimentally, the effects of substituents are difficult to quantify. In a recent study, the question of stabilization was addressed in the form of a preference of a substituent to be attached to a carbonyl group rather than to a methyl group [158]. Thus, a positive ΔH for the reaction

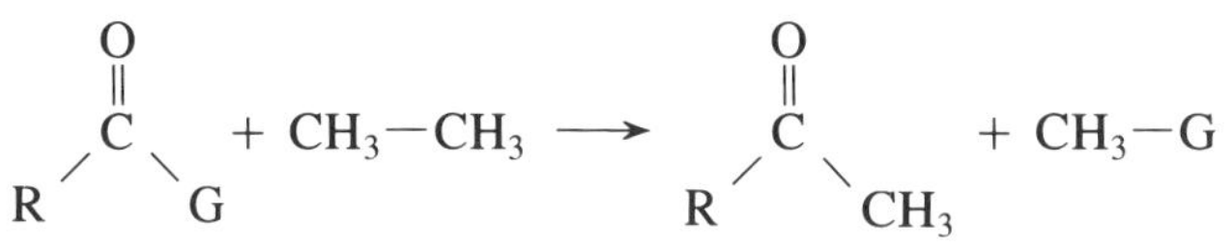

$$\underset{R}{\overset{O}{\underset{\|}{C}}}\diagdown_G + CH_3-CH_3 \longrightarrow \underset{R}{\overset{O}{\underset{\|}{C}}}\diagdown_{CH_3} + CH_3-G$$

is taken as evidence of net stabilization of the carbonyl group by the group "G." Theoretically derived ΔH values were compared to some which could be derived from experimental data. The calculated values agreed quite closely with the experimental values where both were available. We present here a list of the calculated ΔH values for the above reaction as "G," ΔH (kJ mol^{-1}) and refer the reader to the original reference for a detailed discussion [158]:

H, −43; CH$_3$, −39; NH$_2$, +77; OH, +93; F, +67; SiH$_2$, −53; PH$_2$, −16; S, +23; Cl, +28; CN, −46; CH=CH, −2; C≡CH, −23; CF$_3$, −52. Thus, electropositive groups (SiH$_3$), and both Z (CN, CF$_3$) and "C" (CH=CH$_2$, C≡CH) substituents prefer attachment to methyl rather than carbonyl, possibly because more stabilization is available to them from π-type (hyperconjugative) interactions with the methyl group orbitals than with the carbonyl group orbitals.

CHAPTER 9

NUCLEOPHILIC SUBSTITUTION REACTIONS

CASE STUDY 4: NUCLEOPHILIC SUBSTITUTION AT SATURATED CARBON

The prototypical nucleophilic substitution reaction is that of an alkyl halide, although the "leaving group" may be any group which forms a polar bond to carbon and yields a weak Lewis base. The tosyl group (p-toluenesulfonyl) is a common leaving group displaced in a nucleophilic substitution reaction. Two mechanistic routes are distinguished, S_N1 and S_N2. These are discussed separately below although it should be recognized that the two mechanisms may compete or even merge.

Unimolecular Nucleophilic Substitution S_N1

The S_N1 mechanism involves a rate-determining heterolytic cleavage of the alkyl-halide bond to yield an intermediate carbocation which undergoes rapid reaction with available electron donors, including solvent. The typical reactions of carbocation intermediates were discussed in Chapter 7. The solvolysis of alkyl halides is an example of the involvement of carbocations in the S_N1 mechanism, in other words, where the final outcome is a nucleophilic substitution. The first step is a heterolytic cleavage of the C—X bond. Properties of X which favor heterolytic cleavage, namely electronegativity difference with carbon (the larger, the better), and degree of overlap of X orbital with the sp^n orbital of carbon (the smaller the better), have already been elucidated. The transition state has partial cleavage of the C—X bond and may have substantial carbocationic character. Factors which favor carbocation formation will also lower the energy of the transition state. Thus, X: or "C" substituents will accelerate the rate of the first step. The reaction is reversible. The reverse reaction provides the mechanism for the conversion of alcohols to alkyl halides. Polar solvents are necessary in order to stabilize the resulting ions. However, the direct nucleophilic involvement of solvent in the solvolysis of derivatives of tertiary C is a matter of debate [159]. We noted earlier that heterolytic rupture is easiest for the C—X bond and most difficult for C—C in the series C—C, C—N, C—O, C—F. In fact, in the absence of appropriate substituents at one or both ends of the bond, *none* of these bonds will undergo heterolytic rupture except under extreme conditions. However, as the above reaction sequence implies, the —OH group can be converted to a much better leaving group by the simple device of protonation. Protonation effectively increases the electronegativity of the oxygen atom. The same can be accomplished by

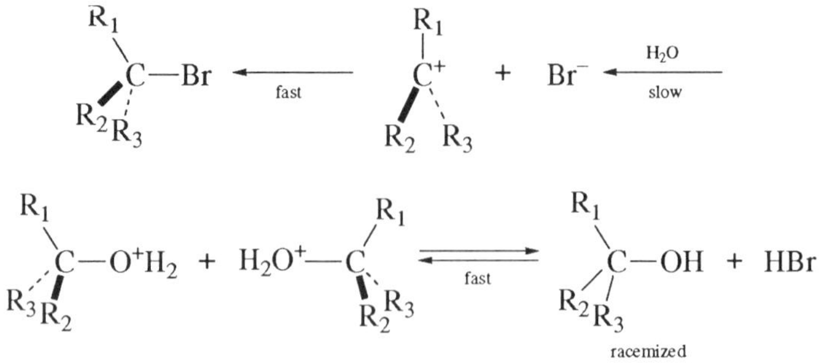

$$\underset{R_2 \; R_3}{\overset{R_1}{C}}-Br \; \underset{fast}{\longleftarrow} \; \underset{R_2 \; R_3}{\overset{R_1}{C^+}} \; + \; Br^- \; \underset{slow}{\overset{H_2O}{\longleftarrow}}$$

$$\underset{R_3 R_2}{\overset{R_1}{C}}-O^+H_2 \; + \; H_2O^+-\underset{R_2 R_3}{\overset{R_1}{C}} \; \underset{fast}{\rightleftarrows} \; \underset{R_2 R_3}{\overset{R_1}{C}}-OH \; + \; HBr$$

racemized

alkylation or by the attachment to the oxygen of other Lewis acids such as metal cations, BF_3, $AlCl_3$, $FeCl_3$, etc. The same is true to a lesser extent for $-NR_2$ and $-F$, for different reasons. The basicity of N makes attachment to Lewis acids an energetically favorable process, but the net difference between the effective electronegativity of N and that of C is not sufficiently large to render $-N$ good leaving group. On the other hand, the low basicity of $-F$ renders attachment to most Lewis acids ineffective in competition with most solvents, especially those which are polar enough to facilitate heterolytic cleavage.

Bimolecular Nucleophilic Substitution S$_N$2

The S$_N$2 mechanism implies a concerted bimolecular reaction which proceeds with inversion of stereochemistry at the central carbon atom. The transition

$$CH_3O^- + \underset{R_2\ R_3}{\overset{R_1}{C}}-Br \longrightarrow CH_3O^{-\frac{1}{2}}\cdots\underset{R_2\ R_3}{\overset{R_1}{C}}\cdots Br^{-\frac{1}{2}} \longrightarrow$$

$$CH_3O-\underset{R_2\ R_3}{\overset{R_1}{C}} + Br^-$$

state has a pentacoordinated carbon atom with a trigonal bipyramidal geometry, the departing and incoming groups occupying the apical positions. The S$_N$2 mechanism is one of the most important in organic chemistry and has been the subject of numerous theoretical and experimental investigations [160]. We wish to examine this reaction using principles of orbital interaction theory to explain (a) why leaving group ability is in the order, Cl > Br > I, (b) to discover the factors which govern *nucleophilicity*, and (c) to determine which substituents may accelerate the rate of reaction. Since nucleophiles are typically Lewis bases, we expect the dominant factors which will govern the interaction of the nucleophile with the alkyl halide will be of the two-electron–two-orbital type involving the HOMO of the nucleophile and the LUMO of the alkyl halide.

The Geometry of Approach. An interaction diagram for the approach of the nucleophile is depicted in Figure 9.1(a). The σ^* LUMO is polarized toward C, so the most favorable overlap and hence, interaction, occurs near the C atom. Front side approach of the nucleophile is *not* favored since the nucleophile would have to approach amidst the nodal surfaces of the σ^* orbital [Figure 9.1(b)]. There is little interaction (which would be repulsive) between the nonbonded orbital and the σ_{CX}-bond orbital since the latter is polarized toward the more electronegative X. The structures of several complexes of halide ion and methyl halides have been studied theoretically and experimentally in the gas phase where the reaction proceeds via a double-welled reaction potential with the trigonal bipyramidal structure as the transition state [161], [162]. Of course, back-side approach accomplishes the observed inversion of configuration at C. Since there are three groups already on the C atom and the incoming nucleophile must squeeze between them, the incoming nucleophile must overcome destabilizing four-electron–two-orbital-type interactions in order to get close enough so that overlap becomes large enough that reaction may occur. Thus, steric effects are expected to be of major concern for operation of

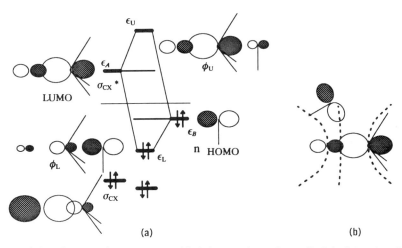

Figure 9.1. The two-electron–two-orbital interaction of an alkyl halide σ* orbital (LUMO) and the nonbonded orbital, n, of a nucleophile: **(a)** most favorable approach; **(b)** unfavorable front-side approach on nodal surface.

the S_N2 mechanism. In practice, the mechanism does not occur at tertiary alkyl centers.

Nucleophilicity. A distinction is usually made between *nucleophilicity* and Lowry–Bronsted *basicity*. The latter involves specifically reaction at a proton which is complexed to a Lewis base (usually H_2O), while the former refers to reactivity at centers other than H. Linear correlations have been shown for gas-phase basicity (proton affinity) and nucleophilicity of nitrogen bases toward CH_3I in solution [163] where the solvent is not strongly involved in charge dispersal.

The ability to overlap at longer range would favor larger atomic centers as opposed to small first row atom centers as nucleophiles; additionally, reactivity is enhanced if the n orbital is higher and therefore closer in energy to the σ* LUMO. Both factors predict the observed relative nucleophilicities, $I^- > Br^- > Cl^- > F^-$, and $-S- > -O-$, and P > N. The n orbital of a given kind of atom may be raised by substitution by X: substituents, a phenomenon which is known as the α effect [146], [164]. Hence hydrazines (NH_2NR_2), hydroxylamines (NH_2OR), peroxides (O^-OR), and similar substances exhibit enhanced nucleophilicity by virtue of their raised, and possibly more extended and polarizable, HOMOs.

Leaving Group Ability. As the picture of ϕ_L [Figure 9.1(a)] implies, the HOMO–LUMO interaction involves some delocalization of charge into the alkyl halide, specifically into an orbital that is *antibonding* between the C and X. This reduces the bond order of the C—X bond and may result in rupture of the bond. Nucleophilic attack on a saturated C atom (the normal case for

CASE STUDY 4: NUCLEOPHILIC SUBSTITUTION AT SATURATED CARBON   155

S_N2 reaction) inevitably leads to rupture of the C—X bond if the incoming nucleophile can get close enough to make a bond. Clearly, the lower the LUMO, the more reactive the substrate (alkyl halide in this case). The nature of the leaving group will certainly affect the position of the LUMO. The expected interaction diagrams for different alkyl-halide bonds are shown in Figure 9.2. Clearly the LUMO energies are in the order consistent with the observed reactivity series. However, as the diagrams of Figure 9.2 suggest, the *polarization* of the LUMO is least for the C—I bond due to the similarity of the electronegativities of C and I. One should then question why attack of the nucleophile with the iodine does take place. However, for the same reason that the C—X bond is weak, there is not much energy to be gained by attempting to form a bond to I.

The Transition State. The MOs of the reacting complex at mid-point in the reaction are shown in Figure 9.3. The pentacoordinated structure is a true *transition structure* for the reaction at a carbon center. However, nucleophilic attack at other centers may lead to stable structures involving three-center–

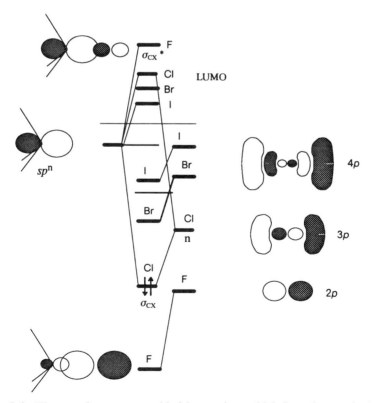

Figure 9.2. The two-electron–two-orbital interactions which form the σ and σ^* orbitals of alkyl halides. The $5p$ orbital of I is not shown.

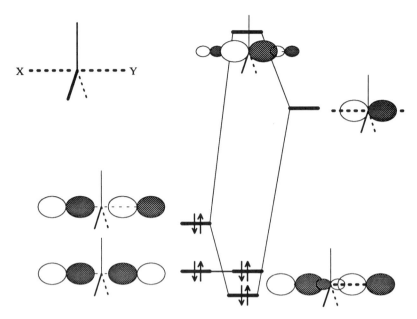

Figure 9.3. Interaction diagram which yields the nature of the bonding in the axial bonds of a trigonal bipyramid. The HOMO is nonbonding. If enough stabilization results from the bonding combination, a stable intermediate may be formed. There are no known examples where this is the case when C is the central atom.

two-electron bonds. The best known example involving H is the anion FHF^-. "Hypervalency" in second or higher row elements in compounds such as PF_5, SF_4, or ClF_3, may also be interpreted in this way. The collinear arrangement of the apical bonds may be attributed to three-center–two-electron bonding without involvement of the $3d$ orbitals.

Substituent Effects. The overwhelming influence of substituents on the rate of S_N2 reactions is steric in nature; bulky groups hinder the reaction by preventing the approach of the nucleophile to the central C atom. Nevertheless, one can examine electronic effects of substituents adjacent to the site of substitution on the rate of reaction. Direct electronic interaction, which may selectively stabilize or destabilize the transition structure, will accelerate or slow down the reaction, respectively. The interaction of the transition structure orbitals with each of the three classes of substituents, X:, Z, and C, are shown in Figure 9.4. In the case of X: and "C" substituents, weak stabilization of the TS is predicted by the simple orbital-interaction theory, while Z substituents are expected to have a stronger stabilizing effect. Experimentally, Z-substituted alkyl halides, e.g., α-haloketones, undergo the greatest acceleration in rate of bimolecular substitution, by up to five orders of magnitude. Alkyl halides with a C substituent, e.g., allylic or benzylic halides, are accel-

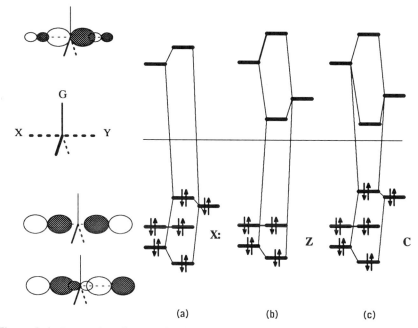

Figure 9.4. Interaction diagram for the MOs of the trigonal bipyramid with a substituent in the equatorial plane α to the site of substitution: **(a)** X:; **(b)** Z; **(c)** C. Only Z substituents are expected to have a strong stabilizing effect on the TS.

erated by up to two orders of magnitudes. Apparently, X: substitution has a mixed effect on the rate of S_N2 reaction, but the rate of S_N1 reaction may be accelerated by up to eight orders of magnitude.

Another Description of the S_N2 Reaction: The VBCM Model

The basic features of the S_N2 reaction, the site of reaction, the stereochemistry (inversion), nucleophilicity, leaving group ability, and the gross effects of α substituents, are readily deduced from orbital-interaction theory. Nevertheless, subtle distinctions having to do with nucleophilicity, leaving group ability, or effects of substituents are beyond the capability of the simple model. Recourse to quantitative electronic structure calculations is often not helpful. The Hartree–Fock procedure fails to provide a reasonable relative energy for the transition state for reasons discussed in Chapter 3, and post-Hartree–Fock calculations yield the correct relative energies, but at the expense of losing the conceptually simple orbital picture. The valence bond configuration mixing (VBCM) model developed by Shaik and co-workers [8], [165], [166], [167], preserves much of the features of the orbital picture while providing a semiquantitative description of the S_N2 reaction. The basic features are given below. The reader is directed to the book by Shaik, Schlegel, and Wolfe [160] for a full description.

158 NUCLEOPHILIC SUBSTITUTION REACTIONS

The general S_N2 reaction may be represented as

$$Nu:^- + R - L \rightarrow Nu - R + :L^-$$

The reactants are represented by two valence bond configurations, $R_1^R = Nu:^- + R - L$ and $R_2^R = Nu\cdot + R\cdot\cdot\cdot L^-$. The second valence-bond configuration, R_2^R, represents the situation where an electron has been transferred from the HOMO of the nucleophile, $Nu:^-$, to the σ^* orbital of the bond involving the leaving group, L. Prior to interaction of the reactants, the reactant condition is described by valence-bond configuration, R_1^R. As the reactants approach each other, the HOMO–σ^* interaction becomes increasingly important and the description of the reactants requires a greater contribution from R_2^R. A parallel situation applies to the products, which are described by valence-bond configurations, $R_1^P = Nu - R + :L^-$ and $R_2^P = Nu\cdot\cdot\cdot R^- + \cdot L$. During the course of the substitution reaction, it is considered that primary reactant configuration, R_1^R, will have a tendency to evolve not into the primary product configuration, R_1^P, but rather into R_2^P. Similarly, the origins of R_1^P are considered to lie in R_2^R. The interplay of the valence-bond configurations is displayed schemati-

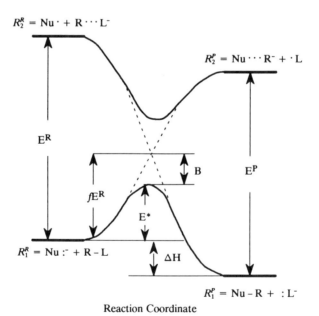

Figure 9.5. A schematic drawing of the S_N2 reaction coordinate according to the VBCM model. The energy gaps, E^R and E^P are identified with the sum of the ionization potential and electron affinities of the appropriate species. The avoided crossing occurs at a fraction of E^R determined by the reaction enthalpy, ΔH, and the expected steepnesses of the descending curves. The activation energy for the reaction is $E^* = fE^R - B$, where B is the energy of the avoided crossing.

cally in Figure 9.5 in which the key feature is the intended and avoided crossing of R_1^R and R_2^R. The diagram in Figure 9.5 may be interpreted semi-quantitatively in as much as the energy change for the reaction may be equated to the enthalpy change for the reaction, ΔH, and the energy gaps between R_1^R and R_2^R, and between R_1^P and R_2^R, may be evaluated from the vertical ionization potentials and vertical electron affinities of the donors and acceptors, respectively. The point of intended crossing is expressed as a fraction, f, of the reactant energy gap, E^R, where f depends on both ΔH and the steepness of the descents of R_2^R and R_2^P. The latter factor is governed principally by the stability of the three-electron bonds (R· · ·L$^-$ and Nu· · ·R$^-$) which in turn may be deduced from electronic characteristics of R, Nu, and L. A strong (delocalized) three-electron bond corresponds to shallow descent and therefore large f, while a weak three-electron bond (localized) corresponds to smaller f. Various means were proposed to estimate B, the extent of the avoided crossing, or B may be assumed to be constant in a series of similar reactions. The activation energy for the reaction is given as $E^* = fE^R - B$, and for a series of similar reactions is largely governed by the two factors which determine the magnitude of f. The VBCM approach predicts that X:-type substituents (specifically halogens, where the involvement of the substituent nonbonded electron pairs cannot be avoided) assist the delocalization of the three-electron bonds and therefore slow down the rate of reaction, the largest retardation occurring when X: and L: are the same. On the other hand, Z-type substituents reduce E^R (increase the electron affinity) without significant delocalization of the three-electron bond and therefore lead to an acceleration of the reaction, but only toward powerful nucleophiles.

CHAPTER 10

BONDS TO HYDROGEN

HYDROGEN BONDS AND PROTON ABSTRACTION REACTIONS

Hydrogen atoms form the exposed outer surface of the majority of organic molecules. A reagent suffering a random collision with a substrate is most likely to encounter a bonded hydrogen atom. The interaction between a substance B and a compound in the vicinity of a hydrogen atom, *bonded to an element which is more electronegative than it is* (which includes C), is shown in Figure 10.1. The important interaction is a two-electron–two-orbital interaction between the HOMO of B and the antibonding σ orbital of the bond to H (σ^*). If the σ^* orbital involves H and C, it probably will not be the LUMO, and if B is not extraordinarily basic, i.e., have a very high energy, localized HOMO, then little consequences will ensue from the interaction other than a weak van der Waals [61] interaction which is always present. If the σ^* orbital *is* the LUMO, the interaction, which involves charge transfer, may be quite strong resulting either in a complex stabilized by a hydrogen bond, or rupture of the σ bond with concomitant transfer of the proton to B, a Lowry–Bronsted acid-base reaction. Bonds to hydrogen from O and F, and probably N, are highly polarized and there will be a substantial electrostatic component to the interaction if B is even partially negatively charged. In Figure 10.1, the HOMOs of the bases, B: and A:, are depicted as generic s orbitals. In fact, these could be nonbonded p or sp^n hybrid orbitals, or also π orbitals as in olefins, enolates or enamines, or strained σ-bonding MOs as in cyclopropanes.

Hydrogen Bonds

Hydrogen bonding interactions are considerably weaker than ionic interactions and covalent bonds, but have a profound effect on many chemical and physical

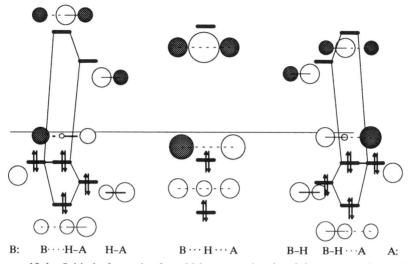

Figure 10.1. Orbitals for a simple acid base reaction involving proton abstraction. Charges are ignored.

properties [168], and determine the shapes of large molecules such as proteins and nucleic acids. Protein secondary structure is determined by H-bonding between the carbonyl oxygen of one amide unit and the N—H bond of another. The two strands of the double helix of nucleic acids are held together by complementary N· · ·H—N and =O· · ·H—N bonds between thymine and adenine, and between cytosine and guanine. Each H—B bond contributes about 20 kJ/mol to the stabilization of the complex. The initial complex, shown on the left-hand side of Figure 10.1 is the diagram for a hydrogen bond. A collinear geometry for B· · ·H—A yields maximum overlap of the HOMO (n_B) and LUMO (σ^*_{HA}) but since the local symmetry of the LUMO is spherical (it is mostly $1s_H$), collinearity is not a strong requirement. The extent of H transfer from A to B in the H-bonded complex, B:· · ·H—A, and the energy associated with formation of the complex is determined by the nature of A: and B:. Numerous hydrogen-bonded complexes have been studied in the gas phase (for a review of complexes of NH_3 and a variety of proton donors, see [169], [170]). Probably the most studied hydrogen bond is that in water dimer, found to be -23 ± 3 kJ/mol experimentally [171] and -19.8 kJ/mol theoretically [172]. The structure of water dimer is as expected on the basis of a dominant HOMO (n_O) − LUMO (σ^*_{OH}) interaction. Optimum hydrogen bonding occurs when A and B are balanced in electronegativity, have maximum electronegativity, and B: is negatively charged. The strongest H-bonded complex is probably FHF^-. Bonding in this symmetrical complex is shown in the middle of Figure 10.1. Symmetrical H bonds of the type $[-OHO-]^-$ have also been found between two oxygen atoms, for example, in monoanions of dicarboxylic acids in nonpolar solvents [173]. The complex $[H_3O^+\cdot\cdot\cdot F^-]$, is responsible for the anomalously low acidity of HF ($pK_a = 3.2$) in aqueous solution. Since the HOMO is nonbonding, simple PMO theory would predict that it makes little difference to the thermodynamic stability whether that MO is occupied or not. Such bonding is observed in electron deficient systems and is discussed below. If B is substantially less electronegative than A:, proton transfer will occur and the H bond will be as depicted on the right-hand side of Figure 10.1.

Proton Abstraction Reactions

Many reaction steps in organic chemistry require the abstraction of a proton by a Lewis base. Interaction diagrams for the elementary stages for the reaction B: + H—A → B—H + A: are shown in Figure 10.1. The reaction parallels the S_N2 reaction; it is a nucleophilic substitution at H. We will restrict our attention to reactions which involve an abstraction of a proton from C. As stated above, the σ^* orbital of a C—H bond is unlikely to be the LUMO of H—A. However, if the σ^* orbital of a C—H bond is not too high in energy, the probability of reaction may still be relatively high due to the exposed nature of the hydrogen, the polarization toward H and the lack of nodes. The first factor makes close approach possible; the other two factors allow large orbital overlap from a wide range of angles of approach.

The energy of σ^*_{CH} may be lowered by two distinct mechanisms. A change in the hybridization of the C sp^n orbital toward smaller n, i.e., more s character, is accompanied by a lowering of the energy (or increase of electronegativity) of the hybrid orbital. The orbital-interaction analysis predicts two consequences, a lowering of the σ^* orbital, and increased polarization of the σ^* orbital toward H. Both factors combine to increase the interaction with the σ^* orbital and increase the reactivity toward a given base. For instance, the pK$_a$ values for H—CH$_2$CH$_3$, H—CH=CH$_2$, and H—C≡CH are 48, 44, and 24, respectively. The second mechanism for lowering the energy of the σ* CH orbital is via admixture into lower energy unoccupied MOs, especially the LUMO. This occurs whenever a Z or C substituent is attached to the carbon atom which bears the C—H bond, as shown in Figure 10.2. The generic p orbital of the Z substituent [Figure 10.2(a)] or C substituent [Figure 10.2(b)] would be the orbital of a π system at the point of attachment of the carbon bearing the H in question. The LUMO of the Z or C substituent is lowered somewhat by in-phase interaction with the $σ^*_{CH}$ orbital. Because the resulting LUMO has some admixture of the $σ^*_{CH}$ orbital, there is an increased probability that the nucleophile will overlap with the hydrogen 1s orbital, resulting in rupture of the C—H bond and yielding a π-delocalized carbanion. Of course, the dominant component of the new LUMO is the LUMO of the substituent and attack of the nucleophile is most likely at that site in the absence of additional considerations such as steric effects, or strong coulomb interactions. The pK$_a$ values of several Z-substituted carbon acids are: H—CH(C(O)CH$_3$)$_2$ 9; H—CH$_2$NO$_2$ 10; H—CH$_2$C(O)CH$_3$ 20; H—CH$_2$C(O)OCH$_3$ 24; H—CH$_2$SO$_2$CH$_3$ 31 (DMSO); H—CH$_2$CN 31 (DMSO). Allylic and benzylic (i.e., C substituted) carbon acids are considerably weaker (C$_6$H$_5$CH$_2$—H 41; CH$_2$=CH—CH$_2$—H 43) unless more than one C substituent is present [(C$_6$H$_5$)$_2$CH—H 33] or the resulting carbanion may be "aromatic" (cyclopentadiene 16).

Two special cases need to be considered: activation by alkyl halide, and by

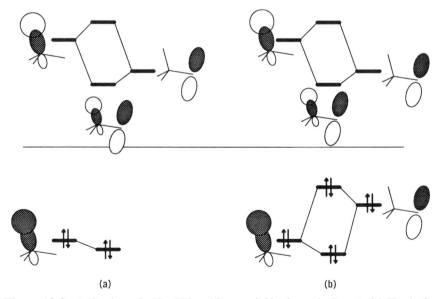

Figure 10.2. Activation of a C—H bond by a neighboring substituent: **(a)** Z substituent; **(b)** C substituent.

164 BONDS TO HYDROGEN

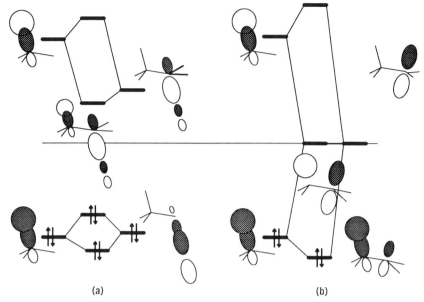

Figure 10.3. Activation of a C—H bond by a neighboring substituent: **(a)** C—X bond; **(b)** carbocationic center.

a formal cationic center. The interaction diagrams are shown in Figure 10.3(a) and (b), respectively. These are involved in elimination reactions of alkyl halides, E2 and E1, respectively.

The E2 Elimination Reaction

The LUMO of a molecule with an alkyl halide bond flanked by a C—H bond is shown in Figure 10.3(a). The LUMO is composed primarily of the σ^*_{CX} with some σ^*_{CH} mixed in *in-phase*. The amount of interaction, and therefore the energy of the LUMO depends on the orientation of the C—H and C—X bonds. Interaction is strongest when the two bonds are coplanar, and slightly better if they are *anti*-coplanar. Interaction of the LUMO with the HOMO of a Lewis base (nucleophile) will direct the base to the points where the best overlap occurs, namely to the backside of the C end of the C—X bond as already discussed, resulting in a nucleophilic substitution by the S_N2 mechanism. However, there is a possibility of attack at the H end of the C—H bond, and this mode may be the most probable if the Lewis base is not a "good nucleophile" or if attack at C is sterically hindered. Attack by Lewis base (i.e., addition of electrons) at the H end of the LUMO is accompanied by a reduction of the σ bond order of the C—H and C—X bonds, and an increase of π-bond order between the two carbon atoms. Thus, the overall course of the reaction is a concerted formation, by *anti* elimination, of a C—C π bond and a B—H σ bond and rupture of the C—H and C—X σ bonds. If the C—H and C—X cannot adopt a coplanar configuration, the LUMO is not lowered in energy, the C—C π bond cannot be formed, and there is no mechanism for breaking the C—X bond if base attack is at the C—H bond. Operation of the E2 mechanism has a very strong stereoelectronic requirement that the C—H and C—X bonds be able to achieve a (nearly) coplanar arrangement. In terms of the intrinsic stabilization, there is not much difference between the *syn* coplanar and *anti* coplanar arrangements. The observation that the E2 reaction proceeds predominantly by *anti* elimination is easily explained on steric grounds. The *anti* arrangement of two bonds at adjacent tetracoordinated centers corresponds to a minimum in the potential function for rotation about the single bond, whereas the *syn* (or eclipsed) arrangements correspond to a maximum. In cyclic systems, where adjacent C—H and C—X bonds are forced into a *syn* coplanar arrangement by ring constraints, the E2 elimination still proceeds, albeit at a reduced efficiency probably due to steric shielding of the C—H bond by the adjacent halogen.

The gas phase E2 reaction of CH_3CH_2Cl with F^- and PH_2^- (proton affinities 1554 and 1552 kJ/mol) has been investigated by high level *ab initio* computations [174]. With F^- as the nucleophile, a small difference of 4 kJ/mol was found favoring the S_N2 pathway over the E2 (*anti*) pathway. However, the E2 (*anti*) pathway was preferred over the E2 (*syn*) route by 53 kJ/mol. Fluoride was predicted to be considerably more reactive than PH_2^-, for which relative transition state energies of 0, 49, and 84 kJ/mol were found for the S_N2, E2 (*anti*), and E2 (*syn*) transition states, respectively.

The E1cB Mechanism Reaction

The E1cB mechanism has the same features as the E2 mechanism except that proton abstraction by the base proceeds essentially to completion prior to departure of the leaving group. A variant of this mechanism may intervene whenever the leaving group is a poor leaving group, or an exceptionally stable carbanion may be formed (i.e., due to the presence of Z substituents in addition to the polar σ bond and/or a hybridization effect). The factors which lead to stabilization of carbanions have been discussed in Chapter 7.

The E1 Elimination Reaction

The rate determining step of the E1 elimination reaction is precisely the same as previously discussed for the S_N1 reaction. The interaction diagram for the C—H bond and an adjacent carbocationic center is shown in Figure 10.3(b). Because the σ and σ^* C—H bond orbitals are equally spaced relative to the energy of the *p* orbital at the cationic site, the LUMO energy is approximately the same as the energy of the unperturbed cationic *p* orbital. Reactivity with Lewis bases remains very high but is reduced somewhat by delocalization of the orbital (smaller coefficient on the *p* orbital). Notice that the presence of the adjacent C—H bond results in stabilization of the carbocation by a lowering of the energy of the σ_{CH} orbital. Concomitant delocalization of the C—H bonding electrons is accompanied by weakening of the C—H bond and partial bond formation between the H and the C at the cationic site. Hydride transfer may result if this is energetically favorable. The most probable course of the reaction with a Lewis base, is formation of a σ bond at the cationic site. However, there is a possibility of attack at the H end of the C—H bond, and this mode may be enhanced if the base is a "good" Lowry-Bronsted base (forms a strong bond to H). Both addition of the nucleophile to C and proton abstraction are reversible. The equilibrium may often be channeled toward proton abstraction by removal of the more volatile olefin by distillation.

REACTION WITH ELECTROPHILES: HYDRIDE ABSTRACTION AND HYDRIDE BRIDGING

The C—H bond is normally not very basic and will not interact with Lewis acids as a rule. However, in the presence of very powerful Lewis acids, such as carbocations, or if substituted by powerful π-electron donors (X: or C substituents), hydride abstraction from a carbon atom may be accomplished, corresponding to an *oxidation* of the C atom.

Activation by π Donors (X: and C Substituents)

Abstraction of a hydride from carbon is almost invariably an endothermic process. The rate of the reaction depends on the stability of the transition structure which closely resembles the product carbocation and is expected to be stabilized by the same factors, among them, substitution by X: and C substituents. Nevertheless, initial interactions set the trajectory for the hydride abstraction reaction. The interaction of a C—H bond with a C substituent is shown in Figure 10.4(b). The feature relevant to the present discussion is that the HOMO which involves some admixture of the C—H bond, has been raised in energy. Therefore, attack by electrophiles, while most likely at the π bond of the "C" substituent, is also possible at the C—H bond. The interaction of an X: substituent with a CH bond is shown in Figure 10.4(a). In general, a single X: or "C" substituent is not sufficient to activate the C—H bond toward hydride abstraction.

Hydride Abstraction

The interaction of a C—H bond with a strong Lewis acid (low energy LUMO) is shown in Figure 10.5(a). The *p* orbital of a carbocation as the LUMO is

REACTION WITH ELECTROPHILES 167

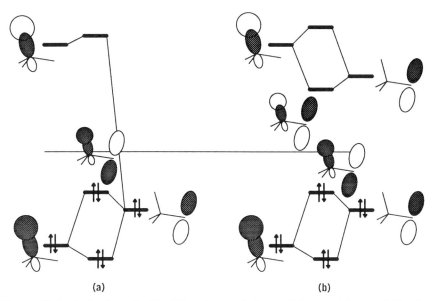

Figure 10.4. Activation of a C—H bond toward electrophilic attack by a neighboring substitutent: **(a)** X: substitutent; **(b)** "C" substituent (only the adjacent p orbital is shown).

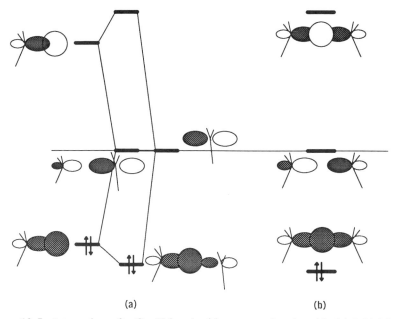

Figure 10.5. Interaction of a C—H bond with a strong Lewis acid: **(a)** initial interaction; **(b)** bonding MOs of a hydride bridge.

shown by way of example. Examples of hydride abstraction reactions are shown in the following Scheme 1.

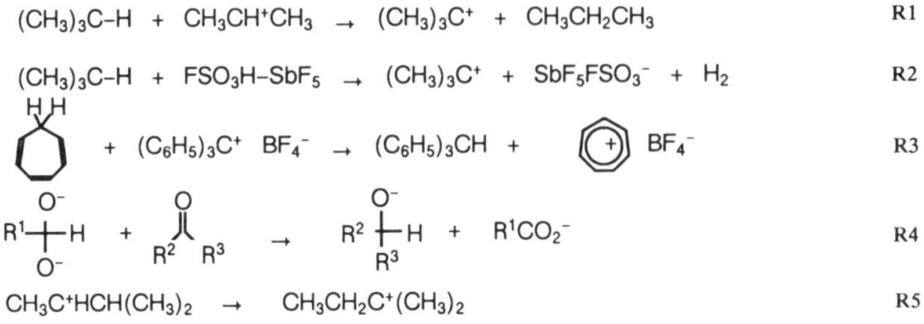

$$(CH_3)_3C-H + CH_3CH^+CH_3 \rightarrow (CH_3)_3C^+ + CH_3CH_2CH_3 \qquad R1$$

$$(CH_3)_3C-H + FSO_3H-SbF_5 \rightarrow (CH_3)_3C^+ + SbF_5FSO_3^- + H_2 \qquad R2$$

[cyclic structure with H,H] + $(C_6H_5)_3C^+$ BF_4^- → $(C_6H_5)_3CH$ + [tropylium cation]$^+$ BF_4^- R3

$$R^1\underset{O^-}{\overset{O^-}{+}}H + \underset{R^2 \quad R^3}{\overset{O}{\jjlarge}} \rightarrow R^2\underset{R^3}{+}H + R^1CO_2^- \qquad R4$$

$$CH_3C^+HCH(CH_3)_2 \rightarrow CH_3CH_2C^+(CH_3)_2 \qquad R5$$

Scheme 1

The first reaction provides a route for the reduction of alkyl halides since the carbocation (isopropyl, in R1) may be prepared from action of $AlCl_3$ on the corresponding alkyl halide. Reactions of the type R1 are also important in the process, *catalytic cracking*, in the manufacture of gasoline. They have also been studied in mass spectrometric experiments [175]. Reaction R2 is one route to the preparation of carbocations under stable ion conditions. Reaction R3 is employed in the laboratory synthesis of the tropylium cation. Reaction R4, the (crossed) Cannizzaro reaction is unusual in that it takes place under strongly basic conditions. The oxy dianion is an intermediate in the reaction of concentrated hydroxide with the aldehyde, R^1CHO. None of R^1, R^2, or R^3 may have hydrogen atoms α to the carbonyl groups. Formaldehyde ($R^1 = H$) is readily oxidized and is useful in the reduction of other ketones or aldehydes. Reaction R5, a 1,2 migration of hydride may be regarded as a special case of this class of reaction although we will see it again in connection with the Wagner–Meerwein rearrangement as a thermally allowed [1,2] sigmatropic rearrangement in connection with our discussion of pericyclic reactions (*vide infra*).

Hydride Bridges

Hydride bridge bonding is common in boron compounds, the simplest example of which is B_2H_6, and in transition metal complexes. We restrict our discussion to instances where hydride bridging occurs between carbon atoms. The MOs of a hydride bridged carbocation are shown in Figure 10.5(b). These are entirely analogous to the MOs previously shown for two-electron–three-center bonding (middle of Figure 10.4), except that the nonbonding orbital is higher in energy and unoccupied. One of the isomers of protonated ethane, $C_2H_7^+$ **1**, has precisely the bonding shown in Figure 10.2. The C—H—C bond is not linear, the angle being about 170° according to high level MO calculations. Several bridged cycloalkyl carbocations of the type **2** have been prepared [176].

$$\begin{bmatrix} H & H \\ \diagdown & \diagup \\ \diagup & H & \diagdown \\ H H & H H \end{bmatrix}^+ \mathbf{1} \quad \begin{bmatrix} H \\ \bigcirc \\ H \end{bmatrix}^+ \mathbf{2} \quad \begin{bmatrix} H \\ H \diagup \triangle \diagdown H \\ H & H \end{bmatrix}^+ \mathbf{3}$$

Complexes between a number of alkyl cations and alkanes have been detected in mass spectrometric experiments [175]. The "nonclassical" structure of the ethyl cation, **3,** may be cited as another example of hydride bridging (for a discussion, see [44]).

REACTION WITH FREE RADICALS: HYDROGEN ATOM ABSTRACTION AND ONE- OR THREE-ELECTRON BONDING

The C—H bond is normally not very polar. As a result the σ_{CH} and σ_{CH}^* orbitals are widely separated and more or less symmetrically disposed relative to "α." Sluggish reaction is expected with carbon-free radicals, but rapid reaction may be anticipated with both electrophilic- and nucleophilic-free radicals. Examples of both kinds of reaction are ubiquitous in organic chemistry. An *ab initio* investigation of the former, involving oxygen centered free radicals has been carried out [177]. The reactivity spectrum may be modified by substitution on the carbon bearing the hydrogen atom. As we have seen in Chapter 7, all three kinds of substituents stabilize the carbon centered free radical intermediate.

HYDROGEN BRIDGED RADICALS

The observation of negative activation energies in a number of reactions between carbon centered free radicals and HX (X = Br, I) [178], [179] has been interpreted as evidence of intermediate complex formation. The existence of the complex between methyl radical and HCl or HBr as a hydrogen bridged species has been established by high level *ab initio* calculations [58]. In the complex, the C—H bond is considerably elongated compared to the H—X bond.

CHAPTER 11

AROMATIC COMPOUNDS

CASE STUDY 4: REACTIONS OF AROMATIC COMPOUNDS

The Cyclic π Systems by Simple Hückel MO Theory

The molecular orbital energy level diagrams for the three- to seven-ring π systems are shown in Figure 11.1. The pattern may be extended. For any regular n-gon, there is a unique MO at $\alpha - 2|\beta|$ which will describe the distribution of two electrons. The remaining orbitals are degenerate in *pairs*, and able to accommodate four electrons, therefore, the origin of the Hückel

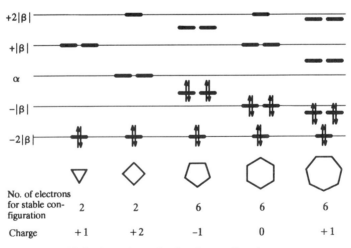

Figure 11.1. SHMO results for the smaller ring π systems.

4n + 2 rule for aromaticity. All of the cyclic π systems have exceptional stability if they have 4n + 2 electrons (two for the three- and four-membered rings, and six for the others shown in Figure 11.1). Examples of each type are known experimentally. However, only benzene is stable in an absolute sense, having a relatively low pair of HOMOs, a relatively high pair of LUMOs, and a large HOMO–LUMO gap. It is electrically neutral, and unstrained in the σ framework.

Aromaticity in σ-Bonded Arrays?

Two novel forms of carbon, with formulas C_{18} [180] and C_{60} [181] owe their stability to aromaticity with cyclic arrays of p orbitals which do not fall strictly into the class of cyclic systems discussed above. Theoretical studies indicate

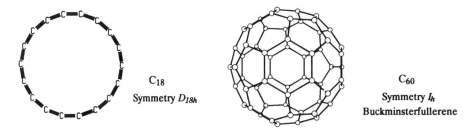

C_{18}
Symmetry D_{18h}

C_{60}
Symmetry I_h
Buckminsterfullerene

that the allotrope, C_{18}, is a planar cyclic structure and therefore has one planar cyclic array of 18 electrons of the above type. It also has a *second* cyclic 18 electron array in the σ framework, albeit the orbitals overlap in a π fashion [182]. The allotrope, C_{60} [183], has a cyclic polyaromatic three-dimensional structure which has also been argued to be "aromatic" [184].

As we have already seen, the Hückel rules do not require that the orbitals be p orbitals overlapping in a π fashion. In fact, cyclic arrays of σ bonds will have substantially the same pattern of orbital energies as shown in Figure 11.1 and the "4n + 2" rule will apply. This observation will become especially relevant when we examine pericyclic reactions, reactions in which more than two bonds are being made and broken at the same time. Cyclic transition structures in which 4n + 2 electrons are involved in the bond reorganizations will be exceptionally stabilized (i.e., "aromatic") permitting relatively low activation energies for the reactions. The familiar Diels–Alder reaction is an example of this kind of reaction. Conjugation through space or through intervening cyclopropyl rings may also exhibit "aromatic" character, as has been established by theoretical studies on the homotropenylium cation [185].

homotropenylium cation

REACTIONS OF SUBSTITUTED BENZENES

The intrinsic stability of the aromatic π system has two major consequences for the course of reactions involving it directly. First, the aromatic ring is less susceptible to electrophilic, nucleophilic, and free radical attack compared to molecules containing acyclic conjugated π systems. Thus, reaction conditions are usually more severe than would normally be required for parallel reactions of simple olefins. Secondly, there is a propensity to eject a substituent from the tetrahedral center of the intermediate in such a way as to reestablish the neutral $4n + 2$ electron π system. Thus, the reaction is two step—an endothermic first step resulting in a four-coordinate carbon atom, and an exothermic second step, mechanistically the reverse of the first, in which a group is ejected. The dominant course is therefore a *substitution* reaction rather than an addition.

ELECTROPHILIC SUBSTITUTIONS

We will restrict our consideration to reactions of substituted benzenes and to nitrogen heteroaromatic systems in which the reaction takes place first with the

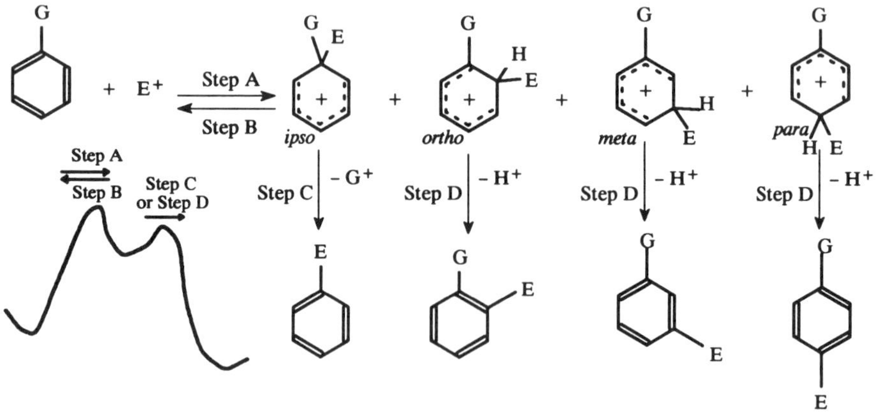

Scheme 1

π system. The simplest example of reaction of a monosubstituted benzene with an electrophile (Lewis acid) is shown in Scheme 1. The electrophile may attach itself to the π system (Step A) in four distinct modes, *ipso*, *ortho*, *meta*, and *para*. The reactivity of the aromatic ring, and the mode of attachment of the electrophile will be influenced by the specific nature of the substituent group, which may be of X:, Z, or C type. Detachment of the electrophile (Step B) is most probable from the *ipso* intermediate unless the group G itself is such that it can detach as a stable electrophile (Step C). The most likely second step is abstraction of the proton by a weak base (Step D) yielding a mixture of *ortho*-, *meta*-, and *para*-substitution products.

Effect of Substituents on Substrate Reactivity

The overall rate of reaction is governed by the activation energy, or more properly, the free energy of activation, $\Delta G^{\ddagger}$. Since the first step is rate-determining and endothermic, or endergonic, it is expected that the effect of the substituent on the kinetics and regioselectivity of the reaction would be greatest on the transition structure which resembles the reactive intermediate. Nevertheless, as previously argued, the course of the reaction, that is to say, which of the four distinct reaction channels is favored, may well be determined by initial interactions between the electrophile and the aromatic substrate. The electronic characteristics of X:-, Z-, and C-substituted benzenes, using aniline, benzaldehyde, and styrene, respectively as models, have been derived from SHMO calculations and are shown in Figure 11.2. While the relative energy levels may readily be deduced from simple orbital interaction considerations, the polarization of the HOMO and LUMO is best derived by actual calculation.

Electrophilic Attack on X: Substituted Benzenes

The X:-substituted benzene (aniline, Figure 11.2) is activated toward electrophilic attack since the HOMO is raised significantly. The electrophile would be directed to the *ortho*, *para*, and *ipso* positions of the ring, and to the X: substituent itself. The *ipso* channel is usually nonproductive since the common heteroatom-based X: substituents are not easily displaced as Lewis acids. Loss of substituent is frequently observed with tertiary alkyl substituted benzenes. Attachment of the electrophile to the X substituent is most likely if X: is NR^1R^2. The substituent is converted to a Z substituent via the low-lying σ^* orbital,

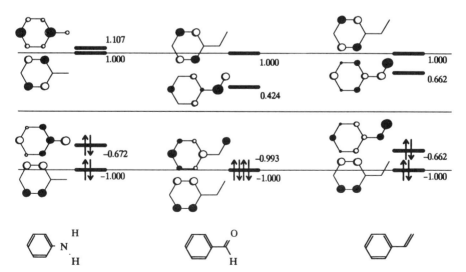

Figure 11.2. The SHMO frontier orbitals of aniline, benzaldehyde, and styrene as prototypes of X:, Z, and C substituted benzenes.

174 AROMATIC COMPOUNDS

and the ring is deactivated toward further electrophilic attack. The *ortho* and *para* channels lead to products. The interaction diagrams for an X:-substituted pentadienyl cation, substituted in the 1, 2, and 3 positions, as models of the transition states for the *ortho*, *meta*, and *para* channels, are too complex to draw unambiguous conclusions. The HOMO and LUMO of the three pentadienyl cations with an amino substituent are shown in Figure 11.3. Notice that the LUMO of each is suitable to activate the C—H bond at the saturated site toward abstraction by base. Curiously, the *meta* cation has the lowest LUMO and should most readily eliminate the proton. The stabilities of the transition states should be in the order of the Hückel π energies. These are $6\alpha - 8.762|\beta|$, $6\alpha - 8.499|\beta|$, and $6\alpha - 8.718|\beta|$, respectively. Thus, the *ortho* and *para* channels are favored over the *meta* channel, and the *ortho* route is slightly preferred over the *para*. Experimentally, *para* substitution products are often the major ones in spite of there being two *ortho* pathways. The predominance of *para* products is usually attributed to steric effects.

Electrophilic Attack on Z-Substituted Benzenes

The Z-substituted benzene (benzaldehyde, Figure 11.2) is not activated toward electrophilic attack since the HOMO of benzene is scarcely affected. No preferred site for attack of the electrophile can be deduced from inspection of the HOMOs. The interaction diagrams for a Z-substituted pentadienyl cation, substituted in the 1, 2, and 3 positions, as models of the transition states for the *ortho*, *meta*, and *para* channels, are complex. The HOMO and LUMO of the three pentadienyl cations with a formyl substituent are shown in Figure 11.4. The stabilities of the transition states should be in the order of the Hückel π

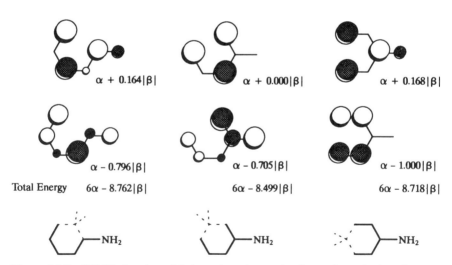

Figure 11.3. SHMO frontier orbitals and total energies for amino-substituted pentadienyl cations.

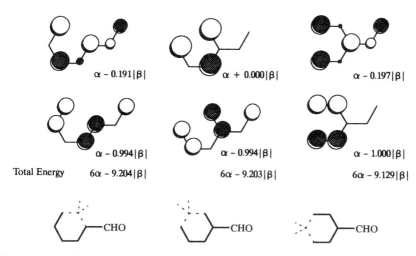

Figure 11.4. SHMO frontier orbitals and total energies for formyl-substituted pentadienyl cations.

energies. These are $6\alpha - 9.204|\beta|$, $6\alpha - 9.203|\beta|$, and $6\alpha - 9.129|\beta|$, respectively. Thus, by SHMO, the *ortho* and *meta* channels are favored over the *para* channel, with no distinction between the *ortho* and *meta* pathways. Experimentally, *meta*-substitution products are usually the major ones, contrary to the SHMO predictions. Either the SHMO method fails in this case, or the predominance of *meta* products may be attributed to steric effects.

Electrophilic Attack on C-Substituted Benzenes

The C-substituted benzene (styrene, Figure 11.2) is activated toward electrophilic attack since the HOMO is raised significantly. The electrophile would be directed to the *ortho*, *para*, and *ipso* positions of the ring, and to the "C" substituent itself. In fact, if the substituent is a simple olefin, electrophilic attack is almost exclusively on the external double bond. Electrophilic substitution of the ring is observed only if the "C" substituent is another aryl group. The *ipso* channel is nonproductive since an aryl group cannot depart as a Lewis acid. The *ortho* and *para* channels lead to products. As in the previous two cases, the interaction diagrams for a C-substituted pentadienyl cation, substituted in the 1, 2, and 3 positions, as models of the transition states for the *ortho*, *meta*, and *para* channels, are too complex to draw simple conclusions. The HOMO and LUMO of the three pentadienyl cations with a vinyl substituent are shown in Figure 11.5. The stabilities of the transition states should be in order of the Hückel π energies. These are $6\alpha - 8.055|\beta|$, $6\alpha - 7.878|\beta|$, and $6\alpha - 8.000|\beta|$, respectively. The situation is entirely analogous to the X:-substituted case. The *ortho* and *para* channels are favored over the *meta* channel, and the *ortho* route is slightly preferred over the *para*. Experi-

176 AROMATIC COMPOUNDS

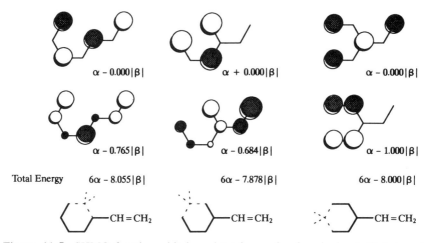

Figure 11.5. SHMO frontier orbitals and total energies for vinyl substituted pentadienyl cations.

mentally, the predominance of *para* products is usually attributed to steric effects.

Electrophilic Attack on N Aromatics: Pyrrole and Pyridine

The SHMO orbitals of pyrrole, pyridine, and pyridinium are shown in Figure 11.6. The HOMO of pyrrole is the same as that of butadiene. Thus pyrrole is more reactive than benzene toward electrophilic attack. Attack, leading to substitution occurs mainly at the 2 and 5 positions where the electron density of the HOMO is concentrated. In the case of pyridine [Figure 11.6(b)], the HOMO is not the π orbital, but the nonbonded MO, n_N, which would be situated at approximately $\alpha - 0.5|\beta|$. Thus, it is not pyridine but pyridinium [Figure 11.6(c)] which undergoes electrophilic attack and substitution. The reactivity is much less than that of benzene, although this could not be deduced directly from the SHMO calculation. Neither does the calculation suggest the reason that electrophilic substitution occurs mainly at the 3 and 5 positions, since the π HOMO is equally distributed between the 2, 3, 5, and 6 positions. It is probable that preference for the 3 and 5 positions originates from the lower activation energy leading to the 2-azapentadienyl dication intermediate. The 2-, and 1-azapentadienyl dications serve as models for the transition states. Their SHMO π energies are $4\alpha - 6.908|\beta|$ and $4\alpha - 6.810|\beta|$, respectively. Substitution into the 4 position of pyridine may be accomplished directly via the N-oxide, whose HOMO is shown in Figure 11.6(d). On the basis of the SHMO calculation, one would expect greatly enhanced reactivity for the N-oxide, although polarization of the HOMO indicates that electrophilic attack would occur predominantly on the oxygen if it is not protonated under the reaction conditions.

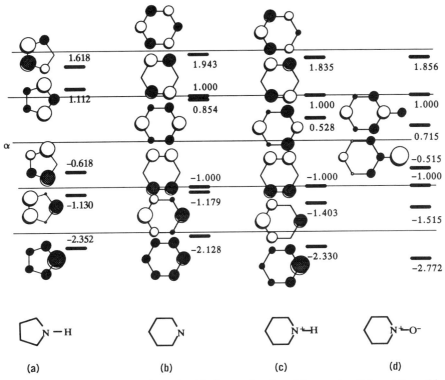

Figure 11.6. SHMO orbitals and orbital energies for: **(a)** pyrrole; **(b)** pyridine; **(c)** pyridinium; **(d)** pyridine-N-oxide (HOMO and LUMO only shown).

NUCLEOPHILIC SUBSTITUTIONS

Nucleophilic substitution in aromatic systems occurs by two very different mechanisms, by initial proton abstraction, and by initial addition to the π system. The former will be discussed separately in the next "Case Study." In the present section, we will restrict our consideration to reactions of substituted benzenes and to nitrogen heteroaromatic systems in which the reaction takes place first with the π system. The simplest example of reaction of a substituted benzene with a nucleophile (Lewis base) is shown in Scheme 2. The scheme is entirely analogous to Scheme 1 which described electrophilic attack, with the exception that hydrogen elimination (as hydride in Step D) is not normally feasible. The nucleophile may attach itself to the π system (Step A) in four distinct modes relative to a reference substituent, G, namely *ipso*, *ortho*, *meta*, and *para*. The reactivity of the aromatic ring, and the mode of attachment of the nucleophile will be influenced by the specific nature of the substituent group, which may be of X:, Z, or C type. Detachment of the nucleophile (Step

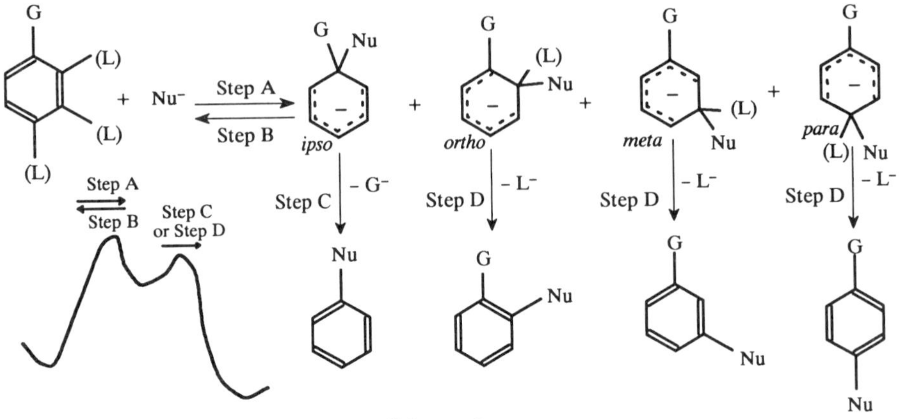

Scheme 2

B) is most probable from *all* intermediates unless the group G itself is such that it can detach as a stable nucleophile (Step C), or there is such a group (L) situated at the site of attack. In the event that a "leaving group," L, is present at the site of attack, the most likely second step is loss of that group (Step D) yielding either an *ortho-*, *meta-*, or *para*-substitution product, as the case may be.

Effect of Substituents on Substrate Reactivity

The overall rate of reaction is governed by the activation energy. Since the first step is rate-determining and endothermic, it is expected that the effect of the substituent on the kinetics and regioselectivity of the reaction would be greatest on the transition structure which resembles the reactive intermediate. Nevertheless, as previously argued, the course of the reaction, that is to say, which of the four distinct reaction channels is favored, may be determined by initial interactions between the nucleophile and the aromatic substrate. The electronic characteristics of X:-, Z-, and C-substituted benzenes, using aniline, benzaldehyde, and styrene, respectively as models, have been derived from SHMO calculations and are shown in Figure 11.2. The SHMO calculations on the model systems suggest that X:-substituted benzenes would not be activated toward nucleophilic attack on the π system, but that Z- and C-substituted benzenes should be reactive in this respect, Z-substituted benzenes more so than C-substituted benzenes since the LUMO is lower in energy. In practice, neither X:- nor C-substituted benzenes undergo nucleophilic attack except by proton abstraction.

Nucleophilic Attack on Z-Substituted Benzenes

The Z-substituted benzene (benzaldehyde, Figure 11.2) is strongly activated toward nucleophilic attack since the LUMO of benzene is substantially lowered. According to the distribution of the LUMO of benzaldehyde, nucleophilic attack is directed to the *ortho* and *para* positions in the ring and to the Z substituent also. The interaction diagrams for a Z-substituted pentadienyl anion, substituted in the 1, 2, and 3 positions, as models of the transition states for the *ortho*, *meta*, and *para* channels, are too complicated to be useful. The HOMOs of the three pentadienyl anions with a formyl substituent are identical to the LUMOs of the corresponding cations shown in Figure 11.4. The stabilities of the transition states should be in the order of the Hückel π energies. These are $8\alpha - 9.585|\beta|$, $8\alpha - 9.203|\beta|$, and $8\alpha - 9.513|\beta|$, respectively. Thus, by SHMO, the *ortho* and *para* channels are favored over the *meta* channel, with a slight preference for the *ortho* pathways. Experimentally, Z substituents *ortho* or *para* to the site of substitution accelerate the reaction.

Nucleophilic Attack on N Aromatics: Pyrrole and Pyridine

The SHMO orbitals of pyrrole, pyridine, and pyridinium are shown in Figure 11.6. The LUMO of pyrrole is higher in energy than that of benzene. As a consequence, pyrrole does not undergo nucleophilic attack in the π system. Just the reverse is true for pyridine, whose LUMO [Figure 11.6(b)] is somewhat lower than that of benzene. The observed reactivity of pyridine is much higher than is implied by the modest lowering of the LUMO. It is likely that the active substrate in the nucleophilic addition is not pyridine but rather metallated pyridine which will resemble pyridinium ion [Figure 11.6(c)]. Complexation of the lone pair of electrons substantially lowers the energy of the LUMO which is concurrently polarized into the 2, 4, and 6 positions. Even hydride can be eliminated in the second step is some cases. The 1-, 2- and 3-azapentadienes serve as models for the metallated transition states for the pathways involving addition of the nucleophile to the 2, 3, and 4 positions, respectively. Their SHMO π energies are $6\alpha - 7.582|\beta|$, $6\alpha - 6.908|\beta|$ and $6\alpha - 7.408|\beta|$, respectively, showing a clear preference for nucleophilic attack in the 2 (or 6), and 4 positions.

CASE STUDY 5: NUCLEOPHILIC SUBSTITUTION BY PROTON ABSTRACTION

In the absence of Z substituents on the aromatic ring and aliphatic sites for nucleophilic attack, including acidic hydrogens, elsewhere in the molecule, nucleophilic aromatic substitution may be accomplished by the mechanism shown in Scheme 3. A strong base abstracts a proton (Step A) to yield an intermediate carbanion localized to the sp^2 hybrid orbital of the aryl ring. The carbanion may reabstract the proton (Step B) or eliminate a leaving group (Step C) (or G itself if it can support a negative charge) to yield a benzyne intermediate. For this reason, this mechanism is usually referred to as the *benzyne* mechanism. The benzyne may undergo attack by the nucleophile (Step D) to

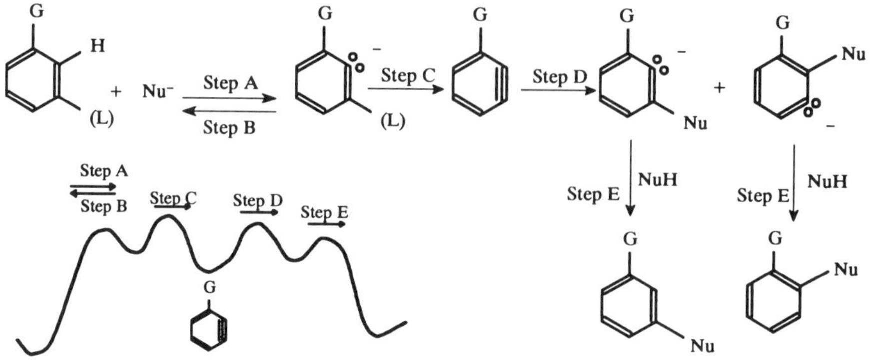

180 AROMATIC COMPOUNDS

Scheme 3

yield either or both of two intermediate carbanions which are subsequently protonated (Step E). Steps D and E are the microscopic reverse of Steps C and A, respectively. Steps A and B may be considered to constitute a special case of the E1cB mechanism discussed earlier. That carbanion which is most stabilized by adjacent low-lying σ^*_{CX} orbitals will be most likely to be formed. In Step C, the better of the two leaving groups (assumed to be L in Scheme 3) departs. Nucleophilic addition to the intermediate benzyne (Step D) is readily explained by PMO arguments. The "extra" π and π^* orbitals of benzyne are compared to those of ethylene in Figure 11.7. *The aromatic π system is not involved in the special properties of benzyne.* The extra benzyne π bond is due to the overlap in π fashion of the two sp^2 hybrid orbitals which lie in the nodal plane of the intact 6π electron system. Two factors contribute to a very low LUMO for benzyne. The sp^2 hybrid orbitals are lower in energy than the $2p$

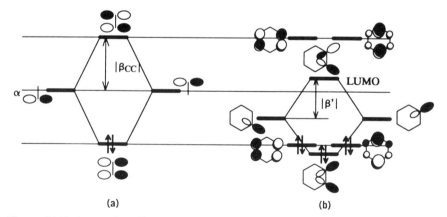

Figure 11.7. Interaction diagrams comparing **(a)** ethylene to **(b)** benzyne. Note that $|\beta'| < |\beta_{CC}|$. The "aromatic" π system is largely unperturbed with nearly degenerate orbitals at $\alpha - |\beta_{CC}|$ and $\alpha + |\beta_{CC}|$.

orbitals from which the ethylene π orbitals are constructed. Secondly, the intrinsic interaction between the two sp^2 orbitals is less than the normal β_{CC} since the orbitals have less *p*-character and are tipped away from each other. The low LUMO of benzyne makes the molecule a strong Lewis acid, susceptible to attack by bases, and a reactive dienophile in Diels–Alder reactions, as we shall see later.

CHAPTER 12

PERICYCLIC REACTIONS

GENERAL CONSIDERATIONS

In a chemical reaction, one or more bonds are broken and/or formed. Most chemical reactions in which a bond is ruptured proceed at rates which suggest that the activation energy is much less than that required to homolytically rupture the bond. Energy must be gained from the formation of one or more new bonds in order to offset that required to break bonds. The course of a typical reaction of the type A: + B—C → A—B + C: may be followed by examination of the interaction diagrams for the reactants and products, and the orbital characteristics of the TS, as shown in Figure 12.1. Repulsive four-electron–two-orbital interactions are always present but may be offset by a particularly favorable two-electron–two-orbital interaction. Often, in order to achieve a sufficiently large two-electron–two-orbital interaction, very stringent stereoelectronic requirements must be met. Thus, a collinear, backside approach is necessary for an S_N2 reaction, and the E2 elimination reaction requires a coplanar, preferably *anti* arrangement of incoming base, and the C—H and C—X bonds. Pericyclic reactions are reactions in which *more* than one bond is being made or broken at the same time and which have a cyclic transition state. As might be expected, even higher stereoelectronic constraints must be met in order that the reaction can proceed under moderate conditions. Stringent stereoelectronic requirements are invariably accompanied by a high degree of *diastereoselectivity*. Secondary interactions of orbitals will be applied to explain observed regioselectivity as well [186]. Pericyclic reactions are usually subclassified as *Cycloadditions*, *Electrocyclic Reactions*, or *Sigmatropic Rearrangements*. These are shown in general form in Figure 12.2 and are discussed separately below. The definitions and terminology follow closely the work of

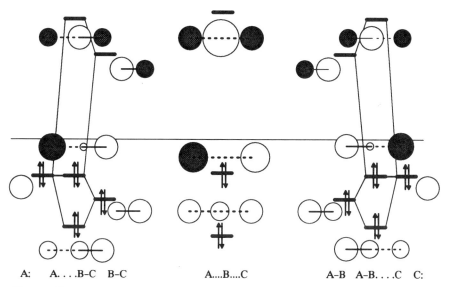

Figure 12.1. Orbitals for a simple substitution reaction: A + B—C → A—B + C.

Woodward and Hoffmann [3] in which more detail may be found. Treatment of the electronic effects via orbital interaction theory closely parallels the frontier orbital method of Fukui [1].

CYCLOADDITIONS AND CYCLOREVERSIONS

A general cycloaddition reaction is shown in Figure 12.2. A molecule with a conjugated system of m π electrons reacts with another molecule with n π electrons to form a cyclic molecule by the formation of two new σ bonds leaving conjugated systems of $(m - 2)$ and $(n - 2)$ π electrons adjacent to the new σ bonds. Here m and n are even, nonzero integers. In the Diels–Alder reaction, $m = 4$ and $n = 2$. Either or both π systems may be part of a more extended conjugated π system. For the purposes of the definition, the *active* π system of each reactant is that bridging the ends of the new σ bonds. In order for the formation of the two σ bonds to be feasible, the active π systems must be able to approach each other in such a way as to enable the terminal p orbitals to overlap simultaneously. In the Diels–Alder reaction, for example, the diene must be able to adopt, or be already in an s-*cis* conformation so that the p orbitals of the 1 and 4 carbon atoms can separately but simultaneously overlap the p orbitals at each end of the olefin.

Stereochemical Considerations

In a cycloaddition reaction, the two active π systems may approach each other in either of two orientations e.g., head to head or head to tail. If one combi-

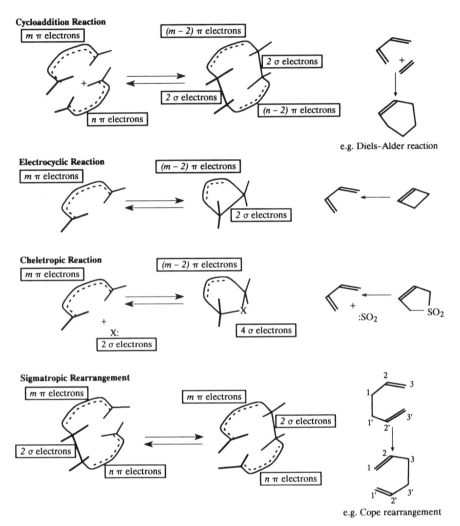

Figure 12.2. The four classifications of pericyclic reactions, with examples of thermally allowed reactions.

nation dominates, the reaction is said to be regioselective. In the course of the reaction, four new saturated centers are formed. With maximum labeling, a total of 16 ($=2^4$) stereoisomeric forms, consisting of eight enantiomeric pairs of diastereomers (if neither polyene is chiral), may be formed. In pericyclic reactions, the stereochemistry is determined by specifying the stereochemical mode in which each component reacts. Each of the two components of a cycloaddition reaction is a conjugated, quasiplanar π system for which two *faces* (top and bottom) may be distinguished. At each end of the π system, a new σ bond terminates. Each σ bond terminates on a line more or less perpendicular to a *face* of the π system. If both σ bonds terminate to the same face (both to the top or both to the bottom), the π system is said to react *suprafacially*, and the mode designation is *s*. If the two σ bonds are formed to opposite faces (one to the top and the other to the bottom, or vice versa), the π system is said to react *antarafacially*, and the mode designation is *a*. The two terminal points may be ends of two separate σ bonds or the opposite ends of the *same* σ bond (in this case, one has an *electrocyclic* reaction which is discussed below). For example, the component designation, $_\pi 4_s$, would indicate that a conjugated π system consisting of four electrons is reacting in a suprafacial manner. Examples of suprafacial and antarafacial reactions of π systems are given in Figure 12.3. All of the reactions illustrated in Figure 12.3 actually give the products shown under the specified reaction conditions. Reactions (4)–(7) are pericyclic reactions.

The reverse process is also useful in synthesis. The active components of the cycloreversion reaction are the two σ bonds which will be broken, and any π systems to which both σ bonds are allylic (i.e., at least one of *m*, *n* is an even number greater than two). The stereochemistry of the reaction may also be specified in terms of a stereochemical mode of reaction of the active components of the reaction.

ELECTROCYCLIC REACTIONS

A general electrocyclic reaction is shown in Figure 12.2. A molecule with a conjugated system of m π electrons cyclizes to form a cyclic molecule. One new σ bond is formed, leaving a conjugated system of $(m-2)$ π electrons in the ring. Here m is an even, nonzero integer. The π system may be part of a more extended conjugated π system. The *active* π system is that bridging the ends of the new ring-forming σ bond. In order for the formation of the σ bond to be feasible, the two ends of the active π system must be able to approach each other in such a way as to enable the terminal *p* orbitals to overlap.

Stereochemical Considerations

In an electrocyclic reaction, a ring is formed from a nominally planar π system. The local plane of the carbon atom involved at each end must rotate through 90° in order that the *p* orbitals may overlap in a σ fashion. The active π system may cyclize by either of two distinct modes, *conrotatory* or *disrotatory*. *Conrotatory* ring closure occurs when the sense of rotation of the two termini is the same, i.e., both clockwise or both counterclockwise. *Disrotatory* ring closure occurs when the sense of rotation of the two termini is opposite, i.e., one end clockwise and the other counterclockwise, or vice versa. If one mode dominates, the reaction is said to be diastereoselective. In the course of the reaction, two new saturated centers are formed. With maximum labeling, a total of 4 $(=2^2)$ stereoisomeric forms, consisting of two enantiomeric pairs of dia-

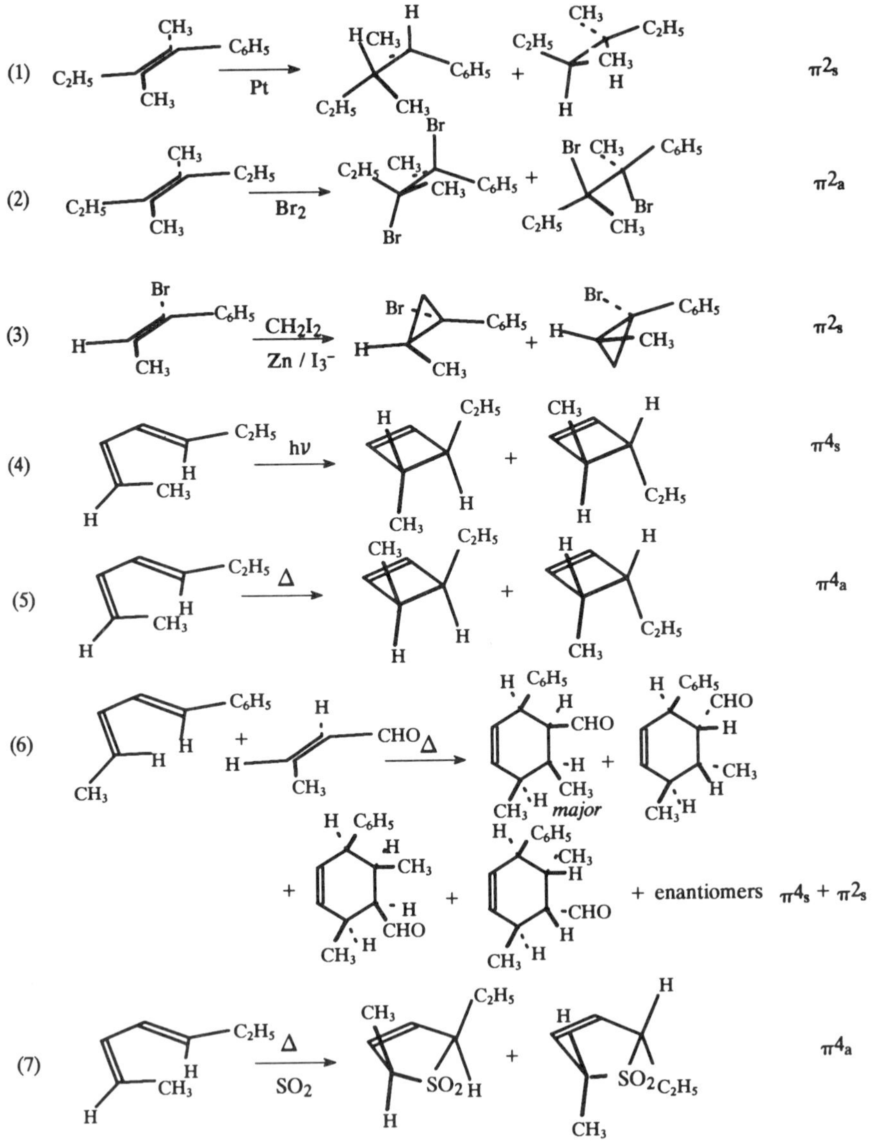

Figure 12.3. Examples of suprafacial and antarafacial reactions of π systems. Notice that the examples serve to describe the stereochemical course of the reaction only. No mechanism is implied by these examples.

stereomers (if the polyene is achiral), may be formed. Reactions (4) illustrates disrotatory closure of the ring, yielding a pair of enantiomeric 2-methyl-3-ethylcyclobutenes. Reaction (5) illustrates conrotatory closure of the ring, also yielding a pair of enantiomeric 2-methyl-3-ethylcyclobutenes which are diastereoisomers of the products of reaction (4). The active m-electron π system may be considered as a single component ($m = 4$ in this case). Disrotatory ring closure corresponds to *suprafacial* reaction of the π system, the conrotatory mode being designated *antarafacial*. A component analysis of the reverse process involves two components, the residual π system, and the σ bond which is to be lost.

CHELETROPIC REACTIONS

A general cheletropic reaction is shown in Figure 12.2. This reaction involves the addition to, or extrusion from, a conjugated system of a group bound through a single atom. The reaction usually involves the elimination of simple stable molecules such as SO_2, CO, or N_2. The atom to which there were two σ bonds carries away a pair of electrons, usually in a sp^n hybrid orbital. The addition of a carbene to a simple olefin to form a cyclopropane is also a cheletropic reaction which, as discussed in Chapter 13, is not predicted to be concerted. Cheletropic reactions incorporate features of both cycloaddition and electrocyclic reactions.

Stereochemical Considerations

In a cheletropic reaction, a ring is formed from a nominally planar π system via bridging of a single atom. The local plane of the carbon atom involved at each end must rotate through 90° in order that the p orbitals may overlap in a σ fashion. As with electrocyclic reactions, the active π system may cyclize by either of two distinct modes, *conrotatory*, or *disrotatory*. If one mode dominates, the reaction is said to be diastereoselective. In the course of the reaction, two new saturated centers are formed. With maximum labeling, a total of 4 $(=2^2)$ stereoisomeric forms, consisting of two enantiomeric pairs of diastereomers (if the polyene is achiral), may be formed. Reactions (3) and (7) of Figure 12.3 illustrate cheletropic reactions of an olefin and a carbene to form a cyclopropane, and a substituted butadiene to form a 3,3-dioxo-3-thiacyclopent-1-ene. Reaction (3) illustrates suprafacial addition of [CH_2] to an olefin, yielding a pair of enantiomeric cyclopropanes. Reaction (7) illustrates conrotatory closure of the diene, yielding a pair of enantiomeric thiacyclopentenes. The forward reaction, (3) or (7), has two components. Disrotatory ring closure corresponds to *suprafacial* reaction of the π system, the conrotatory mode being designated *antarafacial*. A component analysis of the reverse process involves three components, the residual π system, and the two σ bonds which are to be lost.

SIGMATROPIC REARRANGEMENTS

A general sigmatropic rearrangement is shown in Figure 12.2. A molecule with two conjugated systems of m and n π electrons rearranges in such a way that the σ bond appears to migrate across each π system, forming a molecule with

188 PERICYCLIC REACTIONS

the same description but with the π bonds shifted by one carbon atom and the σ bond in a new position. Here m and n are even integers. One or *both* may be zero. A "zero" may indicate either that there is no active π system adjacent to that end of the σ bond, or that there is an active π system which consists of a single empty p orbital (a trivial π system with no electrons). In order for a sigmatropic rearrangement to take place, there must be at least *one* active π system. Sigmatropic rearrangements are sometimes described by a pair of numbers [i, j]. These count the number of *atoms* over which the two ends of the σ bond migrate, counting the atom at each end as "1." Thus the Cope rearrangement illustrated in Figure 12.2 is a [3, 3] rearrangement since each end of the migrating σ bond moves over three atoms. For the formation of the new (rearranged) σ bond to be feasible, the two ends of the active π systems must be able to approach each other in such a way as to enable the terminal p orbitals to overlap, or, in the case of a [1, j] rearrangement (i.e., only one end of the σ bond migrates), the nonmigrating end of the σ bond must be able to approach the other end of the active π system closely enough so that new bond formation can be initiated.

Stereochemical Considerations

In a sigmatropic rearrangement, with maximum labeling, the number of elements of stereochemistry is preserved. There are four, chirality at each end of the σ bond, and geometric isomerism at the far end of each active π system. After the rearrangement, the roles are reversed. A maximum of 16 ($=2^4$) stereoisomeric products may be formed. Even if one end of the σ bond does not migrate but is an asymmetric center, the stereochemistry of that center may change.

The stereochemical mode of reaction of a σ bond (as a component) may be specified as *suprafacial* or *antarafacial* in a fashion parallel to the π components. A sigma component of the reaction always has two electrons (a π component may have any even number, including 0). A σ bond component is regarded as reacting *suprafacially* if the stereochemical fate at each end is the same, i.e., the configuration is retained or inverted at each end. A σ component is regarded as reacting *antarafacially* if the stereochemical fate at the two ends is different, i.e., the configuration is retained at one end and inverted at the other, or vice versa. If one end of the bond is a hydrogen atom, that end is considered to react with retention of configuration. The stereochemical mode of reaction of that bond is then determined by whether configuration is retained (s) or inverted (a) at the other end.

COMPONENT ANALYSIS (*ALLOWED* OR *FORBIDDEN*?)

A component analysis of pericyclic reactions proceeds by the following steps which are illustrated for each of the classes of reaction in Figure 12.4:

COMPONENT ANALYSIS (*ALLOWED* OR *FORBIDDEN*?)

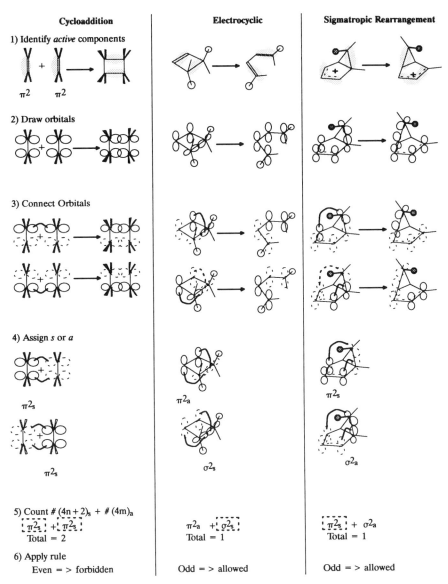

Figure 12.4. Procedure for general component analysis illustrated for each of the three types of pericyclic reactions.

1. Examine the reactants and products identifying the *active* components of each. The active components are those σ and/or π bonds which are broken in the reactants, and formed in the products.
2. Draw the constituent orbitals for each component. These are p orbitals at each end of the π system, and (usually) sp^3 hybrid orbitals at each end

of σ components (if the σ bond terminates at H, the orbital there is an *s* orbital). *Disregard orbital phases.*

3. Identify which orbitals of the reactants must overlap to form the newly formed bonds of the products and connect these with a curved line. When orbitals overlap in a π fashion, one may *arbitrarily* choose the pair of lobes (top or bottom) when making the connection.
4. Examine the components of the reactants. Each component should have *two* curved lines entering it, one to each end. The curved lines indicate the stereochemical mode of reaction of each component. For a π component, if the two curved lines are connected to the same face, the mode is *s*, if to opposite faces, *a*. For a σ component, the mode is *s* if both curved lines are connected to the larger (inner) lobes of the sp^3 orbitals, or both to the smaller (backside) lobes. Otherwise the mode is *a* as in the illustrated sigmatropic rearrangement in Figure 12.4. If the component is a single sp^n orbital, it is called ω. If both curved lines terminate to the same lobe, that component is reacting as *s*, if to different lobes, then *a*.

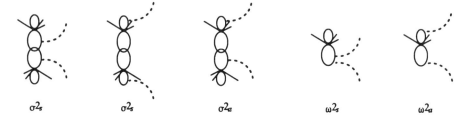

$\sigma 2_s$ $\sigma 2_s$ $\sigma 2_a$ $\omega 2_s$ $\omega 2_a$

5. Add up the components, counting only $(4n + 2)$ components if they are *s*, and $4m$ components if they are *a*. Remember that $(4n + 2)$ and $4m$ are the numbers of electrons. Thus the allylic π system in Figure 12.4 is a two-electron component, i.e. $(4n + 2)$ with $n = 0$. A component may have no electrons, in which case it is a $4m$ component ($m = 0$).
6. Apply the rule (see below).

The Rule for Component Analysis

The general rule for all pericyclic reactions was formulated by Woodward and Hoffmann ([3], p. 169). *A ground state pericyclic change is symmetry-allowed if the total number of* $(4n + 2)_s$ *and* $(4m)_a$ *components is* odd.

The Diels–Alder Reaction

The Diels–Alder reaction, shown in Figure 12.3, Reaction (6), is probably the best known and synthetically the most useful of all pericyclic reactions. The diene must be in, or be able to achieve, a quasi-*s-cis* geometry. The Diels–Alder reaction of the parent molecules, 1,3-butadiene and ethene, is difficult, requiring high temperature and pressure (e.g., 36 h at 185°C at 1800 psi [187]) but has been shown to proceed in a concerted synchronous fashion both experimentally [187] and theoretically [187], [188]. The reaction is accelerated by substitution by electron donors (X: or C substituents) on the diene moiety, and by the presence of electron withdrawing substituents (Z type) on the alkene, which then becomes a "good dienophile." Many examples of the Diels–Alder reaction with the reverse substitution pattern (reverse-demand Diels–Alder) are also known. The reaction proceeds with a high degree of stereospecificity at both diene and dienophile moieties as expected from its component analysis, $_\pi 4_s + {_\pi 2_s}$, and often in the case of unsymmetrically substituted components a high degree of regioselectivity (head-to-head versus head-to-tail selectivity). Additionally, the longitudinal orientation of the Z-substituted dienophile in the reaction is consistent with a degree of preference for the more crowded *endo* transition state leading to *endo* diastereoselectivity. All aspects of the Diels–Alder reaction are readily understood in terms of the orbital interaction diagram.

Orbital Interaction Analysis. An orbital interaction diagram for the Diels–Alder reaction is shown in Figure 12.5(a). The geometry of approach of the two reagents which ensures a maximum favorable interaction between the frontier MOs (dashed lines) preserves a plane of symmetry at all separations. The MOs are labeled according to whether they are symmetric (S) or antisymmetric

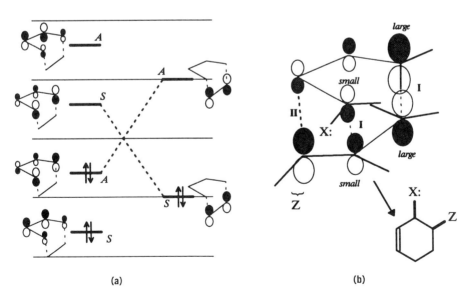

Figure 12.5. (a) Orbital interaction diagram for an *s-cis* diene and an olefin in the orientation for maximum overlap; (b) the interacting frontier MOs of an X:-substituted diene and a Z-substituted dienophile showing primary (I) and secondary (II) interactions.

(A) with respect to reflection in the plane. Simultaneous overlap of *both* HOMO–LUMO pairs is a necessary feature of all pericyclic reactions, permitting charge transfer in both directions in the multiple bond-forming process. The charge transfer and back transfer is required to avoid excessive charge separation in the reaction but is not necessarily synchronous in the sense that the extent of charge transfer in each direction is the same. In the normal course of the Diels–Alder reaction, the reactivity is controlled by a dominant HOMO (diene)–LUMO (dienophile) interaction. The polarization and energies of unsymmetrically substituted frontier MOs of olefins and dienes was discussed in Chapter 6. The orbitals are shown in Figure 12.5(b) in the orientation which provides maximum overlap (the *large–large small–small* orientation) in the primary interactions (I). The reaction is therefore regioselective, producing 3X,4Z-cyclohex-1-enes. The secondary interaction (II) [186] leads to *endo* diastereoselectivity and *cis* diastereomers, namely an enantiomeric pair of *cis*-3X,4Z-cyclohex-1-enes [Figure 12.5(b)]. Substantial progress has been achieved in enantioselective Diels–Alder reactions where one or both of the components are substituted by chiral groups or complexed to chiral organometallic groups.

The Cope Rearrangement

The Cope rearrangement, illustrated in Figure 12.6, and variations of it, are synthetically important reactions ([99], p. 1021). Reaction of the parent 1,5-hexadiene has an activation energy of 140 kJ/mol [189] and has been shown

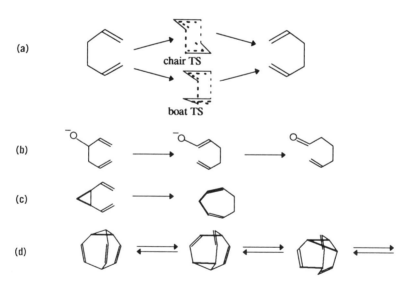

Figure 12.6. (a) The Cope rearrangement of 1,5-hexadiene; (b) the oxy-Cope rearrangement; (c) the divinylcyclopropane rearrangement; (d) the degenerate rearrangements of bullvalene.

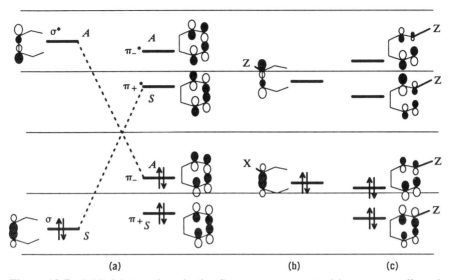

Figure 12.7. Orbital interactions in the Cope rearrangement: **(a)** symmetry allowed interactions of the σ and π orbitals; **(b)** effect of X: (or C) and Z (or C) substituents in the 3 position; **(c)** effect of Z substituents at the 1 position.

by *ab initio* computations [190], [191] to proceed in a synchronous concerted manner. The chair-like transition state is slightly preferred over the boat-like form [Figure 12.6(a)]. Variations include the anionic oxy-Cope rearrangements shown in Figure 12.6(b) [192], and the rearrangement of divinyl cyclopropanes to 1,4-cycloheptadienes [Figure 12.6(c)]. The 1209600-fold degenerate rearrangement of bullvalene [Figure 12.6(d)] is rapid at room temperature, displaying only a single peak in the NMR spectrum. The structure has been shown to have C_{3v} symmetry by neutron diffraction studies [193] and by temperature-dependent solid-state NMR [194].

Orbital Interaction Analysis. An orbital interaction diagram for the Cope arrangement is shown in Figure 12.7(a). The reaction may be initiated by electron donation from the σ bonding orbital into π_+^*, or from π_- into σ^* (dashed lines). The effect of a Z-type substituent at the 3 position [Figure 12.7(b)] is to lower the energy of the σ^* orbital thereby increasing the interaction with the π_- orbital and facilitating the reaction. After rearrangement the Z substituent would occupy the 1 position of the resulting 1,5 diene [Figure 12.7(c)]. While the LUMO energy is lowered by the Z substituent, the concomitant polarization of the adjacent π^* orbital and energy separation of the two π^* orbitals act to reduce the bonding interaction (positive overlap) between the two π^* orbitals. These factors, together with the energy lowering of one of the π bonding orbitals combine to shift the equilibrium of the Cope rearrangement so as to leave the Z substituent in the 1 position. The effect of an X:-type substituent in the 3 position is also shown in Figure 12.7(b). The raised σ bonding orbital also serves to facilitate the reaction, placing the X: substituent in the 1 position. The effect of the X: substituent in this position (not shown) is the inverse of that of the Z substituent, but the result is the same; the equilibrium is shifted in this direction. The oxy-Cope rearrangement shown in Figure 12.6(b) is an example.

The σ bonding orbital may also be raised by incorporation into a three- or four-membered ring. The divinylcyclopropane to cyclohepta-1,4-diene [Figure 12.6(c)] is an example, as is the rapid degenerate rearrangement of bullvalene [Figure 12.6(d)] and related compounds.

1,3-DIPOLAR CYCLOADDITION REACTIONS

A general 1,3-dipolar cycloaddition reaction is shown below. The "dipole" is a triad of atoms which has a π system of four electrons and for which a dipolar

$$\begin{array}{c} X \\ \| \\ Y \end{array} + \begin{array}{c} A^+ \\ \diagdown \\ B \\ \diagup \\ C^- \end{array} \longrightarrow \begin{array}{c} X{-}A \\ | \quad \diagdown \\ \quad \quad B \\ Y{-}C \diagup \end{array}$$

resonance form provides an important component of its bonding description. The requirements have also been described as "Atom A must have a sextet of outer electrons and C has an octet with at least one unshared pair" ([99], p. 743). The substrate is an olefin and an alkyne or a carbonyl. The reaction therefore is of $_\pi 4 + {_\pi}2$ type like the related Diels–Alder reaction and has proved to be very useful synthetically for the construction of five-membered ring heterocycles [195], [196]. Evidence suggests the reaction is concerted and regioselective. "Dipoles" fall into two general categories [99].

1. *16-electron:* Those for which the dipolar canonical form has a double bond on the sextet atom and the other resonance structure has a triple bond. Examples are azides $(R-N_3)$, diazoalkanes $(R_2C=N=N)$, and nitriloxides $(R-C\equiv N-O)$.
2. *18-electron:* Those in which the dipolar canonical form has a single bond on the sextet atom and the other form a double bond. Examples are ketocarbenes $[R-C:-C(O)R]$, azoxy compounds $[R-N=N(O)R]$, and nitrones $(R_2C=N(O)R)$.

If all three atoms of the triad belong to the first row, a total of 18 different uncharged species, six of the *16-electron* kind, and 12 of the *18-electron* kind, are possible.

The reactivity and regioselectivity of 1,3-dipolar cycloadditions have been discussed in terms of the frontier orbitals [197]. Most of the features may be understood on the basis of simple Hückel MO theory. The HOMO and LUMO π orbitals and π orbital energies for all 18 combinations of the parent dipoles are shown in Figure 12.8. The frontier orbitals of many of the 1,3-dipoles have

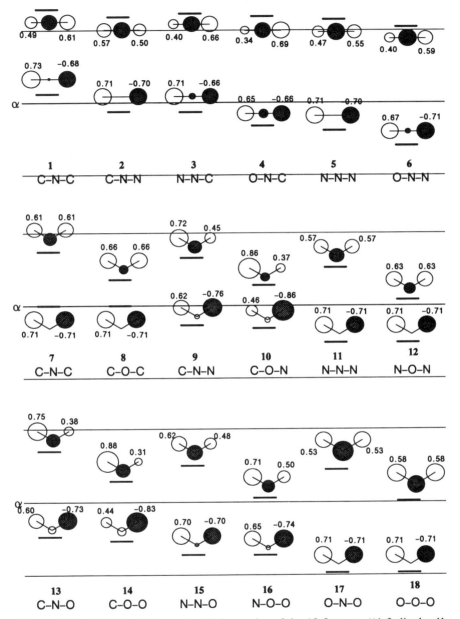

Figure 12.8. SHMO orbitals and orbital energies of the 18 first row "1,3-dipoles."

previously been derived by CNDO/2 and extended Hückel theory [198]. The first six structures, all of *16-electron* type, are shown in greater detail below. The SHMO parameters for the dicoordinated C atom of **1**, **3**, and **4**, and the monocoordinated N atom of **2**, **5**, and **6**, were derived by the addition of 0.25 $|\beta|$ to the normal coulomb integrals (α's) of the tricoordinated C and dicoor-

196 PERICYCLIC REACTIONS

$$\overset{R_1}{\underset{R_2}{\diagdown}}\overset{-}{C}-N^+\equiv C-R_3$$

1
Nitrile ylides

$$\overset{R_1}{\underset{R_2}{\diagdown}}C=N^+=N^-$$

2
Diazoalkanes

$$\overset{R_1}{\diagdown}N-N^+\equiv C-R_2$$

3
Nitrile imines

$$O^--N^+\equiv C-R$$

4
Nitrile oxides

$$\overset{R}{\diagdown}N=N^+=N^-$$

5
Alkyl azides

$$O^--N^+\equiv N$$

6
Nitrous oxide

dinated N atoms respectively, as discussed in Chapter 7. It is apparent that each of the first six parent or alkyl-substituted dipoles has a very high LUMO and so will not function in the first instance as an electrophile. All have quite a high HOMO which decreases rapidly across the series 1–6. The nucleophilicity is predicted to be relatively high for the first stable dipole, **2** and lower for the others, **4, 5,** and **6**. In fact, N_2O has been shown to react with "C"-substituted and X-substituted olefins but does not react with Z-substituted olefins [199]. The orbital energies of N_2O appear to be seriously overestimated in SHMO theory, perhaps because no account is taken of substitution by strongly electronegative groups. The experimental ionization potential of N_2O is quite high, 12.9 eV [200]. For the rest of these dipoles, a good dipolarophile is analogous to a good dienophile in the normal Diels–Alder reaction, in other words, a Z-substituted olefin.

The HOMO of none of the first six parent dipoles is strongly polarized. Little regioselectivity is expected due to the primary interaction, HOMO (dipole)–LUMO (dipolarophile). For dipoles **4–6**, the LUMO is relatively strongly polarized but it should not be expected that significant regioselectivity be observed due to the secondary interaction, LUMO (dipole)–HOMO (dipolarophile), since the HOMOs of good dipolarophiles for these dipoles are not strongly polarized themselves. While the HOMO orbitals of the parent dipoles are not strongly polarized, this will probably not be the case for substituted dipoles. Substituents in the terminal positions will perturb the π system in the same direction as already discussed in Chapter 6 in connection with olefin reactivity. Substituents of the X: and "C" type will raise the HOMO further and polarize it away from the substituent. Substituents of the Z and "C" type will lower the LUMO and also polarize it away from the substituent. Substitution at the middle position is not possible for neutral dipoles of the 16-electron type.

The remaining 12 dipoles, which are of the *18-electron* type, are shown explicitly below. Imine ylides (**7**) and carbonyl ylides (**8**) are not stable but may be generated *in situ* by pyrolysis of suitably substituted aziridines and oxiranes. The energy of the HOMO, and therefore the nucleophilicity of the parent *18-electron* dipoles decreases from very high to very low across the

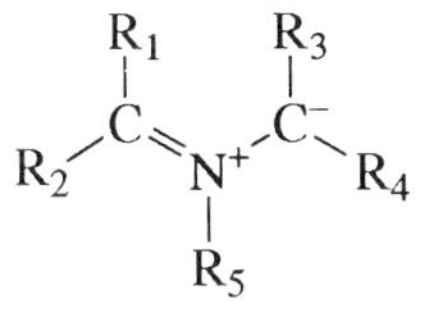

7 Imine ylide

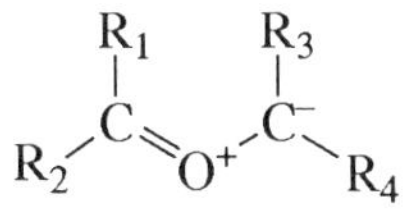

8 Carbonyl ylide

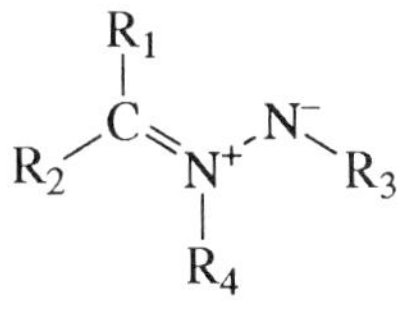

9

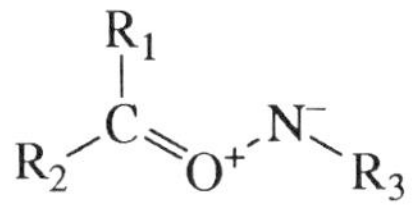

10 Carbonyl imine

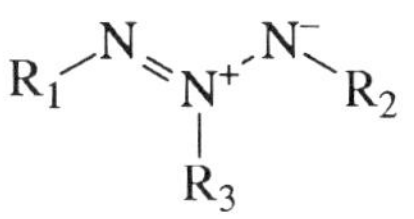

11

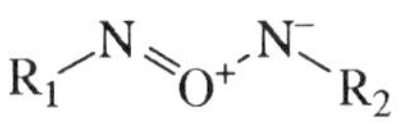

12 Nitrosyl imine

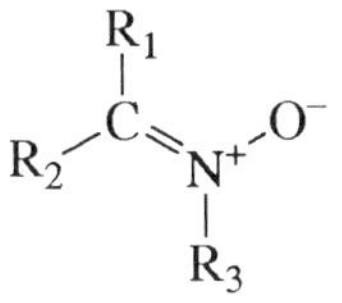

13 Nitrone

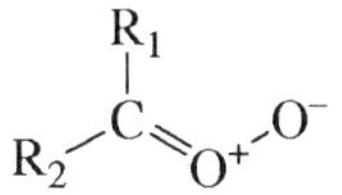

14 Carbonyl oxide

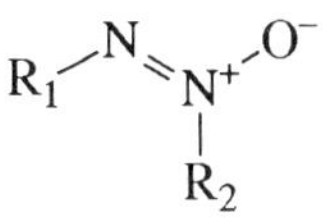

15 Azoxy

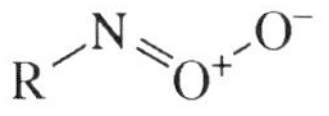

16 Nitrosyl oxide

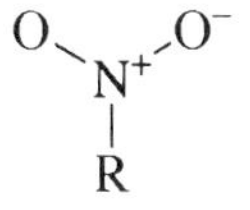

17 Nitro compound

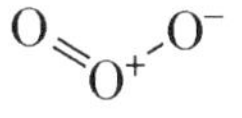

18 Ozone

series **7–18**. In the same series, the electrophilicity increases from moderate to high, being consistently higher when the central atom is oxygen.

The ozonolysis of olefins may be analyzed as a sequence of two 1,3-dipolar cycloadditions, initial electrophilic attack by ozone (**18**) to form the first intermediate which decomposes into a carbonyl compound and a carbonyl oxide (**14**), followed by nucleophilic 1,3-dipolar addition of the carbonyl oxide (**14**) to the ketone yielding the molozonide.

CHAPTER 13

ORBITAL AND STATE CORRELATION DIAGRAMS

GENERAL PRINCIPLES

Orbital interaction theory as introduced and utilized in the previous chapters has served to point us in a direction along a reaction coordinate. We have been able to find the most favorable geometry in which two molecules can approach each other by considering the interactions of the frontier orbitals, and we have seen that *carried to the logical extreme*, the initial interactions imply making and breaking of bonds and the formation of new molecules. Indeed, the initial trajectory implied from the orbital interaction theory very often leads to the expected products when the reaction is exothermic. For endothermic reactions, information with respect to the structure and stability of the transition state is necessary to decide upon the reaction pathway and this could be deduced from examination of the reactive intermediates (or products if the reaction is concerted). The purpose of *correlation* diagrams is to follow a system *continuously* from one description to another, assuming that material balance is maintained, for example, a chemical reaction from reactants to products, or a molecule from one conformation to another. A correlation diagram, or a comparison of several correlation diagrams, would permit one to identify preferred reaction pathways, or more stable conformations, and offer an explanation for experimentally observed preferences.

What should be correlated? In an *orbital* correlation diagram, the shapes and energies of orbitals are examined to see if the electronic structure of the reactants could be smoothly converted into the electronic structure of the products, each defined by the structures and occupancies of their respective orbitals. The nodal characteristics of orbitals are very resistant even to rather large perturbations and will tend to be conserved in chemical reactions. If an element of symmetry, for example, a mirror plane, is maintained during the course of the reaction, the nodal characteristics separate the orbitals into two sets, the members of one set being symmetric with respect to reflection in the symmetry plane, the others being antisymmetric with respect to reflection. Orbital symmetries are conserved in the reaction. If the number of occupied orbitals of each symmetry type is not the same in the reactants as in the products, that reaction will not take place readily. Even if symmetry is not present, *local* symmetry usually is sufficient to distinguish the orbitals, hence the principle of *Conservation of Orbital Symmetry* which formed the title of the classic text by Woodward and Hoffmann [3], or the *Orbital Correlation Analysis using Maximum Symmetry* (OCAMS) approach of Halevi [9].

WOODWARD–HOFFMANN ORBITAL CORRELATION DIAGRAMS

An abbreviated description of the generation of Woodward–Hoffmann orbital correlation diagrams is presented here. A full description is available in numerous sources, most notably [3]. To begin with, a reaction coordinate is proposed. The active components of the reactants and products are maximally symmetrized by ignoring substituents and the presence of embedded heteroatoms. Symmetry elements conserved in the course of the reaction (at least those which intersect bonds being formed or broken) are identified and group orbitals are combined to provide symmetry adapted MOs if necessary. In- and out-of-phase combinations of like-symmetry reactant orbitals are taken if they are close in energy. The orbitals of products are treated the same way. Inspection of the reactant and product orbitals may suggest obvious correlations. However, orbitals of like symmetry will not cross, while orbitals of different symmetry may. The lowest energy MO of the reactants is connected to the lowest energy product MO which has the same symmetry. The next higher reactant MO is similarly correlated with a product MO of the same symmetry type, and so on until the highest energy reactant MO has been correlated. The numbers and symmetry types of reactant and product orbitals must be the same, so that all orbitals are connected. If all reactant orbitals which are occupied in the ground-state electronic configuration correlate with ground-state occupied MOs of the product(s), the reaction is thermally allowed by the postulated mechanism. If one or more reactant MOs correlate with product orbitals which are empty in the ground state, the reaction is not allowed.

Orbital correlation diagrams are useful for cycloadditions and electrocyclic reactions but not for sigmatropic rearrangements since no element of symmetry is preserved.

Cycloaddition Reactions

The orbital correlation diagrams for the attempted face-to-face dimerization of two olefins ($_\pi 2_s + {}_\pi 2_s$) to form a cyclobutane, and for the Diels–Alder reaction are shown in Figure 13.1(a) and (b), respectively. Figure 13.1(a) displays a

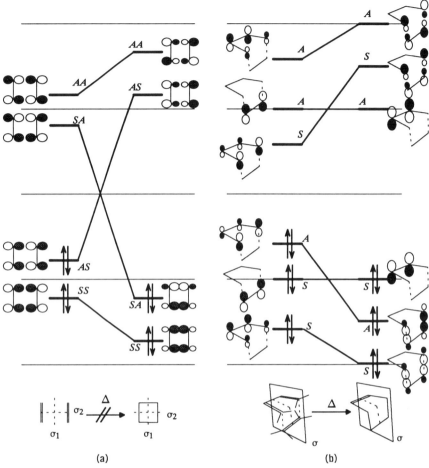

Figure 13.1. (a) Orbital correlation diagram for $_\pi 2_s + {}_\pi 2_s$ dimerization of two olefins to form a cyclobutane; (b) orbital correlation diagram for $_\pi 4_s + {}_\pi 2_s$ cycloaddition of a diene and an olefin (the Diels–Alder reaction).

typical correlation diagram for a reaction which is orbital-symmetry forbidden. The HOMO of the olefin pair (the out-of-phase combination of the two π bonds) correlates with an unoccupied MO of cyclobutane. Therefore, an attempt to squeeze two olefins together in a face-to-face geometry (the least motion pathway to cyclobutane) would tend to yield a highly excited state of the cyclobutane. The reverse is also true. Rupturing a cyclobutane into two olefins would produce one or both olefins in a highly excited state. On the other hand, the orbital correlation diagram for the Diels–Alder reaction [Figure 13.1(b)] is typical of thermally allowed pericyclic reactions. Ground-state reactants would yield ground-state product(s) directly.

Electrocyclic Reactions

As discussed in Chapter 12, electrocyclic reactions may proceed in a conrotatory or disrotatory fashion, that is, the π system cyclizes in an antarafacial or suprafacial manner, respectively. Since there is only a single component, it should be *counted* according to the general component analysis for a thermally allowed reaction, i.e., $_\pi(4m)_a$ = conrotatory and $_\pi(4n + 2)_s$ = disrotatory. Thus, the cyclization of butadienes to cyclobutenes (or more likely the reverse reaction) should proceed in a conrotatory fashion ($m = 1$), and the reaction of 1,3,5-hexatrienes should proceed in a disrotatory manner ($n = 1$). As illustrated in Figure 13.2, using butadiene as an example, a two-fold axis of symmetry is maintained throughout a conrotatory electrocyclic closure or opening, while a mirror plane of symmetry is preserved during disrotatory ring closure. The orbital correlation diagrams for electrocyclic reactions of butadiene and hexatriene are shown in Figure 13.3(a) and (b), respectively. In each case, the solid lines connecting reactant and product MOs show the correlation for *con*rotatory ring closure, the MOs being classified according to their properties, **S** or **A**, with respect to a preserved C_2 axis of symmetry. The dashed lines show the correlation for *dis*rotatory ring closure, the MOs being classified S or A with respect to a preserved *mirror plane* of symmetry. The thermally allowed conrotatory opening of cyclobutenes has been confirmed experimentally. Likewise, the disrotatory mode of ring closure of (Z)-1,3,5-hexatriene has been demonstrated.

The electrocyclic reactions of three-membered rings, cyclopropyl cation and cyclopropyl anion, may be treated as special cases of the general reaction. Thus, cyclopropyl cation opens to allyl cation in a disrotatory manner (i.e., allyl cation, $n = 0$), and cyclopropyl anion opens thermally to allyl anion in a conrotatory manner (i.e., allyl anion, $m = 1$). Heterocyclic systems isoelectronic to cyclopropyl anion, namely oxiranes, thiiranes, and aziridines have also been shown experimentally and theoretically to open in a conrotatory manner [201].

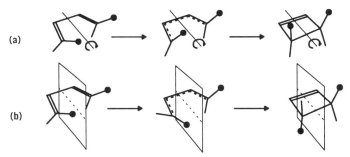

Figure 13.2. (**a**) Conrotatory electrocyclic reaction showing preservation of a C_2 axis of symmetry; (**b**) disrotatory ring closure showing preservation of a mirror plane of symmetry.

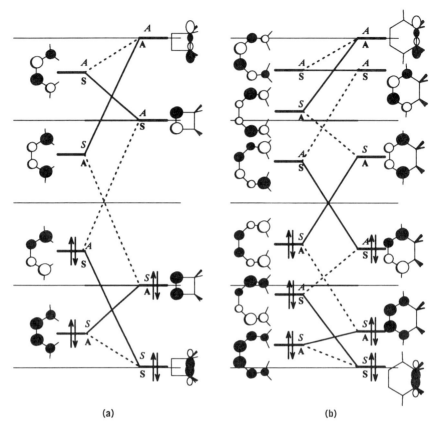

Figure 13.3. (a) Orbital correlation diagram for electrocyclic reaction of butadienes; (b) orbital correlation diagram for electrocyclic reaction of hexatrienes. Solid lines and S, A denote correlation for *conrotatory* motion; dashed lines and S, A denote correlation for disrotatory motion.

Cheletropic Reactions

Cheletropic reactions, in which a single atom is added or extruded, comprise a special case of cycloaddition reactions. Figure 13.4 illustrates correlation diagrams for two typical cheletropic reactions, the loss of SO_2 from a thiirane dioxide [Figure 13.4(a)], and loss of CO from a norbornadienone [Figure 13.4(b)]. The addition of a carbene to an olefin is another example which is discussed below (Figure 13.9).

Photochemistry from Orbital Correlation Diagrams

Photochemical processes are best examined using *state* correlation diagrams, as shown below, and in Chapter 14. Nevertheless, some information may be

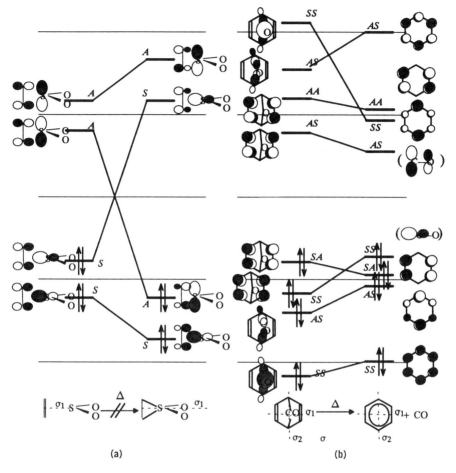

Figure 13.4. (a) Orbital correlation diagram for $_\pi 2_s + _\omega 2_s$ cheletropic addition of SO_2 to an olefin. The symmetry element preserved is a vertical mirror plane; (b) orbital correlation diagram for elimination of CO from a norbornadienone. Two vertical planes of symmetry are preserved.

derived from orbital correlation diagrams. The orbital correlation diagrams for cycloaddition reactions shown in Figure 13.1, are repeated in Figure 13.5, showing the reactants in their lowest excited states. In simple Hückel theory, this corresponds to a single electron jump from the HOMO to the LUMO. The photochemical correlation shown in Figure 13.5(a) suggests that if the reactants are in their lowest excited state, then the product cyclobutane would also be formed in its lowest excited state. This is considered to be a photochemically *allowed* reaction. This should be contrasted with the situation depicted in Figure 13.5(b) where the lowest excited state of reactants correlates with a highly excited state of the product cyclohexene. Thus, the Diels–Alder reac-

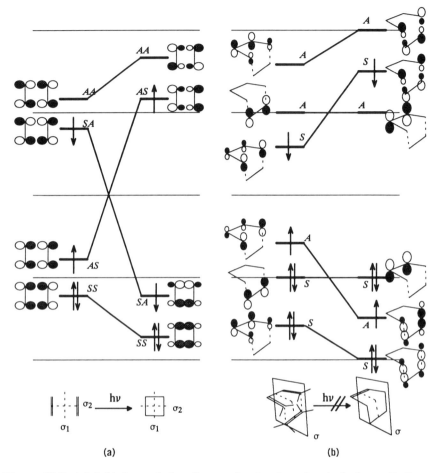

Figure 13.5. (a) Orbital correlation diagram for the photochemical $_\pi 2_s + {}_\pi 2_s$ dimerization of two olefins to form a cyclobutane; (b) orbital correlation diagram for the photochemical $_\pi 4_s + {}_\pi 2_s$ cycloaddition of a diene and an olefin.

tion would not proceed with the observed stereochemistry ($_\pi 4_s + {}_\pi 2_s$) under photolysis conditions. This does not imply that a diene and an olefin will not react under these conditions, but it does suggest that the mechanism will be other than the thermally allowed route. The stereospecific $_\pi 2_s + {}_\pi 2_s$ *unsensitized* photodimerization of *cis*- and *trans*-2-butene has been carried out [202].

Another well-studied case involves the $_\pi 2_s + {}_\pi 2_s$ photochemical conversion

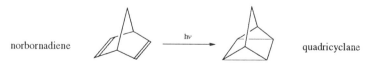

of norbornadiene to quadricyclane [203], possibly as a system for the chemical storage of solar energy.

The orbital correlation diagrams for electrocyclic reactions shown in Figure 13.3, are repeated in Figure 13.6, with the reactants in their lowest excited states. In the case of butadienes [Figure 13.6(a)], the reactant in its first excited state would yield product in its first excited state if the disrotatory pathway (correlation by dashed lines) were followed. The situation is reversed for hexatrienes [Figure 13.6(b)]. In that case, reactant in its first excited state would yield product in its first excited state if the conrotatory pathway (correlation by solid lines) were to be followed. Thus, the expected stereochemistry upon photolysis is the opposite of that expected and observed for the thermal (ground state) reactions. In fact, the photolytic ring opening of simple alkylcyclobutenes has been shown *not* to be stereospecific [204]. One of the

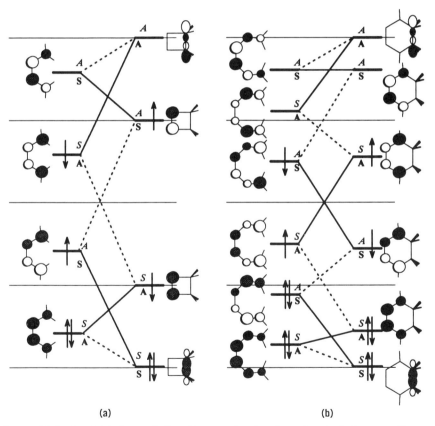

Figure 13.6. (a) Orbital correlation diagram for the photochemical electrocyclic reaction of butadienes; (b) orbital correlation diagram for the photochemical electrocyclic reaction of hexatrienes. Solid lines and **S, A** denote correlation for *conrotatory* motion; dashed lines and *S, A* denote correlation for *disrotatory* motion.

possible reasons for the lack of observed stereoselectivity is competitive radiationless transition to the vibrationally excited ground state and subsequent thermal ring opening in the allowed conrotatory fashion. Partial orbital symmetry control is observed in the photochemical ring opening of a constrained cyclobutane [205]. An alternative mechanism for the loss of stereospecificity in the photolysis of cyclobutenes has been proposed on the basis of theoretical calculations [206] which suggested the existence of three stereochemically distinct transition points for de-excitation of the excited ring-opening cyclobutane. By contrast with the photochemical cyclobutane to butadiene conversion, the photochemical conversion of 1,3-cyclohexadiene has been shown to proceed in a highly stereospecific conrotatory manner to yield the *cis*-hexatriene [207].

LIMITATIONS OF ORBITAL CORRELATION DIAGRAMS

Although very useful in the vast majority of cases, orbital correlation diagrams are known to yield incorrect, or at least suspect, results in a few cases, and may be difficult to interpret in others. The rearrangement of benzvalene to benzene provides an example. The most direct path for the rearrangement preserves a C_2 axis of symmetry. The correlation diagram for the reaction, shown in Figure 13.7, suggests the reaction is thermally allowed. On the contrary, a moderately high barrier is observed for this highly exothermic reaction. In a similar vein, the direct thermal conversion of cyclooctatetraene to cubane is not observed, and was considered by Woodward and Hoffmann to be forbidden on the basis of *intended* orbital correlation. Recent experimental results reveal that in the *retro* process, which is exothermic, cubane may be rearranged to cyclooctatetraene with a relatively low activation barrier, 180 ± 4 kJ mol^{-1} [208]. The rearrangement is thought to involve a number of intermediates which were not observed [208].

STATE CORRELATION DIAGRAMS

Although orbital correlation diagrams often give a reliable indication of the course of a reaction, it is not possible to follow a reaction in any quantitative fashion in this way, even if highly accurate *ab initio* MO energies are available at all points along a reaction coordinate. The ground state and electronically excited states are separated from each other within the Born–Oppenheimer approximation, and by total spin angular momentum (since the usual electronic Hamiltonian does not have terms derived from electron and nuclear spins). *Ab initio* CI or other post-Hartree–Fock techniques are able in principle to produce quantitatively correct correlations of the ground and lower excited states of reactant(s) and trace these accurately along the reaction coordinate to products. However, since, only the few lower electronic states are of chemical interest,

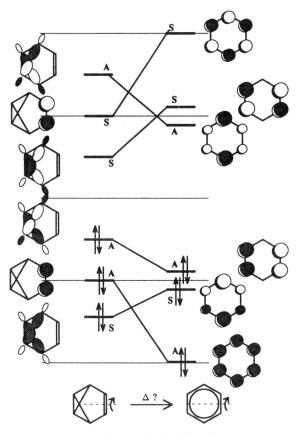

Figure 13.7. Orbital correlation diagram for the thermal rearrangement of benzvalene to benzene. A C_2 axis of symmetry is preserved. See the text for a discussion.

quite adequate *state* correlation diagrams may usually be derived from the simple orbital correlation diagrams which we have examined above. Different electronic states may be generated from the set of occupied and empty MOs by redistributing the electrons. Relative energies of the states may be estimated by simple summation of the electron (MO) energies, and state symmetries may be deduced from the symmetries of the electron distributions (the state symmetry is the direct product of the individual electron symmetries). Singlet and triplet spin configurations are simply represented by the artifice of showing electrons in singly occupied MOs as spin antiparallel (singlet) or parallel (triplet). Triplet states derived from the same MOs as singlet states have the same spatial symmetry but are lower in energy by a magnitude which depends on the magnitude of the exchange integral. Qualitatively, if the SOMOs are spatially close (e.g., π and π^*), the singlet triplet spacing will be large, whereas for spatially remote MOs or orbitals which lie in each others nodal planes (e.g.,

208 ORBITAL AND STATE CORRELATION DIAGRAMS

n_O and π^* of carbonyls), the singlet triplet splitting will be small. Once the states of reactants and products are qualitatively ordered in energy and their spin and symmetry properties assigned, the states may be correlated by a few simple rules which parallel the rules laid out above for orbital correlation diagrams.

Electronic States from MOs

The procedure for obtaining electronic state wave functions from the MOs for a molecule is illustrated for a carbonyl group in Figure 13.8. The frontier MOs are sketched, placed on an energy scale and, if applicable, their symmetry properties designated relative to the molecular point group or the local point group. Thus, the important MOs of carbonyls are the occupied π and n_O (HOMO) MOs and the unoccupied π^* (LUMO) as were developed in Chapter 3 or 8. The local symmetry in the absence of strongly conjugating substituents is C_{2v}. In Figure 13.8(a), the MOs are classified as S or A with respect to the three symmetry operations, or given the appropriate labels of the irreducible representations. In Figure 13.8(b), electronic configurations of the lower elec-

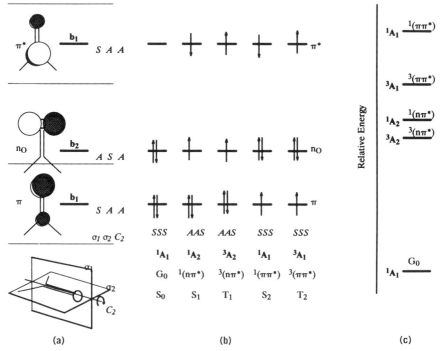

Figure 13.8. (a) The MOs of the carbonyl group. Symmetry classification is with respect to the local symmetry group C_{2v}; (b) the electronic states which can be constructed from the three frontier MOs; (c) the electronic states ranked approximately in relative energy.

tronic states are obtained by redistributing the electrons among the MOs. The spatial composition of the state may be specified by the *configuration*, i.e., a statement of the occupancies of the MOs. The ground state, designated S_0 or G, is obtained when all the lowest energy MOs are doubly occupied in most cases (we have seen an exception in the case of carbenes in Chapter 7). Thus the *configuration* of S_0 is $\cdots (\phi_{HOMO-1})^2(\phi_{HOMO})^2$. Excited *singlet* states are obtained by single (or multiple) electron excitations from the occupied MOs. The first (lowest energy) excited singlet state is usually that obtained from a HOMO to LUMO excitation, and is designated S_1, with configuration $\cdots (\phi_{HOMO-1})^2(\phi_{HOMO})^1(\phi_{LUMO})^1$, or more descriptively, by the MOs involved in the transition (if the MOs are known), e.g., $^1(n_O\pi^*)$. States with doubly occupied MOs only must be singlet states. Where "open shells" are involved, additional states must be considered from reversal of electron spin(s). When two electrons have the same spin, the state is a *triplet*, designated $T_1, T_2, \cdots$ in order of increasing energy, or by superscript in front of the state descriptor, e.g., $^3(n_O\pi^*)$. The spatial symmetry of the state is, in practice, determined by the cross product of the spatial symmetries of the SOMOs, by the rules: $S \times S \Rightarrow S, A \times A \Rightarrow S, S \times A \Rightarrow A, A \times S \Rightarrow A$. The "closed shell" states, like the ground state are S with respect to every symmetry operation, or transform as the totally symmetric irreducible representation of any point group. The energies of the electronic states may be derived from simple Hückel considerations (MO energy differences). Triplet states are lower than the singlets with the same MO configuration. The separation of $^3(n_O\pi^*)$ and $^1(n_O\pi^*)$ is relatively small, whereas the separation of $^3(\pi\pi^*)$ and $^1(\pi\pi^*)$ is large. Singlet states may be reached directly from the ground state by photoexcitation at the appropriate frequency. The electron rearrangement associated with the excitation process is very rapid (about 10^{-15} sec) and the nuclei do not have time to move to their new equilibrium positions. The dynamics of the excitation process and the fate of excited electronic states is discussed in Chapter 14.

Rules for Correlation of Electronic States

When the electronic states of both the reactants and products have been determined and characterized, a correlation diagram may be constructed by connecting the states according to the following rules:

1. As with orbital correlation diagrams, states are correlated from the lowest to the highest.
2. Reactant states will only correlate with product states of the same spatial symmetry and spin multiplicity.
3. Correlation lines define potential energy "surfaces." Surfaces are allowed to cross (intersect) if the states involved are different in spatial symmetry or spin multiplicity, but may *not* cross if both characteristics are the same.
4. Inspection of the orbital makeup of reactant and product states may im-

ply *intended* correlations which would lead to state crossings. If the states involved may not cross by rule 3, an *avoided* crossing occurs. The intention to cross is often depicted by dashed lines.

Example: Carbene Addition to an Olefin

The MOs and electronic states of carbene have been discussed in Chapter 7. The orbital and state correlation diagrams for addition of :CH_2 to ethylene is shown in Figure 13.9. The Walsh bonding picture for the MOs of cyclopropane requires that the σ and σ^* MOs of the ethylene also be included in the diagram. The σ_2 and σ_3 orbitals are degenerate but are shown separated for clarity. The postulated least-motion pathway preserves a vertical plane of symmetry (as well as the other elements of the C_{2v} point group) and the orbitals of reactants and the product cyclopropane are characterized by their behavior towards reflection in this plane. The orbital correlation is shown [Figure 13.9(a)] with the carbene in its S_0 electronic state. Direct insertion by the least-motion pathway is clearly forbidden on the basis of the diagram. Yet it is well known that singlet carbenes add smoothly and stereospecifically to olefins to yield

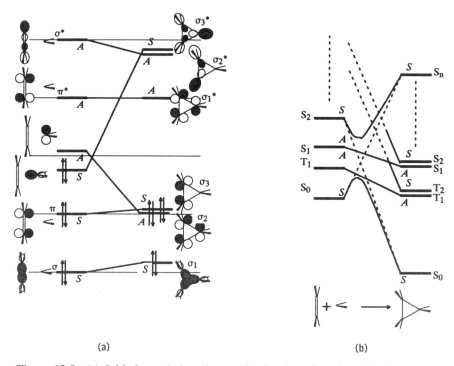

Figure 13.9. (a) Orbital correlation diagram for the direct insertion of carbene into an olefin to form cyclopropane. Symmetry classification is with respect to the vertical bisecting mirror plane; (b) state correlation diagram showing the intended correlations and the avoided crossing of states S_0 and S_2.

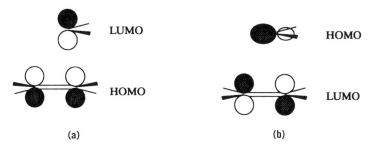

Figure 13.10. Non-least motion approach of a singlet carbene to an olefin showing the most favorable HOMO–LUMO interaction: **(a)** the primary interaction; **(b)** secondary interaction.

cyclopropanes. The apparent contradiction may be resolved by inspection of the state correlation diagram. The electronic configurations of carbene were shown in Figure 7.5. The lower electronic states of the reactants are entirely derived from redistribution of the two valence electrons between the two frontier MOs of the carbene and are labeled as in Figure 7.5: $S_0-(\pi)^2(sp^n)^2$, sym-

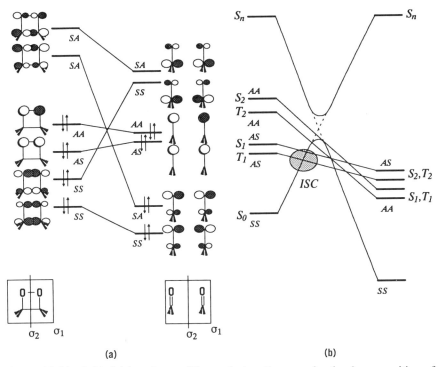

Figure 13.11. Orbital **(a)** and state **(b)** correlation diagrams for the decomposition of dioxetane. The observed chemiluminescence may be due to intersystem crossing (*ISC*) to the triplet manifold in the region denoted by the shaded circle in **(b)**.

metry S; $T_1-(\pi)^2(sp^n)^1(p)^1$, symmetry A; $S_1-(\pi)^2(sp^n)^1(p)^1$, symmetry A; $S_2-(\pi)^2(p)^2$, symmetry S. The lower electronic states of cyclopropane by simple Hückel theory are: $S_0-(\sigma_2)^2(\sigma_3)^2$, symmetry S; T_1, $S_1-(\sigma_2)^2(\sigma_3)^1(\sigma_1^*)^1$, symmetry A; T_2, $S_2-(\sigma_2)^1(\sigma_3)^2(\sigma_1^*)^1$, symmetry S. The pairs of states, S_1, S_2 and T_1, T_2, are degenerate. The high energy cyclopropane excited state $S_n-(\sigma_2)^0(\sigma_3)^2(\sigma_3^*)^2$ is shown as the intended correlation partner of the carbene (+ ethylene) S_0 state. The doubly excited but relatively low energy carbene (+ethylene) state S_2 would correlate directly with the ground state of cyclopropane but is prevented from doing so by the avoided crossing with the ascending S_0 state. The reaction is highly exothermic and the avoided crossing occurs early in the reaction coordinate. The resulting barrier hindering reaction is relatively low, thus explaining the apparent allowedness and concerted character of singlet carbene insertions into olefins. In fact, the barrier may be further lowered if the carbene approaches the olefin in a non-least motion pathway as would be expected from orbital interaction considerations (Figure 13.10) and shown by computation [209], [210]. Similar non-least motion reaction pathways have been theoretically predicted for rapid singlet carbene insertion into C—H bonds [211] and in carbene dimerization to ethylenes [212].

The orbital and state correlation diagrams for the pyrolysis of dioxetanes are shown in Figure 13.11. The formally forbidden $_\sigma 2_s + {_\sigma 2_s}$ reaction proceeds with activation energy in the range, 90–120 kJ mol^{-1}, is highly exothermic, and is accompanied by emission of light. One of the product carbonyl compounds is produced in an electronically excited state. The chemiluminescence is due to phosphorescence with high quantum efficiency [213], [214], indicating that the final product state is a triplet.

CHAPTER 14

PHOTOCHEMISTRY

The ground state of a molecule represents only one of an infinite number of electronic states. Because the electron distribution is different for each state, the bonding, structure, and reactivity will be different from the ground state. Each electronic state is *de facto* a separate chemical species, with its own set of chemical and physical properties. Excited electronic states are higher in energy than the ground state so energy must be introduced in some manner to excite the molecule. In *photo*chemistry the energy is introduced to the molecule by the absorption of a photon of light (hν) which falls in the UV/visible spectral region.

PHOTOEXCITATION

The geometry of a molecule in its ground state represents but one point on a (3N-6)-dimensional potential hypersurface (3N-5 in the case of linear molecules). It is a stationary point (all forces acting on the nuclei are zero) corresponding to a local minimum (all displacements of nuclei from their equilibrium positions lead to a rise in the potential energy). Other stationary points on the ground state potential energy hypersurface may represent different conformations of the same molecule or different molecules (including dissociation fragments). Each excited state has a similar potential hypersurface associated with it. Stationary points on excited state surfaces will not in general coincide with those on the ground-state surface. In Figure 14.1 is shown a one-dimensional cross section of the ground state, S_0, and two excited-state potential surfaces, one of which is *bound* like the ground state, S_1, and the other of which is *unbound*, S_2. In fact, Figure 14.1 may depict a state correlation dia-

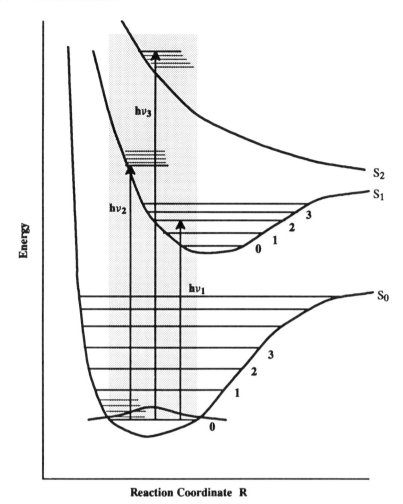

Figure 14.1. Potential energy surfaces for bound states, S_0 and S_1, and a dissociative state, S_2.

gram of the kind discussed in Chapter 13. On *bound* surfaces, the vibrational motion is also quantized. The spacing of the vibrational energy levels (400–4000 cm^{-1} or 4–40 kJ mol^{-1}) is such that at room temperature most of the molecules are in the lowest vibrational state. In addition, *rotational* motion of the molecule about its center of mass is also quantized. The spacing of the rotational energy levels depends inversely on the moment of inertia. For large molecules (more than a few first or higher row atoms) or in solution, the rotational structure is not resolved, although at room temperature, a number of rotational levels will be populated. The range of structures in the ground vibrational state is given by the vibrational wavefunction shown for the lowest vibrational level of S_0 in Figure 14.1. The time required to excite the electron (10^{-15} s) is very short compared to vibrational periods ($> 10^{-13}$ s). The shaded area in Figure 14.1 represents the range of structures from which *vertical* excitation may take place if the energy of the photon corresponds to approximate energy difference between the ground-state lowest vibrational level and some vibrational level of an excited-state potential energy surface. In general, the excited state reached by vertical excitation from the ground state will be *hot* (vibrationally excited) and may dissociate if the vibrational mode of the excited state corresponds to bond stretching. A more detailed representation of the sequence of events after photoexcitation is given by a Jablonski diagram.

THE JABLONSKI DIAGRAM

A generic Jablonski diagram for a molecular system is shown in Figure 14.2. Singlet states and triplet states are shown as separate stacks. Associated with each electronic state is a vibrational/rotational manifold. The vibrational/ro-

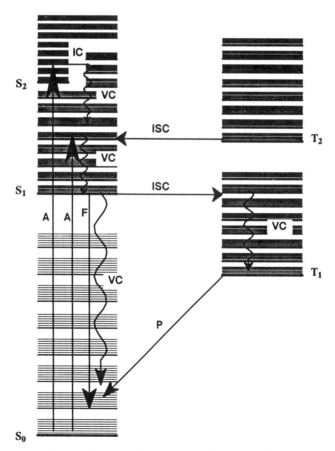

Figure 14.2. A modified Jablonski diagram: A = absorption; F = flourescence; P = phosphorescence; IC = internal conversion; ISC = intersystem crossing; VC = vibrational cascade. See text.

tational manifolds of different electronic states will in general overlap each other. Straight lines represent photon absorption (A) or emission, the latter as *fluorescence* (F) or *phosphorescence* (P). Wavy lines correspond to radiationless transitions which involve vibrational relaxation via a *vibrational cascade* (VC) to the zero vibrational level of the same electronic state. Energy is carried away through collisions with solvent. If vibrational/rotational levels of two electronic states overlap then it is possible to move from one state to the other without a change in energy. If a change in spin multiplicity is not involved (S → S, or T → T), the process is called *internal conversion* (IC); if a change in spin multiplicity *is* involved (S → T, or T → S), the process is called *intersystem crossing* (ISC). IC or ISC are immediately followed by VC.

FATE OF THE EXCITED MOLECULE IN SOLUTION

The fate of an excited molecule depends on the rates (time scales) of competing processes. As already mentioned above, the time scale for photon absorption (A) is very fast, 10^{-15} s. For molecules containing elements with atomic number less than about 50, electron spin is strongly conserved. As a result only singlet states are accessible in the primary excitation process. The time scale for the vibrational cascade and for internal conversion between higher excited states is governed by vibrational motions or collision frequencies in dense medium. Both are about 10^{-12}–10^{-13} s. Internal conversion between S_1 and S_0 may be substantially slower, 10^{-6}–10^{-12} s, of the same magnitude approximately as fluorescence, 10^{-5} s–10^{-9} s. Ultimately, useful photochemistry is limited by the fluorescence lifetime, τ_F, which for absorptions in the near UV may be estimated from the molar extinction coefficient [215] as

$$\tau_F(s) \simeq \frac{10^{-4}}{\epsilon_{max}}$$

A photochemical transformation can be accomplished only if the photoexcited molecule has a chance to do something before it fluoresces. Intersystem crossing time scales may vary considerably, 10–10^{-11} s. Return to the ground state by means of phosphorescence is comparatively slow, 10^2–10^{-4} s. The rates of chemical processes depend on the magnitudes of barriers hindering the change. Intermolecular processes depend additionally on collision frequency between reactant (the photoexcited molecule) and substrate, as well as possible orientational (stereoelectronic) criteria. Effective barriers hindering reaction tend to be lower in photochemical processes partly because bonding in the excited molecule is weaker, and partly because the excited molecule may be vibrationally excited (hot). The collision frequency for intermolecular reactions may be maximized if the substrate can function as the solvent, or if the reaction is intramolecular.

THE DAUBEN–SALEM–TURRO ANALYSIS

The following very useful approach for the analysis of photochemical reactions is due to Dauben, Salem, and Turro [11]. A bond, a—b, made from fragment orbitals, ϕ_a and ϕ_b, of molecular fragments, a and b, is broken in the process under consideration. The relationship to the orbital interaction diagram is transparent (see Figure 14.3). The orbitals, ϕ_a and ϕ_b, are the left- and right-hand sides of the interaction diagram, while the assembly, a—b, is the middle. The "middle" constitutes the reactant. The noninteracting right- and left-hand sides constitute the products. Assume that the orbital, ϕ_b, is lower in energy than ϕ_a. The two electrons originally in the bonding orbital may be distributed in four distinct ways between the two fragment orbitals, ϕ_a and ϕ_b, yielding four states, $Z_1 = \ldots (\phi_b)^2(\phi_a)^0$, $Z_2 = \ldots (\phi_b)^0(\phi_a)^2$, $^1D = \ldots (\phi_b)^1(\phi_a)^1$, and $^3D = \ldots (\phi_b)^1(\phi_a)^1$. The first two states are *zwitterionic*, since the bond dissociates heterolytically. Also, since both electrons occupy the same orbital, both zwitterionic states are *singlet* states and will be totally symmetric with respect to any symmetry operation which may be preserved in the dissociation. By our assumption that $E(\phi_b) < E(\phi_a)$, it is expected that $E(Z_1) < E(Z_2)$. The remaining two states arise from homolytic dissociation of the bond and therefore are *diradical* in character. Both *singlet* and *triplet* states arise. If any residual interaction persists (i.e., if the bond broken was a π bond, or the products are held together in a solvent cage), then the triplet diradical state is

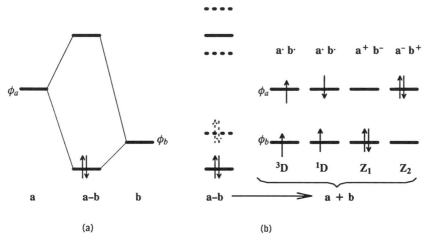

Figure 14.3. (a) Orbital interaction diagram for the formation of a bond between molecule fragments a and b; (b) orbitals for Dauben–Salem–Turro analysis of the rupture of the a—b bond: left-hand side, orbitals of the bond with other occupied and unoccupied MOs of the reactant (molecule a—b); right-hand side, configurations which arise from the orbitals which made up the bond (other orbitals of fragments a and b not shown).

lower than the singlet diradical state. Otherwise the two have the same energy. Since the electrons end up in different orbitals, the spatial symmetry of the diradical states is determined by the symmetry properties of ϕ_a and ϕ_b. The diradical is symmetric (S) if ϕ_a and ϕ_b are both symmetric or both antisymmetric (A) with respect to a preserved symmetry operation. If ϕ_a and ϕ_b have different symmetry properties, one S and the other A, then the spatial symmetry of the diradical states is A. Notice that the latter case could not arise if ϕ_a and ϕ_b are the orbitals involved in the original bond, since they could not have interacted to form the bond if they were of different symmetry. It will generally happen that one or more of the orbitals of the fragments a and b will be comparable in energy to ϕ_a and ϕ_b, so that other diradical and zwitterionic states must be considered and the situation of different symmetries may occur among these. The same holds true for the reactant excited states.

As mentioned earlier, heterolytic cleavage of a bond to form charged species is never observed in the gas phase and is very unlikely in nonpolar solvent. Thus, the expected overall order of the energies of the product states arising from the bond rupture is $E(^3D) \leq E(^1D) < E(Z_1) < E(Z_2)$. If the energy separation of ϕ_a and ϕ_b is large, or if ϕ_a and/or ϕ_b are diffuse, i.e., the coulomb repulsion of two electrons in the same MO (neglected in Hückel theory) is not large, the energy separation of the diradical states from Z_1 may not be very large, and may indeed be reversed if the dissociation is carried out in a highly polar solvent like water. Examination of several reactions of carbonyl compounds will serve to exemplify the principles of the Dauben–Salem–Turro analysis.

NORRISH TYPE II REACTION OF CARBONYL COMPOUNDS

A common photochemical reaction of carbonyl compounds is the transfer of a hydrogen atom to the carbonyl oxygen atom.

The lower electronic states of the reactants (left hand side) are those of the carbonyl group shown in Figure 13.8(c), since the σ_{CH} and σ_{CH}^* orbitals are too far apart to participate in the lower electronic states. In other words, the light will be absorbed by the carbonyl compound. The local group orbitals of the product fragments and the states which arise from them are shown in Figure

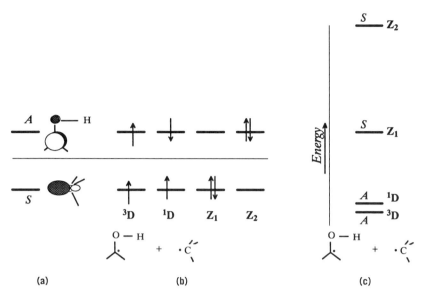

Figure 14.4. (a) Frontier orbitals of the fragments; (b) electronic configurations with two electrons; (c) order of energies of the electronic states.

14.4. The reaction is considered to take place in the plane of the carbonyl group and the orbitals are symmetry typed according to their behavior with respect to reflection in this plane.

A Dauben–Salem–Turro type state correlation diagram for the photochemical step of the Norrish Type II hydrogen abstraction reaction is shown in Figure 14.5. The reactant (carbonyl) states are classified in point group C_s, and also S or A. The placement of the reactant and product states on the same energy diagram need only be approximate. The singlet and triplet $n\pi^*$ states are located by virtue of the observed lowest electronic transition in the UV spectra of carbonyls, about 250 nm or 5 eV relative to the ground state. The $\pi\pi^*$ states are higher, with a larger singlet/triplet gap. Assuming the ground states of reactants and products to be of similar energy, the diradical states may be placed at about the energy required to dissociate a C—C or C—H bond, about 360kJ/mol or 4 eV.

The state correlation diagram indicates that the reaction should proceed via the $n\pi^*$ states which correlate directly with the product states of the respective multiplicity. The $\pi\pi^*$ states of the carbonyl group correlate with higher electronic states of the products. The carbonyl ground state correlates with the Z_1 state of products. Both singlet and triplet $n\pi^*$ states cross with the ascending ground state at the regions marked by shaded circles. At the crossing of singlet states, *internal conversion* (IC) may allow a substantial fraction of the reaction to revert to the reactants. *Intersystem crossing* is usually much less efficient and therefore a higher quantum yield would result if the reaction were carried out on the triplet potential energy surface.

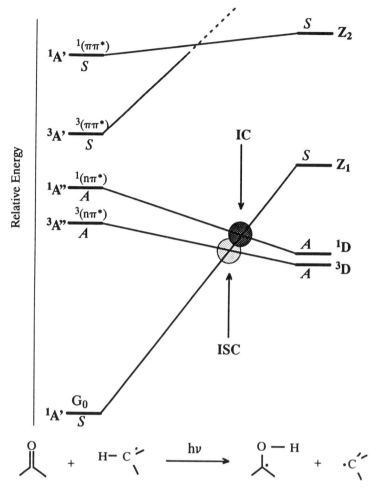

Figure 14.5. Dauben–Salem–Turro analysis of the photochemical step of the Norrish Type II reaction. The reaction is efficient on the $^3(n\pi^*)$ surface to yield triplet diradical products. It is less efficient on the $^1(n\pi^*)$ surface since internal conversion (IC) is relatively efficient.

The efficiency of photon capture to form the $^1(n\pi^*)$ state is very low since the excitation is electric dipole forbidden in the local $C_{2\nu}$ point group of the carbonyl. Direct excitation to the $^3(n\pi^*)$ state is both space and spin forbidden. The triplet state, $^3(n\pi^*)$, may be reached via *sensitization* using compounds which undergo ISC very efficiently and whose lowest triplet states are above the $^3(n\pi^*)$ state of the carbonyl. Conversely, the presence of compounds whose triplet states are below the $^3(n\pi^*)$ state of the carbonyl will result in *quenching* of the reaction.

NORRISH TYPE I CLEAVAGE REACTION OF CARBONYL COMPOUNDS

A second common photochemical reaction of carbonyl compounds is the cleavage of the bond adjacent to the carbonyl group.

$$\text{\Large >\!\!C(=O)\!-\!C<} \xrightarrow{h\nu} \text{\Large >\!\!C(=O)\cdot} + \cdot\text{\Large C<} \longrightarrow \text{(radical reaction products)}$$

As with the Norrish Type II reaction, the lower electronic states of the reactants (left hand side) are those the carbonyl group shown in Figure 13.8(c). The local group orbitals of the product fragments and the states which arise from them are shown in Figure 14.6. The reaction is again considered to take place in the plane of the carbonyl group and the orbitals are symmetry typed according to their behavior with respect to reflection in this plane.

A Dauben–Salem–Turro type state correlation diagram for the photochemical step of the Norrish Type I α cleavage reaction is shown in Figure 14.7. The reactant (carbonyl) states are classified in point group C_s, and also S or A. The state correlation diagram indicates that the reaction should proceed most efficiently via the triplet $\pi\pi^*$ state which correlates directly with the ground state of the products. The $^3(n\pi^*)$ state of the carbonyl group also cleaves relatively efficiently via internal conversion at the crossing of the triplet surfaces.

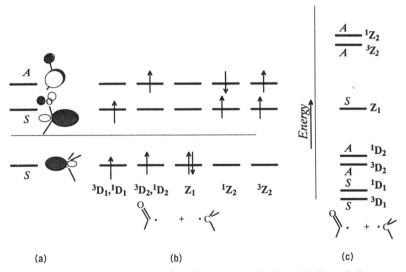

Figure 14.6. (a) Frontier orbitals of the fragments of a Norrish Type I cleavage; (b) electronic configurations with two electrons; (c) order of energies of the electronic states.

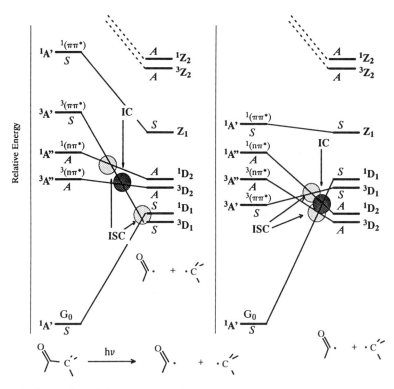

Figure 14.7. Dauben–Salem–Turro analysis of the photochemical step of the Norrish Type I reaction for saturated carbonyls (**a**) and for conjugated carbonyls (**b**). The reaction is most efficient on the $^3(\pi\pi^*)$ surface to yield triplet diradical products. It is also efficient on the $^3(n\pi^*)$ surface since internal conversion (IC) permits formation of products in their ground state.

Reaction via $^1(n\pi^*)$ also occurs since the $^1(n\pi^*)$ state correlates with a higher singlet electronic state of the products and can then undergo internal conversion and vibrational cascade to the lowest singlet state of products. However, in this case, recombination of the radical pair is very fast if the radicals have been held together in a solvent cage or within the same fragment, as in the case of α-cleavage of cyclic ketones. The carbonyl ground state correlates with the $^1D_1$ state of products. This reaction corresponds to thermal cleavage of the α C—C bond.

APPENDIX A

AN INPUT DESCRIPTION FOR GAUSSIAN XX

The following section is extracted from the *Gaussian 92® User's Guide* [216].

The *required* sections of a Gaussian XX input file are listed here. Many other choices are available. Consult the latest *Gaussian XX User's Guide* (currently *Gaussian 92®*) [216] for details. All of the required input is free-format and insensitive to case. Spaces or commas can be used to delimit fields. Input sections are separated by blank lines.

A Gaussian XX [33] run is usually submitted as a batch job. The required input has the following form:

- Job-Type—flagged by an initial #, e.g.,

   # RHF 6-31G* opt

- Job—Title, e.g.,

   Water–geometry optimization—6-31G* basis set–RHF theory

- Molecule specification (see below)
- Variable specification (see below)

®Gaussian 92 is a registered trademark of GAUSSIAN, INC. The material in this Appendix is copyrighted by GAUSSIAN, INC. and is used with permission.

MOLECULE SPECIFICATION INPUT SECTION

This section is always required. It specifies the nuclear positions and the number of electrons of α- and β-spin. There are several ways in which the nuclear configuration can be specified. By default, a Z matrix is read from the input stream. Note that the Z matrix includes cartesian coordinates as a special case. Alternate methods of molecule specification are requested by the Geom route command, as specified in section 5.2.6 of [216].

Charge and Multiplicity. The first line of the molecule specification section is always present regardless of where the nuclear coordinate information comes from. It specifies the net electric charge (a signed integer) and the spin multiplicity (a positive integer). Thus, for a neutral molecule in a singlet state, the entry "0 1" is appropriate. For a radical anion "-1 2" would be used. This is the only molecule specification section required if Geom = Checkpoint was used.

Z-Matrix Input. The input is free-field; the several items on each line may be separated by either blanks or commas.

The remaining lines in the Z-matrix molecule specification section give the relative positions and identities of the nuclei. Most of these will be real nuclei, used later in the molecular orbital computation. However, it is frequently useful to introduce *dummy nuclei* which help to specify the geometry, but are ignored subsequently.

Each nucleus (including dummies) is numbered sequentially and specified on a single line. This set of data is referred to as the Z matrix. The identity and position of the Nth center is specified by the $(N + 1)$th line (the first was the charge and multiplicity line) in the section and is either given relative to the positions of the $N - 1$ previously specified centers or as explicit cartesian coordinates.

The information about the Nth nucleus is contained in up to eight separated items in the line in either format 1:

Element, N_1, Length, N_2, Alpha, N_3, Beta, J

or format 2:

Element, N_1, X Coordinate, Y Coordinate, Z Coordinate

ELEMENT specifies the chemical nature of the nucleus. It consists of the chemical symbol "H" or "C" for hydrogen or carbon or an integer specifying the atomic number. Alternatively, it may be an alphanumeric string beginning with the chemical symbol, followed immediately by a secondary identifying integer. Thus "C5" can be used to specify a carbon nucleus, identified as the fifth carbon of the molecule. This is sometimes convenient in following conventional chemical numbering. Dummy nuclei are denoted by the symbol "X." This item is required for every nucleus, and for the first atom specified (N = 1), it is the only item on the line in format 1, since the first atom is placed at the origin. Ghost atoms (centers which are not ignored during the calculation but which by default have zero charge and no basis functional on them) are specified by the symbol "Bq."

If N_1 is positive, it specifies the (previously defined) nucleus for which the internuclear distance from center N_1 to center N will be given. This item may be either an integer with $N_1 < N$) giving the number of the line defining the other center or an alphanumeric string, matching the element item on a previous Z-matrix line. If N_1 is 0 or -1 or omitted, the cartesian coordinates of center N are expected to follow on the same line. In this case, the center is fixed and may not be optimized. If N_1 is 0, the forces on the center and force constants for its motion will still be calculated in first- and second-derivative calculations, while if N_1 is -1, the center will be fixed in derivative evaluation as well. The main use of $N_1 = -1$ is in specifying a fixed cluster of atoms on which other atoms will be absorbed.

LENGTH is the internuclear distance $R(N, N_1)$. This may be either a positive floating point number in Å or an alphanumeric string of up to eight characters. In the latter case, the length is represented by a *parameter*, for which a value will be specified in a later input section. If a length is to be varied in an optimization run, it must be specified by a parameter. Note that the same parameter may be used for two distinct lengths, in which case they are constrained to be equal. The items N_1 and LENGTH are required for all nuclei after the first. For the second nucleus, only ELEMENT, N_1, and LENGTH are required in format 1 in which case it is placed on the Z axis.

N_2 specifies the nucleus for which the angle $\alpha(N, N_1, N_2)$ will be given. Again this may be an integer or an alphanumeric string which matches a previous ELEMENT entry. Note that N_1 and N_2 must represent different nuclei.

ALPHA is the internuclear angle $\alpha(N, N_1, N_2)$. This may be a floating point number giving the angle in degrees or an alphanumeric string representing a parameter. The numerical value of ALPHA must lie in the range $0 < \alpha < 180°$. N_2 and ALPHA are required of all nuclei after the second. For the third nucleus, only ELEMENT, N_1, LENGTH, N_2, and ALPHA are required. This nucleus is placed in the XZ plane.

N_3. The significance of N_3 and BETA depends on the value of the last item, J. If $J = 0$ or is omitted, N_3 specifies the nucleus for which the internuclear dihedral angle $\beta(N, N_1, N_2, N_3)$ will be given. As with N_1 and N_2, this may be either an integer or an alphanumeric string matching a previous ELEMENT entry. This center must be distinct from those specified by N_1 and N_2.

BETA is the internuclear dihedral angle $\beta(N, N_1, N_2, N_3)$ (if $J = 0$; see below). This may be a floating point number giving the angle in degrees or an alphanumeric string representing a parameter (or a parameter string preceded by a negative sign). The dihedral angle is defined as the angle between the planes defined by (N, N_1, N_2) and (N_1, N_2, N_3) ($-180° \leq \beta \leq 180°$). The sign is positive if the movement of the directed vector $N_1 \rightarrow N$ towards the directed vector $N_2 \rightarrow N_3$ involves a right-handed screw motion.

J. The above descriptions of N_3 and BETA apply if J is zero or absent. Although it is always possible to specify the nucleus N by a bond length, and bond angle, and a dihedral angle, it is sometimes preferable to use a bond length and two bond angles. This is called for by using $J = \pm 1$. If $J = \pm 1$, N_3 specifies that the internuclear angle $\beta(N, N_1, N_3)$ will be given. As before, it may be a floating point number (value in degrees) or an alphanumeric string representing a parameter. In the event of specification by two internuclear angles α and β, there will be two possible positions for the nucleus N. This is fixed by the sign J. Thus $J = +1$ if the triple product $N_1 N \cdot (N_1 N_2 \times N_1 N_3)$ is positive, and $J = -1$ if the product is negative.

X Coordinate, Y Coordinate, and Z Coordinate are floating point numbers (*not* symbols) giving the coordinates of center N.

The Z matrix is terminated by the blank line which indicates the end of the molecule specification section. If no parameters have been introduced and if no special options have been invoked in the job-type section, the input is complete and Gaussian 92 will perform the requested computation.

ACTIVE AND FROZEN VARIABLE SPECIFICATIONS

These two sections define the values of any parameters introduced into the Z matrix in the preceding molecule specification section. These are not required if there are no parameters (all lengths and angles have been specified by floating point numbers). If parameters are used, they are divided into two sets. The first section contains the *active variables*, which are the parameters that are *varied* in an optimization run. This section contains the *initial* values of the variables. If optimization is not called for, these values will be inserted for a single run as if they had been specified numerically in the Z matrix. The second section of parameters contains *inactive variables* which are parameters that are *not* to be varied in the current optimization. The inactive variables can later be activated only for Berny optimizations, otherwise they have identical effects to entering the values directly into the Z matrix.

Each parameter in these sections is given a value on a separate line, the value following the parameter name in free field format. Blanks, commas, or "=" signs may be used as separators. Each section is terminated by a blank line. Thus if the variables (R1, R2, A) are given values (1.08, 1.36, 105.4), the full variable section consists of the four lines:

R1 = 1.08
R2 = 1.36
A = 105.4
(*blank line*)

If the inactive variable section is empty (i.e., if all parameters have been defined in the variable section), a second blank line should not be provided. The active variables section can be separated from the Z matrix by a line containing the string "Variables:" instead of a blank line, and the inactive variables can be separated from the active variables by the string "Constants:" instead of a completely blank line. This is for compatibility with the format of Z matrices in the output of Gaussian 92 jobs, and the NewZMat input converter.

A typical use of frozen variables would involve optimizing the structure of a large molecule, which sometimes is best done in three steps:

1. Optimize the bond angles and distances.
2. Optimize the dihedral angles.
3. Optimize everything.

This scheme has proved useful for larger molecules (30 or more heavy atoms) having many loose degrees of freedom. A similar three-step approach to locating transition states is also sometimes useful:

1. Locate a (poor) guess for the transition structure using Linear Synchronous Transit.
2. Freeze one or two variables corresponding to the reaction coordinate and optimize the other variables (to a minimum, obviously).
3. Release all variables and optimize to a transition state. To produce the required negative eigenvalue of the Hessian, this step must either begin with analytic second derivatives [i.e., Opt = (No-Freeze, CalcFC, TS)] or GEOM = MODIFY must be used to read in the partially optimized Z matrix and then modify it to request numerical differentiation with respect to the key variables.

APPENDIX B

THE INTERACTIVE PROGRAM, SHMO

Simple Hückel Molecular Orbital (SHMO) theory was introduced in Chapter 5. A copy of a computer program for performing SHMO calculations is included with this book. The program allows interactive generation of structures and assignment of coulomb and resonance integrals (α_X and β_{XY}), and solves the SHMO equations. The results may be displayed by means of an energy level diagram, or as individual MOs. Population analysis as a function of the number of electrons is also performed. SHMO is limited to a **maximum of 24 centers**.

INTERACTIVE STRUCTURE INPUT

SHMO permits structure input from a previously saved structure, or manually. The programs begin by:

Step 1

SHMO: Structure definition ⟨ret⟩ if manual input
or enter file name (.HUC is assumed)?

If you had stored a structure from a previous session with SHMO, it may be recalled by specifying its name. If a filename is entered, SHMO retrieves and builds the structure and proceeds to Step 8.

Entering ⟨ret⟩ (i.e., pressing the "return" or "enter" key) causes the statement

SHMO: Enter molecule by following instructions

The instructions are fairly self evident and input is reasonably "user friendly," i.e., it is not easy to "hang" the program inadvertently. The first query deals with number of rings.

Step 2

SHMO: Number of rings (maximum of two, please)?

If the structure has rings, these should be specified first. A maximum of two rings is permitted at this stage, although more rings may be created by connecting side chains (see below).

A response of 0 (or ⟨ret⟩) implies no rings. An acyclic structure will be generated (see Step 7).

A response of "1" or "2" prompts

Step 3

SHMO: Number of atoms in the first ring? Permitted responses: any number n, $3 \leq n \leq 24$.

A regular polygon of size n will be displayed. SHMO will then loop over the atoms, requesting index numbers. Entering ⟨ret⟩ at any time will cause automatic sequential numbering of any unnumbered centers. The numbers are displayed as they are entered.

If the number of rings entered was 2, SHMO asks about the second ring:

Step 4

SHMO: Does the second ring share a side with the first?

Answering "no" (or ⟨ret⟩) permits generation of structures such as biphenyl or stilbene.

Answering "yes" permits generation of structures such as naphthalene or azulene. In either case, structure generation proceeds in a transparent fashion. After the rings and any intervening chain have been defined, the opportunity to add side chains is presented. A linear side chain of arbitrary length may be added to any previously numbered center. Any number of side chains may be added, consistent with the maximum number of centers (24).

Step 5

SHMO: How many side chains? Permitted responses: any number $j \geq 0$ (or ⟨ret⟩)

Enter the number of side chains. If your response is 0 (or ⟨ret⟩), proceed to Step 8.

If you are uncertain of the number of side chains, enter a large number. The loop over the chains may be terminated at any point by entering ⟨ret⟩ at the request for the point of attachment of the next chain (Step 6).

Note: A carbonyl group is constructed from a side chain of length 1, the center being subsequently changed from C (the default) to O (see Step 9). A

230 THE INTERACTIVE PROGRAM, SHMO

branched group such as amide, $-C(O)N$, would be constructed from two chains, the first of length 2 and the second of length 1 attached to it.

Step 6

SHMO: Side chain—attached to atom? Permitted response: the index number of any previously defined center.

SHMO will loop over the requested number of side chains, asking for the length of each chain, displaying it, and permitting automatic or manual numbering as it goes. The loop over the chains may be terminated at any point by entering ⟨ret⟩ at the request for the point of attachment of the next chain. Proceed to Step 8.

Step 7

If there were no rings, input to SHMO would begin here.

SHMO: Length of longest chain? Permitted responses: any number n, $1 \leq n \leq 24$. Enter the length of the longest chain. SHMO will display a staggered chain of length n, using realistic angles of 120°. SHMO will then loop over the atoms, requesting index numbers. Entering ⟨ret⟩ at any time will cause automatic sequential numbering of any unnumbered centers. The numbers are displayed as they are entered. SHMO then presents the opportunity to add side chains. Proceed to Step 5.

Step 8

At this point, the definition of the skeletal structure is complete. You may wish to make a note of the atomic numbering at this stage. Once the calculation proceeds, the numbers will not be displayed again, although various calculated quantities will refer to the centers by number or will be listed in the numbered sequence.

All atoms are C [i.e., $h_X = h_C = 0$, see equation (5.14)] and all bonds are CC bonds [i.e., $k_{XY} = k_{CC} = -1$, see equation (5.15)]. You are presented with the opportunity to add or delete bonds at this stage. By adding bonds, more rings may be created than the initial maximum of two.

SHMO: Add new bonds or modify existing bonds? Permitted responses: y(es), n(o) (or ⟨ret⟩)

Answer "n(o)" or ⟨ret⟩ if no changes are to be made. Proceed to Step 9.

Answer "y(es)" if bond parameters are to be changed. You will then be asked to specify the index numbers of the two centers whose bonding characteristics you wish to change, and then the value of k_{XY} (called "beta" by SHMO) you wish to assign. A value of -1 is equivalent to a CC double bond. A value of 0 deletes the bond. SHMO loops until you enter "n(o)" or ⟨ret⟩. Proceed to Step 9.

Step 9

SHMO now presents the opportunity to change atoms and another chance to change bond parameters:

SHMO: Include heteroatoms or effects of substituents? Permitted responses: y(es), n(o) (or ⟨ret⟩)

Answer "n(o)" or ⟨ret⟩ if no changes are to be made. Proceed to Step 10.

Answer "y(es)" if heteroatoms are to be added and/or bond parameters are to be changed.

SHMO: Site of substitution or heteroatom? Permitted response: the index number of any defined center, or ⟨ret⟩ which skips out of the loop.

Enter the index of the center at which you wish to make a change. SHMO will loop back to this point until you enter ⟨ret⟩ at which point it proceeds to Step 10.

Some default parameters [see Table VII, Chapter 5 [86]] are incorporated in SHMO. Heteroatoms are identified by atomic symbol. If the symbol is recognized, then the internal parameters in the form of h_X [see equation (5.14)] and k_{XY} [see equation (5.15)] are inserted. These may be kept or changed.

You may be prompted for the number of atoms to which the heteroatom is bonded. This is the correct coordination number and may include atoms which are not part of the π system. The information is used to choose the parameters for the heteroatom which depend on the bonding characteristics, and also to establish the effective *nuclear charge* for the purpose of a population analysis. You are presented with the default value of h_X (called the "coulomb integral alpha" by SHMO) and asked to change it by entering a new value or accepting it by just entering ⟨ret⟩. SHMO then loops over all atoms which are bonded to the center. For each, you are presented with the default value of k_{XY} (called "beta" by SHMO) and asked to change it by entering a new value or accept it by just entering ⟨ret⟩.

Step 10

An opportunity is presented to save the structure.

SHMO: Save Structure? Permitted responses: y(es), n(o) (or ⟨ret⟩)

Answer "n(o)" or ⟨ret⟩ if the structure is not to be saved. Proceed to Step 12.

Answer "y(es)" if the structure is to be saved.

Step 11

SHMO: Enter name of structure. Permitted responses: any valid DOS name up to eight characters. HUC will be added as the suffix, and the file is written. Proceed to Step 12.

Step 12

SHMO begs patience while it solves the secular equation. When finished, it requests further instructions.

THE *OPTIONS* MENU

When the calculation is completed, SHMO displays a single line of *Options* and awaits input. A ⟨ret⟩ replaces the line with a second line. The choices are:

PE (Plot Energies): SHMO presents an energy level diagram of the calculated MO energies, in units of $|\beta|$ relative to α, and returns you to the Menu line.

PM (Plot Molecular Orbital): SHMO prompts for the MO you wish to see. A ⟨ret⟩ returns you to the Menu line. A number displays the particular MO superimposed on the molecular skeleton. Phase differences are displayed as shaded and unshaded circles. The radius of the circle is proportional to the absolute value of the coefficient. The circle should be interpreted as one lobe (say the upper) of the p orbital on the center. The MO energy and coefficients are also presented, the latter as a list ordered according to the numbering of the atoms.

A (Analysis): The analysis depends on the number of electrons, which must be specified first. Upon receiving the number of electrons, SHMO displays a page of useful information derived from the MOs. The charge distribution and dipole moment may be displayed. The plot of the charge distribution resembles the MO plot. Circles are proportional to the absolute magnitude of the net charge; positive and negative charges are represented by shaded and unshaded circles, respectively.

The bond orders may also be calculated and displayed.

The opportunity is presented to redo the analysis for a different number of electrons. Answering "no" or ⟨ret⟩ returns you to the Menu line.

SC (Scale Plot): If it happens that the structure is too large to be displayed on the screen, a scale factor of less than unity will reduce the size.

N (New Calculation): Specifying "N" to the Menu request returns you to Step 1.

Q (Quit): Exits SHMO and returns you to DOS.

KNOWN BUGS AND CAVEATS

Analysis: SHMO will carry out the analysis of Stabilization Energy, Ionization Potential, and Electronic Excitation Energy, for any system. However, the internal parameters are only suitable within SHMO theory for hydrocarbons. The printed quantities for heteroatom-containing systems are essentially meaningless and should be disregarded.

APPENDIX C

EXERCISES GROUPED BY CHAPTER

EXERCISES FROM CHAPTER 1

Q1. The snoutene skeleton is shown below. Locate all of the elements of symmetry (state the point group if you know it), and identify the stereochemical relationship between the specified pairs of groups or faces.

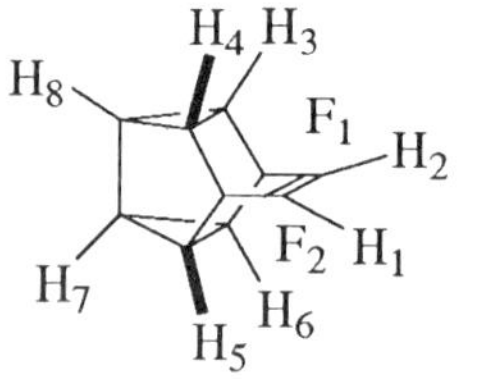

(a) H_1 and H_2
(b) H_3 and H_4
(c) H_3 and H_5
(d) H_3 and H_6
(e) H_3 and H_7
(f) F_1 and F_2 (faces of the π bond)

Q2. Label the pairs of protons shown in boldface in each of the following compounds as *homotopic*, *enantiotopic*, or *diastereotopic*, as required. Assume normal rotational barriers and observations at room temperature.

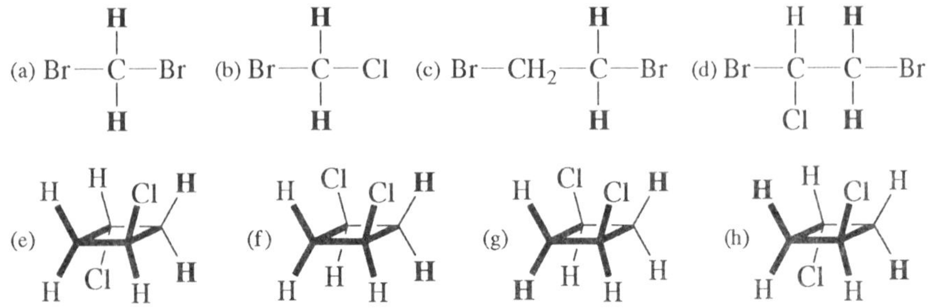

(a) Br—C(H)(H)—Br (b) Br—C(H)(H)—Cl (c) Br—CH₂—C(H)(H)—Br (d) Br—C(H)(Cl)—C(H)(H)—Br

(e), (f), (g), (h)

234

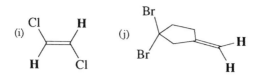

ANSWERS: (a) homotopic, (b) enantiotopic, (c) enantiotopic, (d) diastereotopic, (e) homotopic, (f) diastereotopic, (g) diastereotopic, (h) enantiotopic, (i) homotopic, (j) diastereotopic

Q3. The unsaturated [2.2.2](1,3,5) cyclophane, COMPOUND C was synthesized 17 years ago (Boekelheide, V.; Hollins, R. A. *J. Am. Chem. Soc.*, **1973**, *95*, 3201.).

COMPOUND C

(a) Locate all elements of symmetry.
(b) Analyze the stereotopic relationships among the hydrogen atoms.
(c) Predict, and fully assign the ^{1}H NMR spectrum.
(d) The UV and NMR spectra are unusual in several respects. Can you explain?

Q4. Analyze the stereochemical relationships between the methyl groups of hexamethyl(Dewar benzene). Identify at least one *pair* of methyl groups in which the methyl groups are:
—homotopic
—enantiotopic
—diastereotopic
—heterotopic
Note: If there are no pairs of a given kind, say so.

Q5. The following questions deal with the symmetry and spectroscopic relations of groups within *anti*-sesquinorbornatriene **1**.

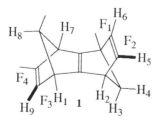

(a) What is the molecular point group? If you cannot remember the name of the point group, locate all elements of symmetry.

(b) In your answers to the questions, use only those hydrogen atoms or faces which are explicitly labeled (if no groups have the specified relationship, say so).
  (i)   Identify a pair of hydrogen atoms which are homotopic.
  (ii) diastereotopic.
  (iii) enantiotopic.
  (iv) constitutionally heterotopic.
  (v)   Identify a pair of faces which are homotopic.
  (vi) diastereotopic.
  (vii) enantiotopic.
  (viii) What is the relationship between H_2 and H_7
  (ix) H_3 and H_8?
  (x) H_5 and H_9?
  (xi) F_2 and F_3?

(c) How many separate ^{13}C nmr chemical shifts should one observe
  (i) under *achiral* conditions?
  (ii) under *chiral* conditions?

(d) Replace **two** H atoms by D in such a way that
  (i)   the only symmetry element is a mirror plane, σ.
  (ii) the only symmetry element is a C_2 axis.
  (iii) the only symmetry element is a center of inversion, i.
  (iv) the molecule is asymmetric.

Q6. Analyze the stereochemical relationships between the groups of cis-2,5-dimethylcyclopentanone **5**. Specifically,

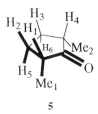

5

Identify at least one *pair* of groups which have the relationship:
  —homotopic
  —enantiotopic
  —diastereotopic
  —constitutionally heterotopic
Note: If there are no pairs of a given kind, say so.

EXERCISES FROM CHAPTER 2

Q1. Chose a small molecule and carry out a HF/STO-3G calculation.

(a) State the name of the compound and draw a clear representation of its structure, correctly oriented in the coordinate system (standard orientation). What is its molecular point group?

(b) Plot the valence MO energy levels, including a few of the unoccupied MOs. Identify the HOMO and LUMO. Make note of the position of the core levels.

(c) Sketch the two highest occupied and two lowest unoccupied molecular orbitals on the basis of the MO coefficients. Note which atoms have the largest coefficients in each MO. Is there any significance to this? Do the MOs transform according to the irreducible representations of the molecular point group?

(d) Comment on the population analysis. Is it reasonable in view of the elemental composition, concepts of electronegativity and resonance, and the molecular symmetry?

(e) Find literature references to previous experimental and theoretical work on the compound you chose. Compare the computed properties (geometry, IP, dipole moment, total energy) with experimental and/or other theoretical studies.

(f) Prepare a 10 min. discourse based on the points of the above questions (with a brief introduction) for presentation before the class. Include in your discourse, discussion of the technical aspects of the calculation as well.

Q2. Write **one** paragraph on the application of the variational principle using each of the following words or phrases in the correct context (not necessarily in the order given; underline each occurrence).

(i) Schrödinger equation
(ii) expectation value
(iii) CI wavefunction
(iv) single determinantal wavefunction
(v) Hamiltonian matrix

Q3. Write **one** paragraph on any aspect of Hartree–Fock theory, but incorporate each of the following words or phrases in the correct context (not necessarily in the order given; underline each occurrence).

(i) single determinantal wavefunction
(ii) Fock operator
(iii) molecular orbital energy
(iv) exchange integrals
(v) basis set

Q4. The following questions deal with a specific application of *ab initio* RHF theory to **ethylene** (C_2H_4).

(a) Using summation notation, write the electronic nonrelativistic Hamiltonian operator for ethylene.

(b) Using any notation which makes it clear, show a single determinantal RHF wavefunction for ethylene.

(c) Assuming that the spatial MOs of part (b) are expanded in a STO-3G basis set,

   (i) What is the size of the basis set?

   (ii) Which basis functions are likely to have the largest coefficients in the lowest *occupied* MO? in the *highest* occupied MO? Explain briefly.

Q5. In Hartree–Fock theory, the many-electron wavefunction is expressed as a single Slater determinant, conveniently abbreviated as

$$\Phi(1, 2, \cdots, N_e) = A[\phi_1(1)\phi_2(2) \cdots \phi_{N_e}(N_e)] \qquad (1)$$

The molecular orbitals ϕ_i, are, by convention, listed in order of increasing energy.

(a) Why is equation (1) a satisfactory representation of a many-electron wavefunction?

(b) What conditions must be satisfied by the MOs in order that equation (1) represents the single determinantal wavefunction of lowest possible energy?

(c) State some mathematical properties of the MOs (involving integration).

(d) State some relationships involving the MOs of equation (1) if equation (1) were a *Restricted* Hartree–Fock wavefunction.

(e) The set of MOs are usually expressed as a coefficient matrix. Why?

(f) Explain the relationship between MO energy, total electronic energy, and total molecular energy (which defines the Born–Oppenheimer potential energy hypersurface).

(g) Write a single determinantal wavefunction for the molecular ion, M^+, in its ground state. What is its energy relative to the energy of equation (1).

(h) What are *virtual* or *unoccupied* MOs?

(i) Write a single determinantal wavefunction for the molecule in its lowest excited state. What is its energy relative to the energy of equation (1).

(j) Express the permanent dipole moment of a molecule in terms of MOs.

EXERCISES FROM CHAPTER 3

Q1. Draw a general two-orbital interaction diagram in which the two unperturbed orbitals are of different energy. Label the diagram and indicate relationships which exist between the labeled components.

Q2. Use a two-orbital interaction diagram to explain or predict some feature of each of the following molecules.

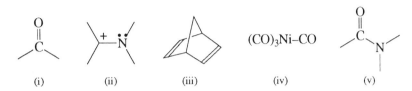

(i)   (ii)   (iii)   (iv)   (v)

ANSWER TO 2(ii): We note that the structure as drawn has one side which is a carbocation, and the other an amine. We should immediately ask whether the presence of the amino group modifies the Lewis acidity of the cationic center, using methyl carbocation as a reference, and whether the properties of the amine, compared to e.g., ammonia, might be affected by the presence of the cationic site. Amines, like ammonia, have a pyramidal (nonplanar) geometry, and are moderately basic. How do the properties of this species compare to these properties which are characteristic of amines? The two ends are connected by a single bond. Single bonds exhibit relatively unhindered rotation, and have characteristic lengths and stretching vibrational frequencies. Do these properties compare with those of methylamine, or are they significantly different? Is the C—N bond longer or shorter? Is rotation more or less hindered? Is the C—N stretching frequency higher or lower? All of these questions may be answered using the simple interaction diagram shown below, constructed from frontier orbitals of the two interacting fragments both of which are tricoordinated centers and therefore have only one valence orbital each. On the left-hand side is the empty $2p$ orbital of the cationic center, and on the right is the nonbonded orbital of the amine, which may be a $2p$ orbital also if the amine is planar, or it may be the sp^n hybrid orbital of a nonplanar tricoordinated center. What does the diagram tell us? We note immediately that there are two electrons and so the interaction is favorable. The geometry will change in such a way as to maximize the interaction. In other words, the N atom will become planar and one end of the C—N bond will rotate relative to the other to maximize the interaction, bringing both ends into coplanarity. There is appreciable π-bond character between the C and the N. Thus the bond will be *shorter* than a typical C—N single bond, as in methylamine, The barrier to torsion about the bond will be substantially *higher*, and the

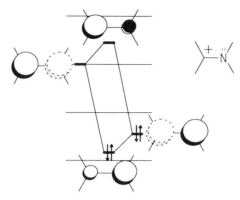

Figure Cl.1. Interaction diagram for [H$_2$C—NH$_2$]$^+$ [answer to question 2(ii)].

stretching frequency will also be **higher**. In terms of basicity, since the nitrogen lone pair has decreased in energy, and the coefficient at N is smaller (it would have been 1 before the interaction), the basicity is less than that of a typical amine. In terms of Lewis acidity, the LUMO is higher in energy than in our reference methyl cation, and the coefficient at C is smaller (it also would have been 1 before the interaction), the Lewis acidity, and electrophilicity will have decreased. It is also more difficult to reduce (add a electron) than a methyl cation.

Q3. Use simple orbital interaction diagrams to explain each of the following:

(a) The bonding in 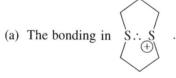 .

(b) Bonazzola and co-workers (Bonazzola, L.; Michaut, J. P.; Roncin, J. *Can. J. Chem.*, **1988,** *66*, 3050) have shown by ESR spectroscopy that singly ionized thiirane cation radical, [thiirane]$^{+\cdot}$, forms a 1:1 complex with thiirane itself. Suggest a structure for the complex. Note: Thiirane is thiacyclopropane $\triangleright_S$.

(c) Hydrazine, N$_2$H$_4$, has a dihedral angle close to 90°.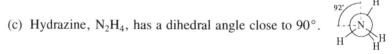

For a discussion of the structures of hydrazines and their ionization potentials, see Klessinger, M.; Rademacher, P. *Angew. Chem. Int. Ed. Engl.*, **1979,** *18*, 826.

ANSWER TO 3(a): We are asked here to provide an interaction diagram which will illustrate the nature of the molecular orbitals which contain the three electrons which are involved in binding together the two sulfur atoms. We note that in the absence of an interaction between them, each sulfur atom is a bent dicoordinated atom and as such has two frontier orbitals, a pure p orbital perpendicular to the plane of the σ bonds to S, and an sp^n hybrid orbital of lower energy directed along the line bisecting the angle made by the σ bonds and in the same plane. In the case of sulfur the p orbital is $3p$, and the sp^n hybrid has rather little "p" character ($n < 1$). In order to set up the interaction diagram, one must have an appreciation of the limitations imposed by the σ framework, which forms an eight-membered ring. Inspection of molecular models quickly reveals that among the possible conformations of the eight-membered ring are two as shown in Figure C1.2(a) and (b) with the valence orbitals superimposed on the molecular skeleton. In the staggered conformation, the sulfur atoms are too far apart to

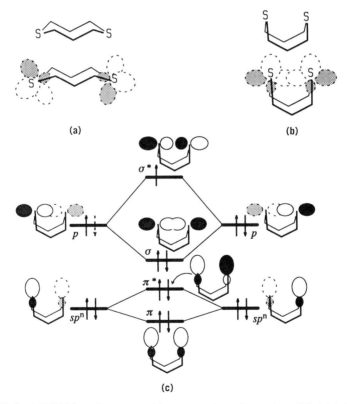

Figure Cl.2. 1,5-Dithiacyclooctane: **(a)** staggered conformation; **(b)** folded conformation; **(c)** interaction diagram in answer to question 3(a) for the bonding in the radical cation.

interact. This is the desirable situation in the neutral molecule since all of the nonbonding orbitals are filled and any interaction would be of the four-electron–two-orbital type, which is repulsive. Maximum interaction occurs in the folded conformation [Figure C1.2(b)]. The interaction diagram is shown in Figure C1.2(c). The sp^n orbitals suffer a weak π-type interaction while the p orbitals interact strongly in a σ fashion. If one, as in this case, or two electrons [see question 4(b)] are missing, then the interaction is favorable and the cation (or dication) will adopt the folded geometry. The three-electron "bond" has two electrons in the σ orbital and one in the σ^* orbital, for a weak net attraction. A suitable answer to this question could have ignored the weaker interactions of the sp^n orbitals.

Q4. Use <u>two-orbital</u> interaction diagrams to explain the **observed** features of the following systems. *Note:* A clear orbital interaction diagram includes pictures of the orbitals before and after the interaction, and shows the disposition of the electrons. A brief verbal explanation of the diagram is also desirable. If more than two orbitals seem to be involved, use your judgement to choose the two most important orbitals.

(a) Hydrogen disulfide S_2H_2, has a dihedral angle close to 90°.

(b) The two sulfur atoms in 1,5-dithiacyclooctyl dication,

are sufficiently close to say that a bond exists between them [see answer to 3(a)].

(c) Ammonia is more basic than phosphine. (*Hint:* Use a proton as the Lewis acid for your answer.)

(d) The NH_2 group of formamide, NH_2CHO, is flat whereas the nitrogen atom of most amines is pyramidal.

(e) Nucleophilic attack on the carbonyl group of ketones, $R_2C=O$ is catalyzed by Lewis acids.

Q5. Rohlfing and Hay recently determined the geometry of FOOF, the fluoro analog of hydrogen peroxide (Rohlfing, C. M.; Hay, P. J. *J. Chem. Phys.*, **1987**, *86*, 4518–4522). The geometry is unusual in three respects, (i) the O—F bonds are unusually long (1.575 Å compared to 1.41 Å in most oxyfluorides), (ii) the O—O bond is unusually short (1.217 Å compared to 1.42 in HOOH), (iii) the FOOF torsion angle is almost 90° (87.5°). Explain all three features.

Q6. The properties of DABCO (1,4-diazabicyclo[2.2.2]octane are of considerable interest. The in- and out-of-phase interaction of the two oc-

DABCO'S HOMO

cupied sp^3 hybridized lone pairs on the two nitrogen atoms may be described by an interaction diagram shown in Figure 3.5. As expected, DABCO is a much stronger base than triethylamine, has two ionization potentials separated by several eV, and is more easily oxidized than trimethylamine. What is unusual about DABCO is that the *in-phase* combination of the nonbonded pairs is the HOMO, and the higher I.P. is *lower* than that of trimethylamine. Explain.

EXERCISES FROM CHAPTER 4

Q1. This question deals with the description of a typical dative bond.
  (a) Draw a two-orbital interaction diagram for the B—N bond, as in $H_3B^- - N^+H_3$. Label the diagram clearly.
  (b) Sketch all of the orbitals. Identify the orbitals (e.g., n, π, π^*, σ, σ^*, etc., as appropriate).
  (c) Comment on the polarity of the B—N bond. Why is the boron atom shown with a formal charge of -1?

Q2. Reaction of alkenes with BrCl proceeds via the intermediate, [alkene-Br]$^+$, rather than the intermediate, [alkene-Cl]$^+$. Explain using a two-orbital interaction diagram.

Q3. Suggest a reason for the rapid expulsion of atomic bromine upon one-electron reduction of N-bromosuccinimide (Lind, J.; Shen, X.; Eriksen, T. E.; Merényi, G.; Eberson, L. *J. Am. Chem. Soc.*, **1991**, *113*, 4629).

Q4. Predict the structures and bonding in donor-acceptor complexes of ammonia with F_2, Cl_2, and ClF. What are the expected relative bond strengths? Are your predictions in accord with the results of high-level theoretical calculations of Røggen and Dahl (Røggen, I.; Dahl, T. *J. Am. Chem. Soc.*, **1992**, *114*, 511)?

Q5. In the dihalogenated ethanes, 1,2-dichloroethane and 1,2-dibromoethane preferentially adopt the conformation in which the two halogens are *anti* to each other as expected on the basis of steric repulsions. On the other hand, 1,2-difluoroethane exists predominantly in the *gauche* conformation (Friesen, D.; Hedberg, K. *J. Am. Chem. Soc.,* **1980**, *102*, 3987). This unexpected behavior has been explained as a manifestation of the *gauche effect* (Wolfe, S. *Acc. Chem. Res.*, **1972**, *5*,

102), from the formation of "bent" bonds (Wiberg, K. B.; Murcko, M. A.; Laidig, K. E.; MacDougall, P. J. *J. Phys. Chem.*, **1990**, *94*, 6956), or as due to hyperconjugative interaction between the C—F and C—H bonds (Radom, L *Prog. Theor. Org. Chem.*, **1982**, *3*, 1). The last explanation relies on orbital interaction theory. Can you reproduce the arguments?

Q6. In a microwave investigation, Lopata and Kuczkowski (Lopata, A. D.; Kuczkowski, R. L. *J. Am. Chem. Soc.*, **1981**, *103*, 3304–3309) determined that the equilibrium geometry of FCH$_2$—O—CHO was as shown below. Explain the near 90° orientation of the CF bond relative

to the plane of the rest of the molecule (the explanation is the same as for the anomeric effect in sugars.)

Q7. Low energy electron impact spectroscopy of [1.1.1]propellane reveals an unusually low energy for electron capture to form the radical anion

(2.04 eV compared to about 6 eV for most alkanes). In addition, appreciable lengthening of the interbridgehead C—C bond is indicated (Schafer, O.; Allan, M.; Szeimies, G.; Sanktjohanser, M. *J. Am. Chem. Soc.*, **1992**, *114*, 8180–8186). Can you explain these observations on the basis of orbital interaction theory? Compare your explanation with that offered in the reference.

Q8. Solve the HMO equations for the σ orbitals and orbital energies of the carbon-carbon and carbon-oxygen bonds: assume that $h(O) = h(C) - |h_{CO}|$, $h_{CC} = h_{CO}$, and $S_{CO} = 0$. Sketch the results in the form of an interaction diagram. Which bond is stronger? Calculate the homolytic bond dissociation energies in units of $|h_{CC}|$. What is the net charge on O, assuming that it arises solely from the polarization of the σ bond?

ANSWER TO Q8: $E(L) = h_C - 1.618|h_{CO}|$; $E(U) = h_C + 0.618|h_{CO}|$; $\phi(L) = 0.851\ sp^n(O) + 0.526\ sp^n(C)$; $\phi(U) = 0.526\ sp^n(O) - 0.851\ sp^n(C)$.

Bond energy: $\alpha(C) + \alpha(O) - 2E(L) = -2.236\beta = 2.236|\beta|$
Ionization potential: $-E(L) = -(\alpha - 0.618\beta) = |\alpha| + 0.618|\beta|$
Electronic excitation ($\pi \rightarrow \pi^*$): $E(U) - E(L) = -2.236\beta = 2.236|\beta|$
Bond order: $2(0.851)(0.526) = 0.895$
Net charges: on O, $-2(0.851)^2 + 1 = -0.447$; on C, $-2(0.526)^2 + 1 = +0.447$

EXERCISES FROM CHAPTER 5

Q1. This question deals with the compound on which you chose to do the SHMO calculation.
  (a) State the name of the compound and draw a clear representation of its structure.
  (b) Do a sketch of the energy levels on the usual α, β energy scale. Identify the HOMO and LUMO.
  (c) Sketch the two highest occupied and two lowest unoccupied molecular orbitals.
  (d) Comment on the Lewis acidity and basicity. Is one expected to dominate in typical chemical reactions? Indicate the probable site(s) of reaction with appropriate reagents (nucleophiles/electrophiles).
  (e) Give the equation of any reaction of the compound which you have found in the literature or a text book (in either case, cite the reference).
  (f) Compare known characteristics (reactivity, polarity, color, etc.) of the compound with what you might expect on the basis of the SHMO calculation.

Q2. The results of a SHMO calculation on (Z)-pentadienone, **1**, are given below (Note: $\alpha_{O:} = \alpha - 0.97|\beta|$, $\beta_{CO} = -1.09|\beta|$):

$\epsilon_1 = a - 1.968|\beta|$ $\phi_1 = 0.57p_1 + 0.53p_2 + 0.45p_3 + 0.35p_4 + 0.24p_5 + 0.12p_6$

$\epsilon_2 = a - 1.505|\beta|$ $\phi_2 = -0.49p_1 - 0.25p_2 + 0.15p_3 + 0.47p_4 + 0.56p_5 + 0.37p_6$

$\epsilon_3 = a - 0.696|\beta|$ $\phi_3 = 0.45p_1 - 0.12p_2 - 0.55p_3 - 0.27p_4 + 0.37p_5 + 0.52p_6$

$\epsilon_4 = a + 0.265|\beta|$ $\phi_4 = -0.38p_1 + 0.44p_2 + 0.28p_3 - 0.51p_4 - 0.15p_5 + 0.55p_6$

$\epsilon_5 = a + 1.156|\beta|$ $\phi_5 = -0.28p_1 + 0.55p_2 - 0.35p_3 - 0.15p_4 + 0.52p_5 - 0.45p_6$

$\epsilon_6 = a + 1.778|\beta|$ $\phi_6 = -0.15p_1 + 0.38p_2 - 0.52p_3 + 0.56p_4 - 0.45p_5 + 0.25p_6$

  (a) Sketch the highest occupied molecular orbital (HOMO) and the lowest unoccupied molecular orbital (LUMO) on the basis of the calculation.

246 EXERCISES GROUPED BY CHAPTER

(b) Assuming that *1* were to react with an electrophile, where is the most likely site of attack? Compare the Lewis basicity of *1* to that of ethylene.

(c) Assuming that *1* were to react with a nucleophile, where is the most likely site of attack? Compare the Lewis acidity of *1* to that of ethylene.

(d) In terms of $|\beta|$, estimate the energy of the lowest $\pi-\pi^*$ electronic excitation of *1*. Is the $\pi-\pi^*$ state likely to be higher or lower than the $n-\pi^*$ state? Explain in less than three lines.

(e) Calculate the π bond order of the C_2-C_3 bond of *1*. What is the net charge on O_1?

(f) Show the most likely product from a Diels–Alder reaction of *1* with 1,3-butadiene. Pay attention to all relevant aspects of stereo- and regioselectivity.

ANSWER TO Q2: There are a number of questions of this type which are primarily concerned with the interpretation of the results of MO calculations, and their relationship with the ideas of orbital interaction theory. The data are derived from a calculation using SHMO.

(a) Pentadienone has six π electons. A "sketch" of the frontier MOs is shown in Figure C1.3. The relative sizes of the atomic orbitals are proportional to the sizes of the MO coefficients and the relative

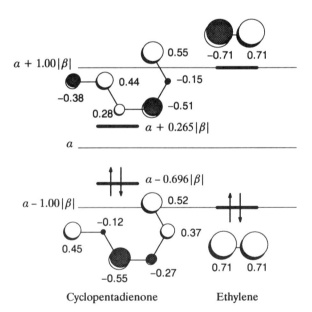

Figure C1.3. The HOMO and LUMO of cyclopentadienone and ethylene (SHMO).

phases are shown by shading. The convention is adopted that a positive signed coefficient corresponds to a *p* orbital with the unshaded lobe up and a negative coefficient is portrayed by a *p* orbital with the shaded lobe up. The actual convention you use is unimportant, but you must be consistent.

(b) Attack by an electrophile will be directed to the part of the HOMO which has the largest coefficient, in this case, atom 3. Atom 6 is a close second. Basicity relative to ethylene is governed by the relative energy of the HOMO, and by the relative magnitudes of the largest coefficients. The HOMO is substantially higher, suggesting greater basicity, but the largest coefficient (0.55) is smaller than in ethylene (0.71), suggesting lower basicity. Since the two factors are in opposition, one might expect that the basicity will be similar, perhaps slightly greater than that of ethylene.

(c) Attack by a nucleophile will be directed to the part of the LUMO which has the largest coefficient, in this case, atom 6. Atom 4 is second. Acidity in the Lewis sense relative to ethylene is governed by the relative energy of the LUMO, and by the relative magnitudes of the largest coefficients. The LUMO is much lower, suggesting significantly greater acidity. Again, the largest coefficient (0.55) is smaller than in ethylene (0.71), suggesting lower acidity. Although the two factors are again in opposition, one might expect that the large energy difference will dominate and that the acidity will be greater than that of ethylene. (In fact, it is extremely difficult to do a nucleophilic attack on ethylene.)

(d) The energy of the $\pi-\pi^*$ excitation in SHMO theory is just the difference of the LUMO and HOMO orbital energies, $0.961|\beta|$. The "n" orbital is a nonbonding orbital on oxygen which lies in the plane of the molecule and is not part of the π system. Its energy is that of a *p* orbital on a monocoordinated oxygen, $\alpha - 0.97|\beta|$. Since the "n" orbital is lower than the highest occupied π orbital, the excitation should have lower energy than the n-π^* excitation so the $\pi-\pi^*$ state will be lower in energy than the n-π^* state.

(e) The bond order of the C2—C3 bond is $2(0.53)(0.45) + 2(-0.25)(0.15) + 2(-0.12)(-0.55) = 0.53$. The total electron population on atom 1 is $2(0.57)^2 + 2(-0.49)^2 + 2(0.45)^2 = 1.53_5$. Since a monocoordinated oxygen atom is neutral with one electron in the π system, the net charge is $1 - 1.53_5 = -0.53_5$

(f) The important orbital interaction for the Diels–Alder reaction is the LUMO (pentadienone)-HOMO (butadiene). The strongest interaction (best overlap) originates if the two molecules are oriented as follows. The interaction is driven by the overlap of the orbitals with the largest coefficients of each MO.

HOMO of butadiene
(above)

LUMO of **1**
(below)

Q3. The results of a SHMO calculation on furan, **1**, are given below (*Note:* $\alpha_O = \alpha_C - 2.09|\beta_{CC}|$, $\beta_{CO} = -0.66|\beta_{CC}|$):

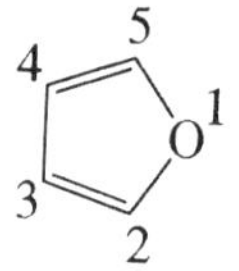

$\epsilon_1 = a - 2.548|\beta|$ $\phi_1 = 0.86p_1 + 0.30p_2 + 0.19p_3 + 0.19p_4 + 0.30p_5$

$\epsilon_2 = a - 1.383|\beta|$ $\phi_2 = 0.43p_1 - 0.23p_2 - 0.60p_3 - 0.60p_4 - 0.23p_5$

$\epsilon_3 = a - 0.618|\beta|$ $\phi_3 = -0.60p_2 - 0.37p_3 + 0.37p_4 + 0.60p_5$

$\epsilon_4 = a + 0.846|\beta|$ $\phi_4 = 0.27p_1 - 0.60p_1 + 0.33p_3 + 0.33p_4 - 0.60p_5$

$\epsilon_5 = a + 1.618|\beta|$ $\phi_5 = -0.37p_2 + 0.60p_3 - 0.60p_4 + 0.37p_5$

(a) Sketch the highest occupied molecular orbital (HOMO) and the lowest unoccupied molecular orbital (LUMO) on the basis of the calculation.

(b) Assuming that **1** were to react with an electrophile, where is the most likely site of attack? Compare the Lewis basicity of **1** to that of ethylene. Show the products of such a reaction (pick a specific eletrophile).

(c) Assuming that **1** were to react with a nucleophile, where is the most likely site of attack? Compare the Lewis acidity of **1** to that of ethylene. Show the products of such a reaction (pick a specific nucleophile).

(d) Consider an equimolar mixture of **1** and 2-cyanofuran, **2**. A Diels–Alder product is obtained from the mixture. Predict the structure of the product. The HOMO and LUMO of **2** are shown below.

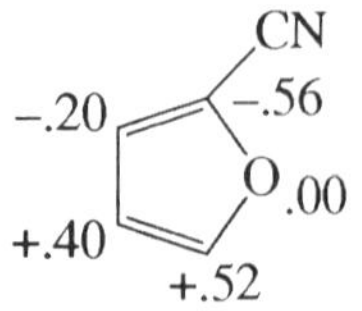

$\epsilon_{HOMO} = \alpha - .796\,|\beta|$

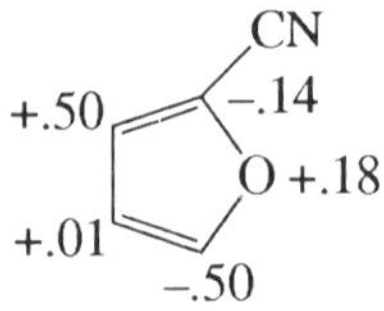

$\epsilon_{LUMO} = \alpha + .254\,|\beta|$

HOMO   LUMO

2

(e) In terms of α and/or $|\beta|$, provide an estimate for the lowest ionization potential of *1* according to the calculation.

(f) In terms of $|\beta|$, estimate the energy of the lowest π–π* electronic excitation of *1*.

(g) Furan *1* belongs to the point group C_{2v}. Indicate whether the HOMO is symmetric or antisymmetric with respect to each of the three symmetry elements of *1*. Do the same for the LUMO.

(h) Calculate the net charge on the oxygen atom of *1*.

(i) Calculate the π bond order of the C—O bond of *1*.

Q4. The results of a SHMO calculation on cyclopentadienone, *1*, are given below (*Note*: $\alpha_{O:} = \alpha - 0.97|\beta|$, $\beta_{CO} = -1.09|\beta|$):

$$\epsilon_1 = a - 2.252|\beta| \quad \phi_1 = 0.46p_1 + 0.56p_2 + 0.38p_3 + 0.31p_4 + 0.31p_5 + 0.38p_6$$

$$\epsilon_2 = a - 1.366|\beta| \quad \phi_2 = 0.64p_1 + 0.24p_2 - 0.18p_3 - 0.48p_4 - 0.48p_5 - 0.18p_6$$

$$\epsilon_3 = a - 0.618|\beta| \quad \phi_3 = -0.60p_3 - 0.37p_4 + 0.37p_5 + 0.60p_6$$

$$\epsilon_4 = a - 0.148|\beta| \quad \phi_4 = 0.56p_1 - 0.43p_2 - 0.33p_3 + 0.38p_4 + 0.38p_5 - 0.33p_6$$

$$\epsilon_5 = a + 1.618|\beta| \quad \phi_5 = -0.37p_3 + 0.60p_4 - 0.60p_5 + 0.37p_6$$

$$\epsilon_6 = a + 1.796|\beta| \quad \phi_6 = 0.26p_1 - 0.67p_2 + 0.46p_3 - 0.17p_4 - 0.17p_5 + 0.46p_6$$

(a) Sketch the highest occupied molecular orbital (HOMO) and the lowest unoccupied molecular orbital (LUMO) on the basis of the calculation.

(b) Assuming that *1* were to react with an electrophile, where is the most likely site of attack? Compare the Lewis basicity of *1* to that of ethylene.

(c) Assuming that *1* were to react with a nucleophile, where is the most likely site of attack? Compare the Lewis acidity of *1* to that of ethylene.

(d) In terms of $|\beta|$, estimate the energy of the lowest π–π* electronic excitation of *1*. Is the π–π* state likely to be higher or lower than the n–π* state? Explain in less than three lines. It is plausible that compounds with this structure might be colored.

(e) Calculate the π bond order of the C_2—C_3 bond of **1**.

(f) Show the most likely product from a Diels–Alder reaction of **1** with 1,3-butadiene. Pay attention to all relevant aspects of stereo- and regioselectivity.

(g) Compound **1** is expected to be very reactive and has never been synthesized. Can you suggest a reason? The tetraphenyl derivative of **1** is stable. Suggest the manner in which the phenyl groups might stabilize the compound.

Q5. The results of a SHMO calculation on fulvene, **1** (R=H), are given below:

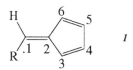

$\epsilon_1 = a - 2.115|\beta| \quad \phi_1 = 0.25p_1 + 0.52p_2 + 0.43p_3$
$\qquad\qquad\qquad\qquad + 0.38p_4 + 0.38p_5 + 0.43p_6$

$\epsilon_2 = a - 1.000|\beta| \quad \phi_2 = 0.50p_1 + 0.50p_2 - 0.50p_4$
$\qquad\qquad\qquad\qquad - 0.50p_5$

$\epsilon_3 = a - 0.618|\beta| \quad \phi_3 = -0.60p_3 - 0.37p_4$
$\qquad\qquad\qquad\qquad + 0.37p_5 + 0.60p_6$

$\epsilon_4 = a + 0.254|\beta| \quad \phi_4 = 0.75p_1 - 0.19p_2 - 0.35p_3$
$\qquad\qquad\qquad\qquad + 0.28p_4 + 0.28p_5 - 0.35p_6$

$\epsilon_5 = a + 1.618|\beta| \quad \phi_5 = -0.37p_3 + 0.60p_4$
$\qquad\qquad\qquad\qquad - 0.60p_5 + 0.37p_6$

$\epsilon_6 = a + 1.861|\beta| \quad \phi_6 = 0.36p_1 - 0.66p_2 + 0.44p_3$
$\qquad\qquad\qquad\qquad - 0.15p_4 - 0.15p_5 + 0.44p_6$

(a) Sketch the highest occupied molecular orbital (HOMO) and the lowest unoccupied molecular orbital (LUMO) on the basis of the calculation.

(b) Assuming that **1** (R=H) were to react with an electrophile, where is the most likely site of attack? Compare the Lewis basicity of **1** (R=H) to that of ethylene.

(c) Assuming that **1** (R=H) were to react with a nucleophile, where is the most likely site of attack? Compare the Lewis acidity of **1** (R=H) to that of ethylene.

(d) In terms of $|\beta|$, estimate the energy of the lowest π–π^* electronic excitation of **1** (R=H). Is it plausible that compounds with this structure might be colored? Explain in three lines or less.

(e) Calculate the π bond order of the C_1—C_2 bond of **1** (R=H).

(f) Calculate the *net charge* of atom C_1 of *1* (R=H).

(g) Choose a suitable reagent and show the most likely product in a reaction between it and fulvene, *1* (R=H).

Q6. Use principles of orbital interaction theory to explain the lower Lewis acidity of allyl cation, $[CH_2CHCH_2]^+$, compared to methyl cation, $[CH_3]^+$.

ANSWER TO Q6: The Lewis acidity depends on the interaction energy ($\Delta\epsilon_L$) from the interaction of the LUMO of the acid with the HOMO of the nucleophile. The interaction is of σ type, with the base HOMO (usually a nonbonded p or sp^n hybrid) interacting end on with the LUMO which for the methyl cation is a single $2p$ orbital, and for the allyl system is a linear combination of $2p$ orbitals. The LUMOs of the two systems are shown below. The magnitude of interaction,

Allyl LUMO
$\epsilon = a + 0.0|\beta|$
$\phi = 0.707p_1 - 0.707p_3$

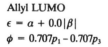

Methyl LUMO
$\epsilon = a + 0.0|\beta|$
$\phi = 1.0p$

$\Delta\epsilon_L \approx h_{AB}^2/(\epsilon_{LUMO} - \epsilon_{HOMO})$, depends inversely on the orbital energy differences and directly on the square of the intrinsic interaction matrix element, which in turn is approximately proportional to the overlap of the two orbitals. The LUMO energies of methyl and allyl are the same. The difference in reactivity, as measured by $\Delta\epsilon_L$ does not arise from the orbital energy differences, but rather from the magnitude of $h_{AB} \approx kS_{AB}$. Since the nucleophile's HOMO can only overlap with one p orbital of the LUMO in each case, the overlap with the allyl LUMO will be smaller because the magnitude of the coefficient (0.707) is smaller (it is 1.0 in the case of methyl).

Q7. Using *simple orbital interaction considerations*, explain the polarity of the N—O bond in amine oxides.

$$R\diagdown\overset{R}{\underset{R}{N^+\!\!-\!\!O^-}}$$

ANSWER TO Q7: This question pertains to the σ bond between a tricoordinated nitrogen atom and a zero-coordinated oxygen atom. It is in *this* chapter rather than the previous one because the student can use the SHMO heteroatom parameters in Table 5.1 as a guide to positioning the interacting orbitals correctly on the same energy scale. ($a_N = a - 1.37|\beta|$) is *lower* in energy than a p orbital of a monocoordinated oxygen ($a_O = a - 0.97|\beta|$). The sp^n hybrid character of the orbital of the N and the *zero* coordination of the oxygen will exaggerate the difference in energies. The correct interaction diagram for

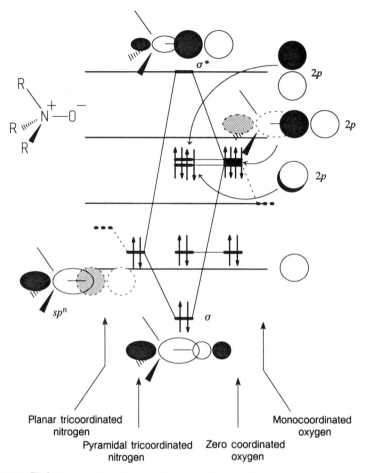

Figure C1.4. Interaction diagram for an amine oxide: answer to question 7.

the N—O bond of the amine oxide is shown in Figure C1.4. Notice that the N—O σ bond is polarized toward nitrogen. Since nitrogen provides both electrons for this dative bond, it loses a fraction of these to the oxygen atom. This is reflected in the formal charges.

EXERCISES FROM CHAPTER 6

Q1. A recent study by Seeman, Grassian, and Bernstein (Seeman, J. I.; Grassian, V. H.; Bernstein, E. R. *J. Am. Chem. Soc.*, **1988**, *110*, 8542) has demonstrated that α-methylstyrene is non-planar in the ground state but is planar in its first excited state. This may be interpreted as evidence that the π bond order of the bond connecting the vinyl group to the ring is greater in the lowest π-π^* state than in the ground state. Can you offer an explanation for this observation by considering the properties of styrene to arise from the interaction of a benzene ring with ethylene?

ANSWER TO Q1: The HOMOs and LUMOs of benzene and ethylene have the same SHMO energies, $\alpha - |\beta|$ and $\alpha + |\beta|$, respectively. These HOMOs interact, the out-of-phase combination becoming the HOMO of styrene. Thus the HOMO of styrene is antibonding across the bond connecting the vinyl group to the ring. Upon photoexcitation, one electron leaves this orbital, resulting in a net increase of the π bond order of this bond. The LUMO of styrene corresponds to the in-phase combination of the LUMOs of ethylene and benzene, and therefore contributes a positive contribution to the π bond order of the connecting bond when it receives the photoexcited electron.

Q2. Use an orbital interaction diagram to explain the observation that tetracyanoethylene is very easily reduced to its radical anion.

Q3. The highly strained olefin, spiropentadiene, has been synthesized (Billups, W. E.; Haley, M. M. *J. Am. Chem. Soc.*, **1991**, *113*, 5084) after more than a decade of theoretical investigations which predicted significant *spiroconjugation*. Discuss the π bonding of this compound.

ANSWER TO Q3: The high symmetry of spiropentadiene (D_{2d}) limits the possibilities for interaction of orbitals of one ring with orbitals of the other. Thus the π bonding orbital of one ring cannot interact by symmetry with either the π or π^* orbitals of the other ring. Only the π^* orbitals of the two rings can interact, but as both are unoccupied, no stabilization ensues from this interaction. However, a plausible interaction which may lead to extra stabilization is the interaction of the occupied π orbital of one ring with the unoccupied Walsh orbital of the other, as shown below, leading to two two-electron–two-orbital stabilizations, a form of homoconjugation.

Spiropentadiene

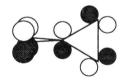

HOMO (π bond)   LUMO (lowest Walsh orbital)

EXERCISE FROM CHAPTER 7

Q1. Use orbital interaction diagrams to explain the fact that tetramethyleneethane exists as a planar species and is triplet in its ground state, while its radical cation exists in a perpendicular conformation. (See

Gerson, F.; de Meijere, A.; Qin, X.-Z. *J. Am. Chem. Soc.*, **1989**, *111*, 1135–1136.)

Q2. The tertiary carbanion, perfluoro-1-methyl-1-cyclobutyl anion is considerably more stable than most carbanions. Explain the role of adjacent fluorine atoms in carbanion stabilization. Use the group orbitals of the trifluoromethyl group in your orbital analysis. What kind (X:, C, or Z) group is $-CF_3$? (See Farnham, W. B.; Dixon, D. A.; Calabrese, J. C. *J. Am. Chem. Soc.*, **1988**, *110*, 2607–2611.)

Q3. Use orbital interaction analysis to explain stabilization of a carbocationic center by a cyclopropyl group. What kind (X:, C, or Z) of a substituent is cyclopropyl? Explain. Predict the orientation of the planar cationic center relative to the cyclopropyl ring.

Q4. A recent study by Abu-Raqabah and Symons (Abu-Raqabah, A.; Symons, M. C. R. *J. Am. Chem. Soc.*, **1990**, *112*, 8614) has characterized the pyridine–chlorine atom three-electron bonded species, Py ∴ Cl, by ESR and UV spectroscopy. Using an orbital interaction diagram, provide a bonding description that explains the red-shifted $\sigma \to \sigma^*$ UV absorption, and the increased length of the N—Cl bond. In an earlier paper, Breslow and co-workers (Breslow, R.; Brandl, M.; Hunger, J.; Adams, A. D. *J. Am. Chem. Soc.*, **1987**, *109*, 3799) considered ring acylated pyridine–chlorine radicals to be π radicals and anticipate special stability for the 4-carboalkoxypyridine-chlorine radical. Show by means of orbital interaction arguments whether Breslow's expectation is justified or not. In other words, would a Z substituent in the 4 position stabilize a pyridine–chlorine π radical?

Q5. By considering the π MOs of the cyclopentadienyl system (C_5H_5) to result from an interaction between *cis*-butadiene π MOs and an sp^2 hybridized C atom, explain the stability of the cyclopentadienyl anion and the instability of the cyclopentadienyl cation.

Q6. Use simple orbital interaction arguments (i.e., orbital correlation diagram) to explain the following:
(a) The stabilization of a carbocation by C—H hyperconjugation.
(b) The stabilization of a carbanion by C—F negative hyperconjugation.
(c) The triplet ground state of phenyl carbene, C_6H_5-C-H.
(d) The singlet nature of cycloheptatrienylidene,
(e) The nucleophilicity of N,N-dimethylaminocarbinyl radical, $(CH_3)_2NCH_2$.

ANSWER TO Q6(d): We can consider the orbitals of cycloheptatrienylidene to arise from the interaction of the π orbitals of hexatriene and the valence orbitals of a dicoordinated carbon atom (a $2p$ orbital and an sp^n hybrid orbital). The π orbitals of hexatriene may be obtained from a SHMO calculation. The interaction diagram is shown in Figure C1.5. The p orbital of the carbene site is raised as a result of the dominant interaction with π_3 of hexatriene. The orbital, π_4, which is closest in energy to the carbene's p orbital does not interact because of symmetry, and π_5 interacts less strongly because the coefficients at the terminal positions of the hexatriene are smaller. The larger HOMO (sp^n)–LUMO (π_4' or π_5') gap permits the ground state to be singlet.

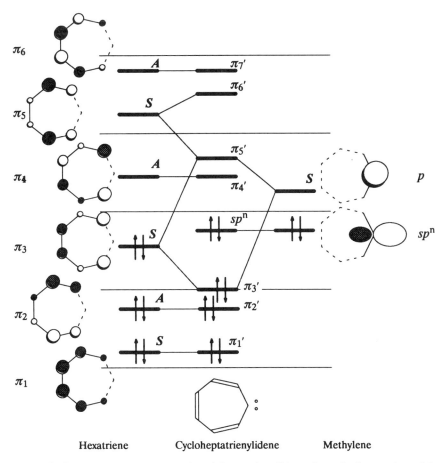

Hexatriene   Cycloheptatrienylidene   Methylene

Figure C1.5. Electronic structure of cycloheptatrienylidene from the interaction of the orbitals of hexatriene and methylene. Symmetry labels refer to the p and π orbitals and a vertical mirror. Answer to question 6(d).

Q7. Use orbital interaction analysis to suggest a reason that tropylium cation $C_7H_7^+$ is such a stable cation. This may be done in either of two ways, i.e., by considering the interaction of a simple carbocation with 1,3,5-hexatriene, or the interaction of an allyl cation with butadiene, both held in the seven-membered planar ring geometry of tropylium. In either case, attention must be paid to orbital symmetry.

Q8. Use orbital interaction analysis to predict the structure of a complex between H_2O and :CH_2. The structure was determined by *ab initio* MO theory by Moreno and co-workers (Moreno, M.; Lluch, J. M.; Oliva, A.; Bertrán, J. *Can, J. Chem.*, **1987**, *65*, 2774–2778).

Q9. Methyl methoxy carbene, $CH_3-C-OCH_3$, was described recently (Sheridan, R. S.; Moss, R. A.; Wilk, B. K.; Shen, S.; Wlostowski, M.; Kesselmayer, M. A.; Subramanian, R.; Kmiecik-Lawrynowicz, G.; Krogh-Jespersen, K. *J. Am. Chem. Soc.*, **1988**, *110*, 7563–7564) as a remarkably selective nucleophile. Explain why the carbene, $CH_3-C-OCH_3$, has a singlet ground state and determine the basis for its "nucleophilicity."

Q10. Benzyne is considered to be the intermediate in strong base-catalyzed nucleophilic aromatic substitution.

(a) Suggest a reason for the ease of formation of the carbanion in step 1.
(b) Suggest a reason for the ease of departure of X^- in step 2.
(c) Suggest a reason for the ease of nucleophilic addition to the double bond in benzyne. (The MW and IR spectra of benzyne have been measured: (MW), Brown, R. D.; Godfrey, P. D.; Rodler, M. *J. Am. Chem. Soc.*, **1986**, *108*, 1296–1297; (IR), Radziszewski, J. G.; Hess, B. A. Jr.; Zahradnik, R. *J. Am. Chem. Soc.*, **1992**, *114*, 52–57.)

Q11. Use orbital interaction analysis to derive the bonding molecular orbitals of ethylbenzenium ion. Consider ethylbenzenium ion to be the result of the interaction of a phenyl group, C_6H_5, and ethylene, C_2H_4, with the appropriate number of electrons. (Direct evidence for the existence of ethylbenzenium ion was obtained by Fornarini, S. and Muraglia, V. *J. Am. Chem. Soc.*, **1989**, *111*, 873.)

ethylbenzenium ion (all substituents are H)

Q12. Use orbital interaction analysis to derive the bonding molecular orbitals of bishomocyclopropenyl cations, such as *4*. The X-ray structure of the hexafluoroantimonate salt of *3* was recently determined by Laube, T. *J. Am. Chem. Soc.*, **1989**, *111*, 9224.

4 (a bishomocyclopropenyl cation)

Q13. Explain Markovnikov's rule, i.e., "In an electrophilic addition of HX to an olefin, the hydrogen goes to the carbon atom which has the most hydrogens."

Q14. Cycloalkanols may be synthesized by free radical intramolecular cyclization (Yadav, V.; Fallis, A. G. *Can. J. Chem.*, **1991**, *69*, 779–789).

(several steps)

Analyze the reaction, paying close attention to substituent effects. Suggest a reason based on orbital interaction criteria for the formation of five-membered rings rather than six-membered rings.

Q15. Discuss stabilization of free radicals by the *captodative* effect. (See for examples (a) Kosower, E. M.; Waits, H. P.; Teuerstein, A.; Butler, L. C. *J. Org. Chem.*, **1978**, *43*, 800. (b) Olson, J. B.; Koch, T. H. *J. Am. Chem. Soc.*, **1986**, *108*, 756. (c) Brook, D. J. R.; Haltiwanger, R. C.; Koch, T. H. *J. Am. Chem. Soc.*, **1991**, *113*, 5910.)

Q16. (a) Offer a rationale for the preferred orientation of an α-sulfonyl carbanion (it is not necessary to invoke the use of *d* orbitals). For a review of α-sulfonyl carbanions, see Boche, G. *Angew. Chem. Int. Ed. Engl.*, **1989**, *28*, 277–297.

(b) The observation of part (a) explains why the equatorial α proton (H_{eq}) in six-membered cyclic sulfones is selectively abstracted under basic conditions. Suggest a reason that the rate of H/D exchange in NaOD/D$_2$O of α-H$_{eq}$ is 36 times faster in **1** than in **2**, and about 10,000 times faster than in **3** (King, J. F.; Rathore, R. *J. Am. Chem. Soc.*, **1990**, *112*, 2001).

preferred conformation   not stable

```
    1                    2                    3
```

EXERCISES FROM CHAPTER 8

Q1. Solve the SHMO equation for the π orbitals and orbitals energies of the carbonyl group: $\alpha(O) = \alpha(C) - |\beta(CC)|$, $\beta(CO) = \beta(CC)$. Sketch the results in the form of an interaction diagram.

ANSWER TO Q1: $E(L) = \alpha + 1.618\beta$; $E(U) = \alpha - 0.618\beta$; $\psi(L) = 0.851p(O) + 0.526p(C)$; $\psi(U) = 0.526p(O) - 0.851p(C)$.
Bond energy: $\alpha(C) + \alpha(O) - 2E(L) = -1.236\beta = 1.236|\beta|$
Ionization potential: $-E(L) = -(\alpha - 0.618\beta) = |\alpha| + 0.618|\beta|$
Electronic excitation ($\pi \to \pi^*$): $E(U) - E(L) = -2.236\beta = 2.236|\beta|$
Bond order: $2(0.851)(0.526) = 0.895$
Net charges: on O, $-2(0.851)^2 + 1 = -0.447$; on C, $-2(0.526)^2 + 1 = +0.477$

Q2. Develop an orbital diagram for the amide group from the interaction of the C=O and N moieties. Explain the positioning of the C=O and N orbitals before and after they interact. (In developing the interactions, consider both the relative energies of the interacting orbitals, and the polarization of the C=O π orbitals. Do not forget the higher nonbonded electron pair on O. What features of the amide functional group are consistent with what you would expect from the analysis?

Q3. 2,5-Dimethylborolane has been shown to be effective in asymmetric reduction of ketones (Imai, T.; Tamura, T.; Yamamuro, A.; Sato, T.; Wollmann, T. A.; Kennedy, R. M.; Masamune, S. *J. Am. Chem. Soc.*, **1986**, *108*, 7402-7404). Using the frontier orbitals of a borolane and a ketone, show the probable course of the initial interaction between the two.

Q4. Metal hydrides are often used to reduce aldehydes to ketones. The rate-determining step is

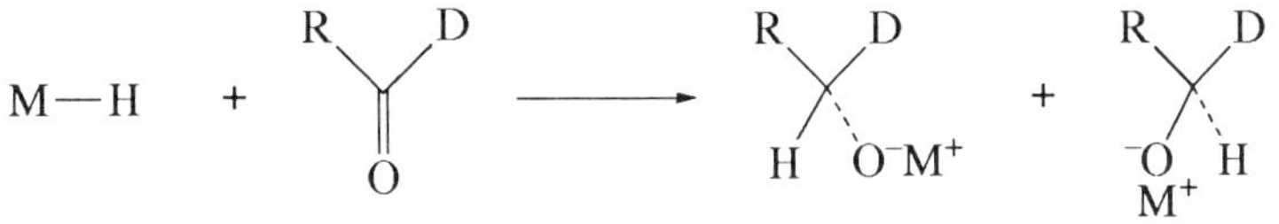

$$M-H \;+\; \underset{O}{\overset{R\;\;D}{\diagdown\!/}} \longrightarrow \underset{H\quad O^-M^+}{\overset{R\;\;D}{\diagup\!\diagdown}} \;+\; \underset{\underset{M^+}{O\;\;H}}{\overset{R\;\;D}{\diagdown\!/}}$$

"M" may be a single metal such as Na or Li, or a complex such as AlH_3 (as in $LiAlH_4$). Using a two-orbital interaction diagram (assume M = Li), show which orbitals are involved in the above step. Predict the most favorable geometry for the approach of the reagents. (For an interesting variation of this reaction using a chiral aluminum hydride to effect enantioselective reduction, see Noyori, and co-workers: Noyori, R.; Tomino, I.; Yamada, M.; Nishizawa, M. *J. Am. Chem. Soc.*, **1984**, *106*, 6717–6725.)

Q5. Explain the higher reactivity of acetone (propanone) compared to methyl acetate ($CH_3CO_2CH_3$) toward reduction by sodium borohydride.

Q6. Singlet carbenes react with carbonyl compounds to produce zwitterionic intermediates which can cyclize to epoxides or react further with another carbonyl to form 1,3-dioxacyclopentanes. Show which orbitals

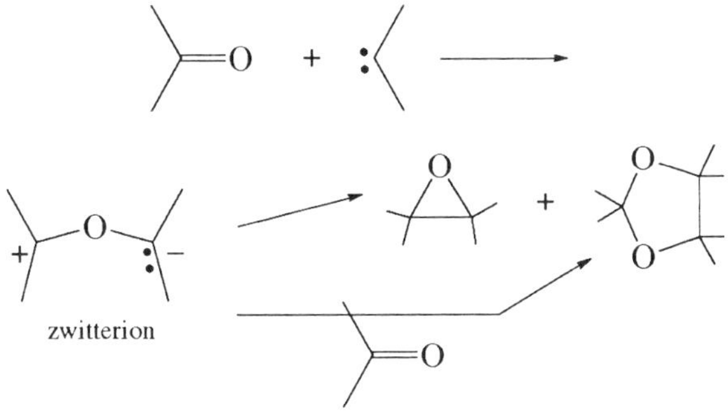

are involved and predict the geometry of approach of the carbene and carbonyl compound in the *initial interaction* that produces the zwitterion.

Q7. Explain the bisected geometry of cyclopropyl aldehyde

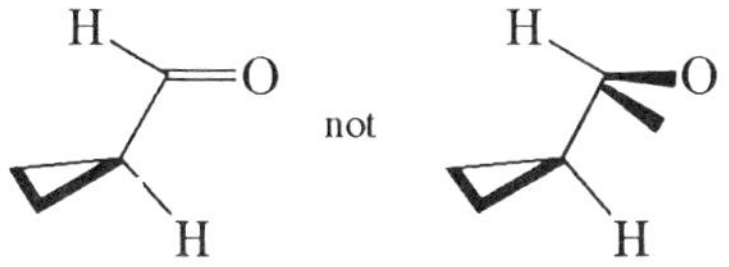

Q8. Alkali metal reduction of ketones may involve the following steps:

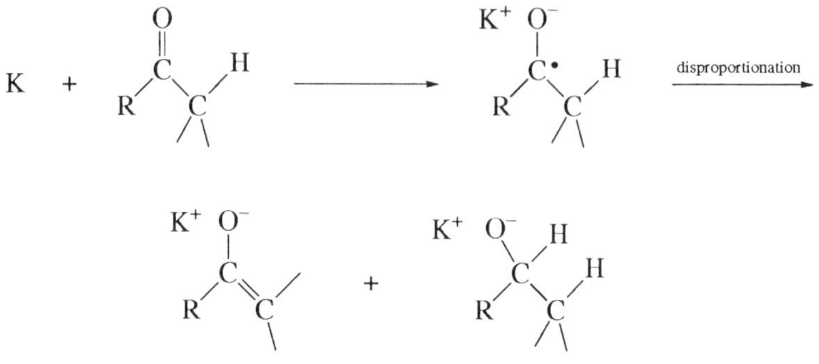

(a) Discuss the electronic structure of each of the intermediates in terms of the orbitals involved.

(b) Show a plausible geometry for the transition state for the disproportionation step, according to orbital interaction theory.

(c) What additional features emerge from the study by Rautenstrauch, et al. (Rautenstrauch, V.; Mégard, P.; Bourdin, B.; Furrer, A. *J. Am. Chem. Soc.*, **1992**, *114*, 1418) on the reduction of camphors with potassium in liquid ammonia?

(d) Present a brief summary of stereochemical considerations in metal reductions of carbonyls as deduced in the theoretical study of Wu and Houk (Wu, Y.-D. and Houk, K. N. *J. Am. Chem. Soc.*, **1992**, *114*, 1656).

Q9. Hahn and le Noble have discovered strongly enhanced stereoselectivity in the reduction of 5-substituted adamantanones where the substitution at C_5 is by positive nitrogen (Hahn, J. M.; le Noble, W. J. *J. Am. Chem. Soc.*, **1992**, *114*, 1916). Thus, E/Z ratios for borohydride reduction of compounds **1–3** where in the range 20–25. Suggest a possible reason for these observations on the basis of orbital interaction theory.

1 $R = CH_3\ I^-$
2 $R = O^-$
3 $R = CH_2CO_2^-$

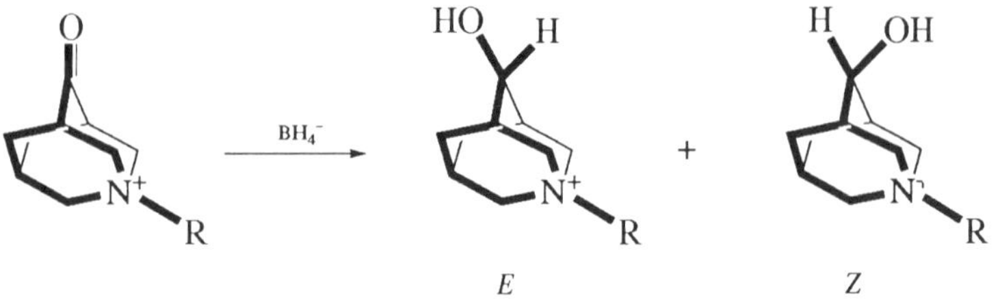

ANSWER TO Q9. A possible explanation for the observed diastereoselectivity of nucleophilic addition to the carbonyl involves a static distortion of the carbonyl group so as to improve π donation of the β σ C—C bonds into the π^* orbital of the carbonyl group, as shown in Figure C1.6. The consequent polarization of the orbital at carbon results in more favorable overlap with the HOMO of the nucleophile (H^- or $H-BH_3^-$, in the present case) and attack from the direction indicated by the arrow. Such a distortion was observed theoretically in 2-adamantyl cation (Dutler, R.; Rauk, A.; Sorensen, T. S.; Whitworth, S. M. *J. Am. Chem. Soc.*, **1989**, *111*, 9024–9029). It may be argued that the static distortion in the ground state of the carbonyl is insufficient to account for the high selectivity observed. According to the theory of Cieplak (Cieplak, A. S.; Tait, B. D.; Johnson, C. R. *J. Am. Chem. Soc.*, **1989**, *111*, 8447–8462), the distortion occurs naturally as one approaches the transition state, one π face being preferred by donation into the σ^*_{CNu} orbital [Figure C1.6(c)].

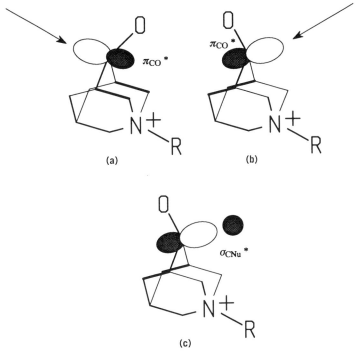

Figure C1.6. Static distortion of carbonyl group to favor overlap with the highlighted bonds. The bonds in **(a)** are poorer π donors because of electrostatic effects of the quaternary N center and interaction with σ^*_{NR}. Thus, interaction **(b)** is favored; **(c)** distortion in the transition state to favor donation into σ^*_{CNu} bond. Answer to question 9.

EXERCISES FROM CHAPTER 9

Q1. Cyclopropanes may be obtained from the Michael condensation of bromo- and chloromalonate carbanions (Le Menn, J.-C.; Tallec, A.; Sarrazin, J. *Can. J. Chem.*, **1991**, *69*, 761–767). The mechanism in-

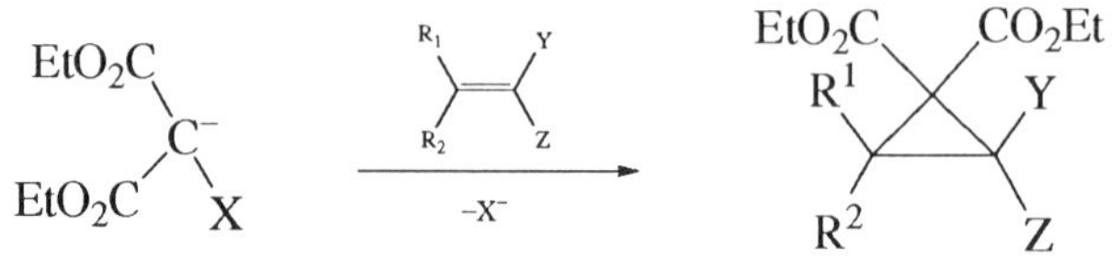

$$\text{EtO}_2\text{C}-\underset{\text{EtO}_2\text{C}}{\overset{|}{\text{C}^-}}-X \quad + \quad \underset{R_2}{\overset{R_1}{>}}=\underset{Z}{\overset{Y}{<}} \quad \xrightarrow{-X^-} \quad \underset{R^2}{\overset{\text{EtO}_2\text{C} \quad \text{CO}_2\text{Et}}{\underset{Z}{\overset{R^1 \qquad Y}{\triangle}}}}$$

volves nucleophilic addition to a Z-substituted olefin followed by an intramolecular bimolecular nucleophilic substitution. Several side reactions also occur. Discuss the chemistry involved in this reaction pointing out substituent effects at each stage.

EXERCISES FROM CHAPTER 10

Q1. Predict the structure of the complex formed between oxirane and HCl. Use orbital interaction arguments and draw an orbital interaction diagram. How does your prediction compare with the experimental structure (Legon, A. C.; Rego, C. A.; Wallwork, A. L. *J. Chem. Phys.*, **1992**, *97*, 3050–3059)?

Q2. In principle, and alkyl halide, $R^1R^2CH-CH-CClR^3R^4$, may react with a nucleophile, Nu^-, by S_N1, S_N2, E1, E1cb, or E2 mechanisms. Choose R^1, R^2, R^3, R^4, and Nu^- so that the course of the reaction may be expected to follow *each* of the mechanisms. The groups and nucleophile should be sufficiently different so that the stereochemical consequences of the reaction are obvious in your answer. Justify your choices using orbital interaction arguments but do not draw orbital interaction diagrams.

Q3. (a) Suggest a reason that fluoboric acid is a very strong acid.
  (b) Why does Ph_3C-OH dehydrate in the presence of HBF_4? Would CH_3OH do the same thing?
  (c) What is the function of acetic anhydride in the Ph_3C^+ synthesis?
  (d) Why does cycloheptatriene lose hydride readily? Would cyclopentadiene do the same?
  (e) Why can one not prepare tropylium tetrafluoborate by the more straightforward route shown below?

cycloheptatriene(H,H) + HBF_4 ⟶ tropylium$^+$ + BF_4^- + H_2

Q4. Suggest an explanation based on orbital interactions for the observed stereochemistry for E2 elimination reactions, i.e., the strong stereoelectronic preference that the C—H and C—X bonds be *anti*-coplanar.

Q5. Using **simple orbital interaction considerations**, explain the following observations. Use orbital interaction diagrams and show the orbitals clearly.
(a) The acidity of the C—H bond in cyclopentadiene ($pK_a = 16$).
(b) The relative acidities of the C—H bond in ethane ($pK_a = 50$), ethene ($pK_a = 44$), and ethyne, ($pK_a = 25$).
(c) Ethanolamine exists in a number of hydrogen bonded conformations, two of which are shown below. One of the two is substantially more stable than the other. Which is more stable? Explain. (Räsänen, M.; Aspiala, A.; Homanen, L.; Murto, J. *J. Mol. Struct.*, **1982**, *96*, 81).

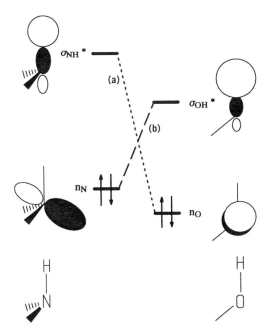

Figure C1.7. The frontier orbitals of an —NH$_2$ group and an —OH group. The stronger hydrogen bonding interaction **(b)** determines the most stable conformation of ethanolamine. Answer to question 5c.

ANSWER TO Q5c: The orbitals and orbital energies of generic $-NH_2$ and $-OH$ groups are shown side by side in Figure C1.7. Conformation **a** and **b** exhibit hydrogen bonding interactions **a** and **b**, respectively. Interaction **b** is favored by the smaller energy gap and additionally by increased polarization of both n_N and σ^*_{OH} orbitals.

EXERCISES FROM CHAPTER 11

Q1. Use orbital interaction diagrams to explain why methoxybenzene (anisole) prefers a conformation in which the methyl group lies in the plane of the aromatic ring.

Q2. Use orbital interaction diagrams to explain why benzyne is an excellent dienophile in Diels–Alder reactions.

Q3. Use orbital interaction diagrams to propose a possible structure of

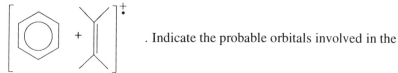

. Indicate the probable orbitals involved in the 680 nm absorption of this species.

Q4. Use orbital interaction diagrams to propose a possible structure of 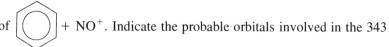 + NO^+. Indicate the probable orbitals involved in the 343 and 500 nm absorptions of this species (see Kim, E. K.; Kochi, J. K. *J. Am. Chem. Soc.*, **1991**, *113*, 4962).

ANSWER TO Q4: We postulate that the structure of the complex will be that which provides the most favorable HOMO–LUMO interaction(s), while minimizing interaction between occupied orbitals. NO^+ is isoelectronic with CO. Its LUMO will be very low lying, a degenerate pair of π^* orbitals polarized toward the nitrogen atom. The benzene HOMOs also are a degenerate pair with a single nodal surface bisecting the ring. Two structures may be postulated for the complex. An attractive possibility involves both HOMOs and both LUMOs, as shown in Figure C1.8(a) analogous to transition metal carbonyl complexes. This complex, which has symmetry C_{6v}, has two serious strikes against it. There is no obvious bonding contribution from the *sp* hybrid orbital on N, as there is in transition metal carbonyl complexes, since the benzene LUMO has the wrong symmetry to interact with it. The *sp* hybrid orbital interactions are destabilizing. The second drawback is that the π^* orbitals of NO^+ are not as highly polarized as they are in CO, leaving more of the orbital not involved in bonding. An alter-

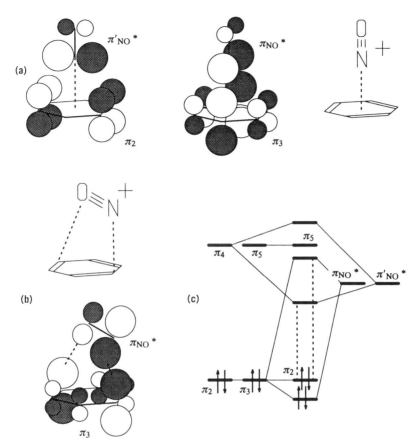

Figure C1.8. (a) The bonding interactions and structure of benzene$-NO^+$ complex with C_{6v} symmetry; (b) the bonding interactions and structure of benzene$-NO^+$ complex with C_s symmetry; (c) interaction diagram for the C_s complex. The dashed lines correspond to possible assignments for the two lowest transitions. Answer to question 4.

native structure which does not have these disadvantages but uses only one of the HOMO–LUMO pairs for bonding, is shown in Figure C1.8(b). The corresponding interaction diagram [Figure C1.8(c)] suggests a possible secondary interaction of the component LUMOs which lowers the LUMO for the system. Proposals for the 343 and 500 nm transitions are shown as dashed lines.

EXERCISES FROM CHAPTER 12

Q1. In a Diels–Alder reaction, when both π systems are polarized, the more favorable overlap, and therefore the stronger interaction occurs when the ends with the larger coefficients get together and the smaller coef-

ficients get together. Predict the major product of each of the following cycloaddition reactions.
(a) (1E)-1-phenyl-1,3-butadiene + acrolein →
(b) (3E)-2-methyl-1,3-pentadiene + formaldehyde →
(c) 1-carbomethoxycyclopentadiene + ethenol acetate →
(d) acrolein + ethenol acetate → (an oxacyclohexene)
(3) methyl 1,3-butadien-1-yl ether + methyl propynoate →

Q2. Provide a **general classification** in terms of "s" and "a" components for each of the reactions given below. Verify, using the "odd sum" rule, that the reaction is *allowed* under the specified reaction conditions.

(a) 2,3-diazabicyclo[2.2.1]hept-2-ene 3 yields bicyclo[2.1.0]pentane 4 upon photolysis.

(b) Compound 5 yields 6 upon heating (hint: there is an enol intermediate).

(c) 2-carbomethoxy-2-cyclohexene-1-one 7 reacts with (E)-1,3-pentadiene to yield the bicyclic product 8 as the major product. What is the role of stannic chloride in the reaction?

(d) Cyclooctatrienone 9 is converted smoothly to the bicyclic dienone 10 upon heating.

(e) 9,10-dideuteriosnoutene 11 is transformed to 7,8-dideuteriosnoutene 12 upon heating.

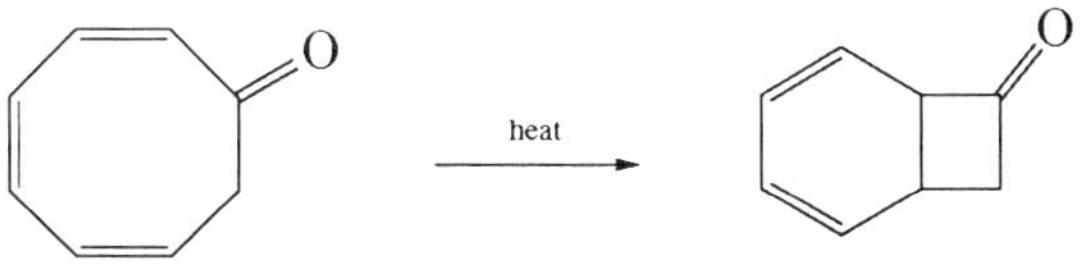

EXERCISES FROM CHAPTER 12   267

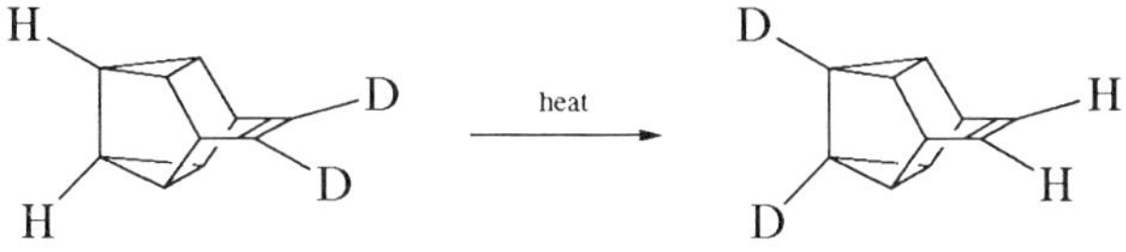

9 → **10** (heat)

11 → **12** (heat)

(f) What are the two optically active products produced upon thermal rearrangement of Feist's ester **5**?

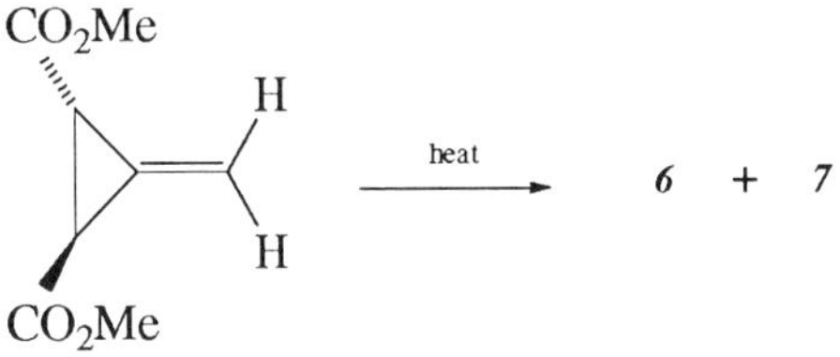

5 —heat→ **6** + **7**

(g) The following rearrangement of cycloheptatrienes is not concerted but proceeds by thermally allowed steps. Suggest a mechanism.

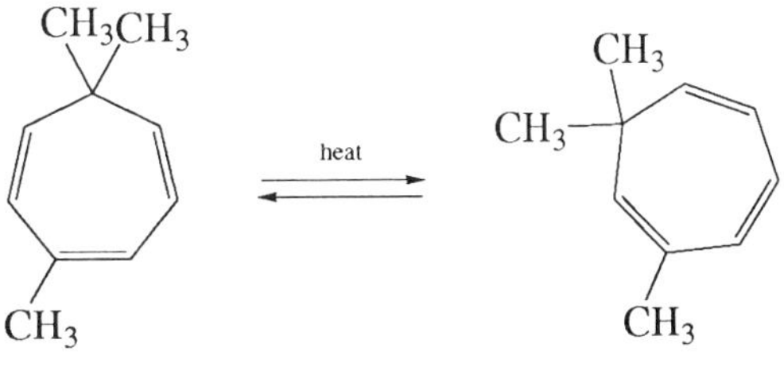

(h) Propose a mechanism for the following reaction:

KO-t-Bu, t-BuOH, 42°

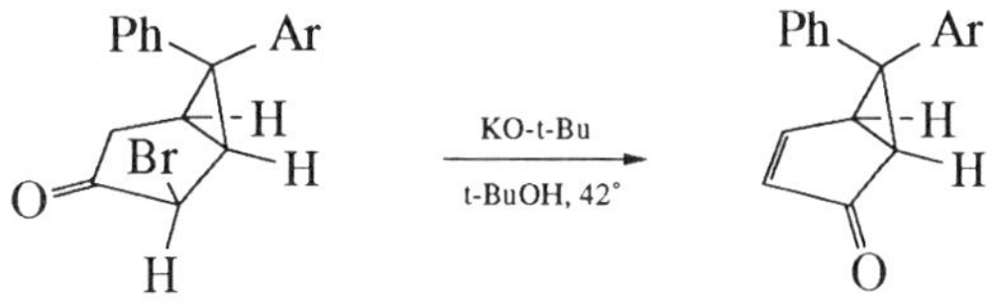

Q3. Classify the following reactions by the *component analysis* method (e.g., $_\pi 4_s + {_\sigma}2_a + \cdots$). Decide whether the reaction as *shown* is thermally allowed. Show clearly the orbitals on which you base your analysis.

(a) The cycloaddition of acetylene to cyclobutadiene to give Dewar benzene:

(b) The rearrangement of Dewar benzene to benzene:

(c) The electrocyclic opening of the steroid-derived cyclohexadiene:

(d) The opening of cyclopropyl cations to allyl cations:

(e) The sigmatropic rearrangement of bicyclo[2.1.1]hex-2-ene to bicyclo[3.1.0]hex-2-ene:

+ enantiomers

Q4. Use orbital interaction diagrams to explain the following observations:
  (a) Benzyne is an excellent dienophile in Diels–Alder reactions.
  (b) A mixture of cyclopentadiene and ethene yields only dicyclopentadiene and not norbornene (bicyclo[2.2.1]hept-2-ene).

(c) 2-methylpropenyl cation adds to cyclopentadiene to form **1**.

Q5. Show the expected products of the following Diels–Alder reactions. Pay careful attention to stereochemistry and regioselectivity, if these considerations are appropriate.

(a)

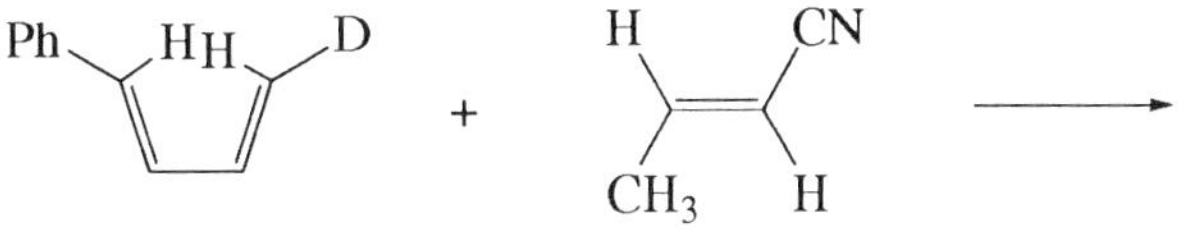

(b)

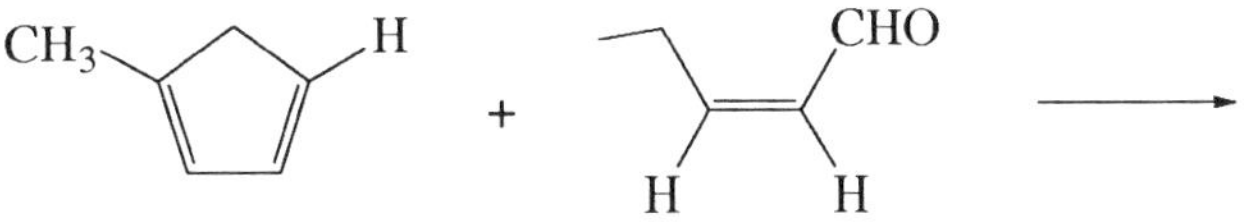

(c)

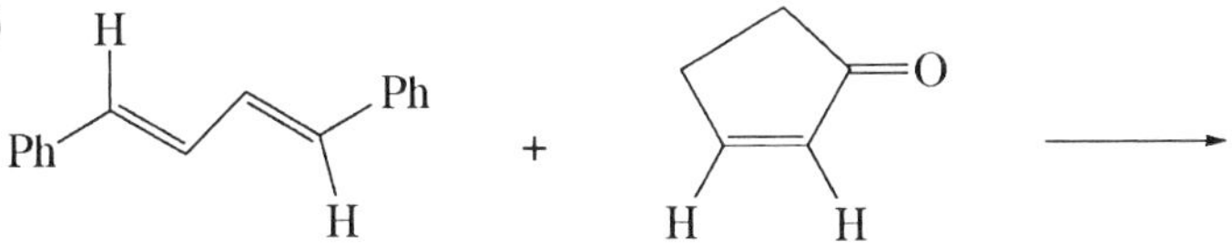

(d)

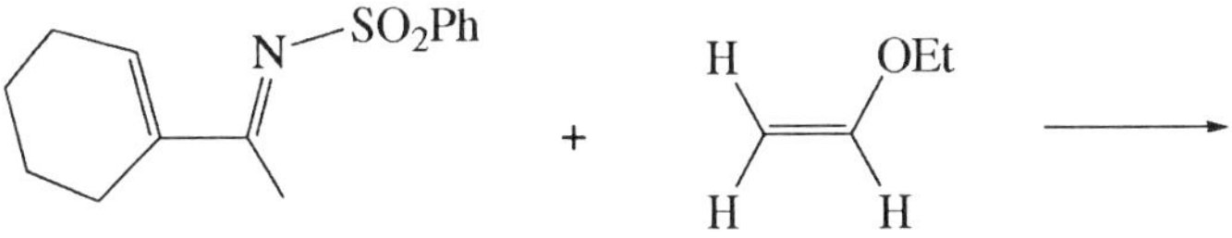

(e)

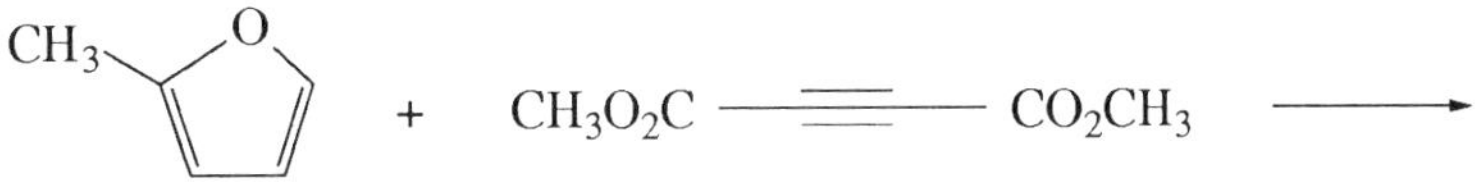

Q6. Predict the product of each of the following reactions.

(a) $H_2C=CH-CO_2H$ + (E)-$H_2C=CH-CH=CHCH_3$ $\xrightarrow{\Delta}$

(b)

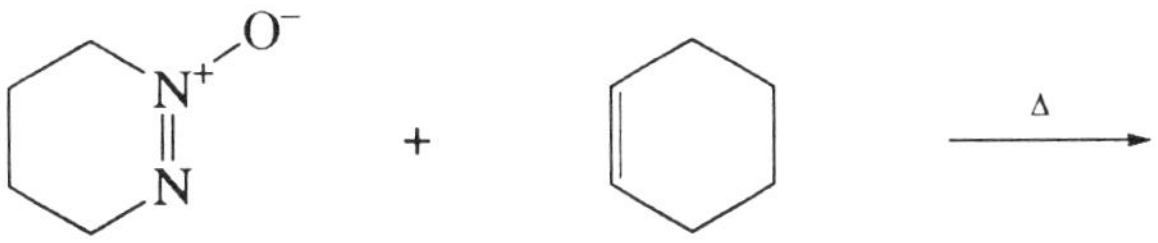

Q7. The origin of the diastereoselectivity found for the Diels–Alder reaction,

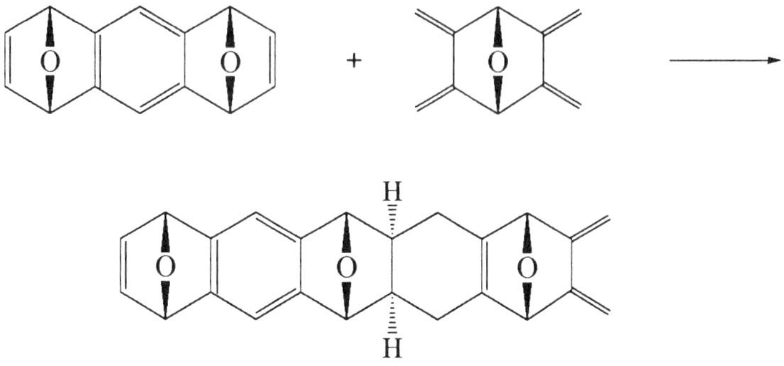

has been attributed to a dominance of interactions between the occupied MOs of the reactants, i.e., to four-electron–two-orbital interactions rather than to the usual secondary aspect of the HOMO–LUMO interaction. Show by suitable orbital interaction diagrams why this may be the case. The relevant references are found in Ashton, P. R.; Brown, G. R.; Isaacs, N. S.; Giuffrida, D.; Kohnke, F. H.; Mathias, J. P.; Slawin, A. M. Z.; Smith, D. R.; Stoddart, J. F.; Williams, D. J. *J. Am. Chem. Soc.*, **1992**, *114*, 6330–6353. *Note*: The process shown in the above reaction forms the basis for the construction of larger stereoregular molecules and has been termed *Molecular* LEGO.

EXERCISES FROM CHAPTER 13

Q1. A key step in one route to the synthesis of hexamethyl Dewar benzene is the cycloaddition of 2-butyne to tetramethylcyclobutadiene (stabilized by Al cation). Using the parent compounds (no methyls) develop a Woodward–Hoffmann orbital correlation diagram for the reaction and determine whether the reaction is thermally allowed.

Q2. This question requires you to construct **orbital correlation** diagrams of the Woodward–Hoffmann type.

(a) Show that the rearrangement of Dewar benzene to benzene is a thermally forbidden process. [Breslow, *et al.* (Breslow, R.; Na-

pierski, J.; Schmidt, A. H. *J. Am. Chem. Soc.*, **1972**, *94*, 5906–5907) have determined that the activation energy for the rearrangement is 96.2 kJ/mol.]

(b) Show that the photochemical electrocyclic ring open of 1,3-cyclohexadiene to *cis*-1,3,5-hexatriene should occur by conrotatory motion. [Although long predicted by the Woodward–Hoffmann

rules, this was first demonstrated experimentally in 1987 (Trulson, M. O.; Dollinger, G. D.; Mathies, R. A. *J. Am. Chem. Soc.*, **1987**, *109*, 586–587.]

(c) Cyclopropyl cations open to allyl cations spontaneously and stereospecifically. Predict the stereochemical course of ring opening, disrotatory or conrotatory?

EXERCISES FROM CHAPTER 14

Q1. In a recent "Accounts of Chemical Research" article on photoexcited ketones, P. J. Wagner reported the following overall reaction (Wagner, P. J. *Acc. Chem. Res.*, **1989**, *22*, 83): The mechanism is thought to involve a diradical intermediate.

(a) Draw the structure of the intermediate diradical and explain how each of the products could be derived from it.
(b) Produce a *state correlation diagram* of the Dauben–Salem–Turro type and analyze the reaction in terms of the carbonyl electronic states which are likely to be involved (you may ignore the effect

of the phenyl group on the carbonyl and radical orbitals, it will not change the relative energies of the states).

Q2. In the article mentioned in the previous question, P. J. Wagner reported the following overall reaction (Wagner, P. *J. Acc. Chem. Res.*, **1989**, *22*, 83): The mechanism is thought to involve two different diradical intermediates.

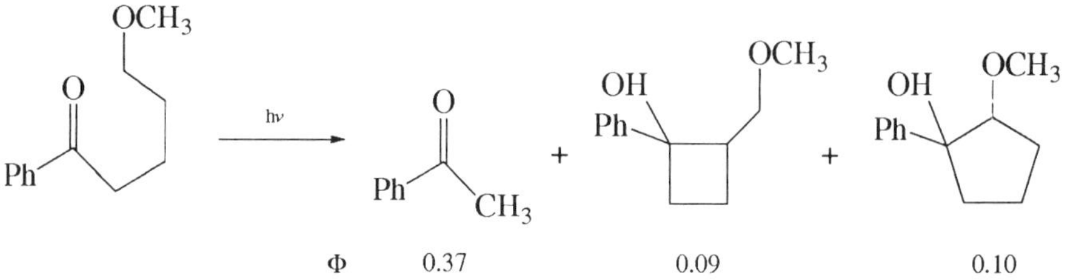

(a) Draw the structures of the intermediate diradicals and explain how each of the products could be derived from them.

(b) Produce a **state correlation diagram** of the Dauben–Salem–Turro type and analyze the reaction in terms of the carbonyl electronic states which are likely to be involved (you may ignore the effect of the phenyl group on the carbonyl and radical orbitals; it will not change the relative energies of the states).

Q3. This question involves the analysis of carbonyl photochemistry using Dauben–Salem–Turro-type correlation diagrams.

(a) Provide an analysis of the photoreduction of benzophenone to benzpinacol in isopropyl alcohol. What is the photoreactive state for the reaction? What is the effect of added naphthalene and why does it have this effect?

(b) A characteristic of the photochemistry of cyclohexadienones is cleavage of the C—C bond next to the carbonyl, as shown below

for 2,2-dimethyl-2,4-cyclohexadienone. Develop a Dauben–Salem–Turro-type state correlation diagram for the photochemical step shown and, *on the basis of your diagram*, discuss the efficiency of the reaction on the singlet and triplet manifold.

Q4. This question involves the analysis of carbonyl photochemistry using Dauben–Salem–Turro-type state correlation diagrams.

(a) 2,4-Di-*t*-butylbenzophenone yields the benzocyclopentanol, **2**, upon photolysis. An intermediate diradical is involved. Show the structure of the intermediate diradical. Develop a Dauben–Salem–Turro-type state correlation diagram for the photochemical step

and, *on the basis of your diagram*, discuss the efficiency of the reaction on the singlet and triplet manifold.

Q5. Photolysis of α-chloro-o-methylacetophenones yields 1-indanones. The mechanism has been studied by laser flash photolysis (Netto-Ferreira, J. C.; Scaiano, J. C. *J. Am. Chem. Soc.*, **1991**, *113*, 5800). Develop

a Dauben–Salem–Turro-type state correlation diagram for the photochemical step and, *on the basis of your diagram*, discuss the efficiency of the reaction on the singlet and triplet manifold. Do the experimental results agree with your analysis?

Q6. The photochemistry of 1,5-hexadien-3-ones has been examined by Dauben and co-workers. A relatively rare dependence of the intramolecular enone–olefin photoaddition of several derivatives has been observed. Discuss the mechanistic possibilities in terms of Dauben–Salem–Turro-type state correlation diagrams. (See Dauben, W. G.; Cogen, J. M.; Ganzer, G. A.; Behar, V. *J. Am. Chem. Soc.*, **1991**, *113*, 5817).

Q7. The mechanism of photoenolization in 1-methylanthraquinone has been studied in detail (Gritsan, N. P.; Khmelinski, I. V.; Usov, O. M. *J. Am. Chem. Soc.* **1991**, *113*, 9615). The reaction was found to occur in both the singlet and triplet (nπ*) states. Develop a Dauben–Salem–Turro-type state correlation diagram for the photochemical step and, *on the basis of your diagram*, discuss the efficiency of the reaction on the singlet and triplet manifold. Do the experimental results agree with your analysis?

Q8. The mechanism of photocyclization of α-(*o*-Tolyl)acetophenones has been explored: (a) Wagner, P. J.; Meador, M. A.; Zhou, B.; Park, B.-S. *J. Am. Chem. Soc.*, **1991**, *113*, 9630. (b) Wagner, P. J; Zhou, B.; Hasegawa, T.; Ward, D. L. *J. Am. Chem. Soc.*, **1991**, *113*, 9640. Develop a Dauben–Salem–Turro-type state correlation diagram for the photochemical step and, *on the basis of your diagram*, discuss the efficiency of the reaction on the singlet and triplet manifold. Do the experimental results agree with your analysis?

MISCELLANEOUS EXERCISES

Q1. Using Orbital Interaction theory wherever appropriate, discuss the chemistry of the diazenium dication of 2,7-diazatetracyclo[6.2.2.2^{3,6}0^{2,7}]tetradecane **1**$^{2+}$. (See Nelsen, S. F. and Wang, Y. *J. Am. Chem. Soc.*, **1991**, *113*, 5905).

Q2. Discuss the mechanism of action of brewer's yeast pyruvate decarboxylase using orbital interaction theory wherever appropriate. For the mechanism, see Zeng, X.; Chung, A.; Haran, M.; Jordan, F. *J. Am. Chem. Soc.*, **1991**, *113*, 5842.

MISCELLANEOUS EXERCISES 275

Q3. Allylic acetals rearrange under acid catalysis to produce tetrahydrofurans. The mechanism has been shown to involve a cyclization step followed by a pinacol rearrangement (Hopkins, M. H.; Overman, L.

E.; Rishton, G. M. *J. Am. Chem. Soc.*, **1991**, *113*, 5354; see also the following two papers, pp. 5365 and 5378). Provide a detailed mechanism for the reaction and discuss features of the steps and intermediates from the point of view of orbital interactions.

Q4. The mechanism of oxidation of amines by hydrogen peroxide has been investigated theoretically (Bach, R. D.; Owensby, A. L.; Gonzalez, C.; Schlegel, H. B.; McDouall, J. J. W. *J. Am. Chem. Soc.*, **1991**, *113*, 6001) and has been shown to be analogous to an S_N2 attack by nitrogen on the O—O bond with simultaneous transfer of hydrogen. Compare the *ab initio* results of Bach, *et al.* to expectations based on orbital interaction theory.

Q5. The mechanism of epoxidation of olefins by peroxyacids has been probed by an "endocyclic restriction test" (Woods, K. W.; Beak, P.

J. Am. Chem. Soc., **1991**, *113*, 6281). It was established that the oxygen transfer takes place via an S_N2-like transition state **A,** rather than a pathway which resembles a 1,3-dipolar addition **B.** Comment on the relative merits of the two transition states using principles of orbital interaction theory. A *spiro* structure similar to **A** for the transition state has been located by *ab initio* calculations (Bach, R. D.; Owensby, A. L.; Gonzalez, C.; Schlegel, H. B.; McDouall, J. J. W. *J. Am. Chem. Soc.*, **1991**, *113*, 2338).

Q6. The photochemical elimination of H_2 from 1,4-cyclohexadiene has been shown experimentally to proceed through a transition state with

C_{2v} or C_2 symmetry (Cromwell, E. F.; Liu, D.-J.; Vrakking, M. J. J.; Kung, A. H.; Lee, Y. T. *J. Chem. Phys.*, **1991**, *95*, 297). Decide on the basis of orbital and state correlation diagrams whether or not the reaction is photolytically allowed.

Q7. Predict the structure of the van der Waals complex between ozone and acetylene. Indicate which frontier orbital interactions should be most important. How does your prediction compare with the experimental structure (Gillies, J. Z.; Gillies, C. W.; Lovas, F. J.; Matsumura, K.; Suenram, R. D.; Kraka, E.; Cremer, D. *J. Am. Chem. Soc.*, **1991**, *113*, 6408)?

Q8. The laboratory preparation of azulene (a) Lemal, D. M. and Goldman, G. D. *J. Chem. Ed.*, **1988**, *65*, 923; (b) Brieger. G. *J. Chem. Ed.*, **1992**, *69*, A262) involves as a key reagent, dimethylaminofulvene, *1* (R = NMe$_2$), which is used it in a reaction with thiophene-

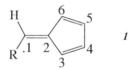

1,1-dioxide to assemble the final azulene skeleton. Unsubstituted fulvene, *1* (R = H), is a very reactive compound. It forms a *dimer* via a cycloaddition reaction in a manner entirely analogous to the reaction between dimethylaminofulvene and thiophene-1,1-dioxide.

(a) Show the most likely product of a reaction of fulvene with itself (i.e., the dimerization). *Hint:* Postulate a transition state for the reaction based on the most favorable interaction of the frontier orbitals, and from it deduce the structure of the most likely product.

(b) Develop a *two*-orbital interaction diagram for the amino-substituted fulvene, *1* (R = NMe$_2$) (for the purposes of the diagram you may treat the methyl groups as if they were H atoms). Show clearly the relative positions of the initial and final orbitals (recall that for a tricoordinated nitrogen atoms, $\alpha_N = \alpha - 1.37|\beta|$), and sketch the initial and final orbitals in the correct orientation.

(c) Use your orbital interaction diagram to discuss the effect of the dimethylamino group on the reactivity of the fulvene.

Q9. "Y" conjugation is often discussed in the literature as a different kind of "aromaticity" because of the prevalence of structures such as NO$_3^-$, CO$_3^{2-}$, and urea [NH$_2$C(O)NH$_2$]. Develop a bonding scheme for the π orbitals of *5*, in which A and B can each donate one *p* orbital to the π system, and A is less electronegative than B. Discuss the optimum number of electrons (2, 4, or 6). Why it is desirable that A is the less electronegative element or group? Do BF$_3$ and *tert*-butyl

$$B-A\overset{B}{\underset{B}{\diagup}} \quad 5$$

cation fit the pattern? *Hint*: The compound has three-fold symmetry and some MOs will be degenerate.

Q10. Use orbital interaction diagrams to explain each of the following (show the orbitals clearly):

(a) Describe the N—N bond in hydrazinium dichloride, $H_3N^+—N^+H_3 \cdot 2Cl^-$.

(b) Account for the fact that *trans*-2,5-dichloro-1,4-dioxane **7** preferentially adopts the diaxial conformation (Koritsánszky, T.; Strumpel, M. K.; Buschmann, J.; Luger, P.; Hansen, N. K.; Pichon-Pesme, V. *J. Am. Chem. Soc.*, **1991**, *113*, 9148). *Hint*: This is an example of the anomeric effect in operation.

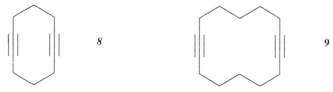

dieq-7   *diax*-7

(c) Offer an explanation for the observation that the lowest ionization potential for the diacetylene **8** (I.P = 8.47 eV) is lower than the corresponding I.P of **9** (I.P = 9.13 eV) (Gleiter, R.; Kratz, D.; Schäfer, W.; Schehlmann, V. *J. Am. Chem. Soc.*, **1991**, *113*, 9258).

(d) Show why C—H bonds next to a carbocationic center are especially acidic, and why proton abstraction yields olefins.

$R_2C^+—CHR_2 + B: \rightarrow$
$\qquad\qquad R_2C=CHR_2 + {}^+B—H$ (B: a weak base)

Q11. The mechanism for the oxidative photofragmentation of α,β-amino alcohols is consistent with a preference for *anti* geometry in the cleavage step (Ci, X.; Kellett, M. A.; Whitten, D. G. *J. Am. Chem. Soc.*, **1991**, *113*, 3893). Provide a rationalization based on the frontier orbitals of the system.

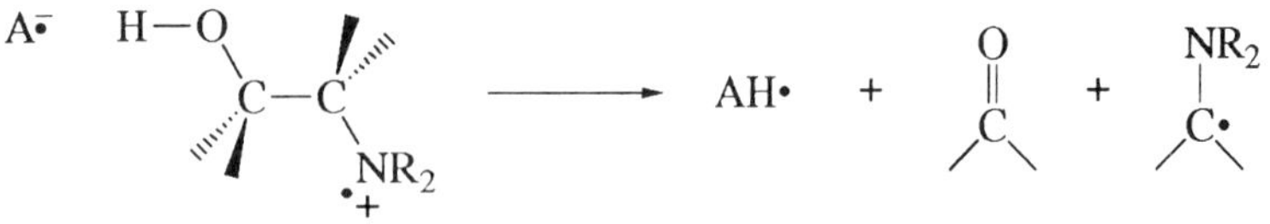

A:

$$\text{H-O}\diagdown\underset{\underset{\overset{+}{\bullet}}{NR_2}}{\overset{}{C-C}} \longrightarrow AH\bullet \;+\; \underset{}{\overset{O}{\underset{}{\|}}}C \;+\; \underset{}{\overset{NR_2}{\underset{}{|}}}C\bullet$$

Q12. Carboxylic acids may be converted to alkyl bromides with the loss of one carbon atom by the Hunsdiecker reaction:

$$R-\overset{O}{\underset{}{\overset{\|}{C}}}-O^- + Ag^+ + Br_2 \xrightarrow{\Delta} R-Br + AgBr + CO_2$$

The mechanism is a free radical chain reaction which is believed to involve the following steps:

Step 1

$$R-\overset{O}{\underset{}{\overset{\|}{C}}}-O^- \;+\; Br_2 \longrightarrow \underset{\mathbf{1}}{R-\overset{O}{\underset{}{\overset{\|}{C}}}-O-Br} \;+\; Br^-$$

Step 2

$$\underset{\mathbf{1}}{R-\overset{O}{\underset{}{\overset{\|}{C}}}-O-Br} \xrightarrow{\Delta} \underset{\mathbf{2}}{R-\overset{O}{\underset{}{\overset{\|}{C}}}-O\bullet} \;+\; \bullet Br$$

Step 3

$$\underset{\mathbf{2}}{R-\overset{O}{\underset{}{\overset{\|}{C}}}-O\bullet} \longrightarrow R\bullet \;+\; CO_2$$

Step 4

$$R\bullet \;+\; \underset{\mathbf{1}}{Br-O-\overset{O}{\underset{}{\overset{\|}{C}}}-R} \longrightarrow R-Br \;+\; \underset{\mathbf{2}}{\bullet O-\overset{O}{\underset{}{\overset{\|}{C}}}-R}$$

The reaction is initiated in Step 2 and propagated by Steps 3 and 4. Analyze each of the steps of the Hunsdiecker reaction in terms of simple orbital interaction theory.

(a) Using a two-orbital interaction, show the interaction which results in the formation of the bromoxy acid, **1**, and displacement of bromide ion in Step 1.

(b) Using an appropriate orbital interaction diagram to describe the O—Br bond of **1**, which ruptures upon heating in Step 2. Why is it feasible that this is the weakest bond in this compound?

(c) Show the electronic structure of the acyloxy free radical, **2**, which is produced in Step 2. It is a σ or π radical? *Hint*: You may wish to construct the MOs of **2** from the interaction of two monocoordinated oxygen atoms, bearing in mind the local symmetry of the carboxylate group.

(d) Use your bonding description from part (b) to explain why the alkyl radical should attack at the Br atom of **1** and not at some other site in Step 4.

ANSWER TO Q12:

(a) This is a simple nucleophilic substitution on Br_2. The Br—Br σ-type interaction is very weak and as a consequence, the LUMO (σ^*_{BrBr}) is low in energy. The interaction is of the two-electron two-orbital type, the other orbital being the HOMO of the carboxylate group, the out-of-phase combination of the in-plane nonbonded orbitals of the oxygen atoms [π_2 will be little lower, see part (c)].

(b) The σ bond of the bromoxy acid is very weak because of the poor intrinsic interaction between a $2p$ orbital and a $4p$ orbital.

(c) The interaction diagram for the carboxylate group is shown in Figure C1.9. Because the in-plane p orbitals (p_σ) overlap more strongly than the out-of-plane orbitals (p_π), the HOMO of the carboxylate anion will be a σ^* orbital, and the neutral carboxylate free radical is predicted to be a σ radical.

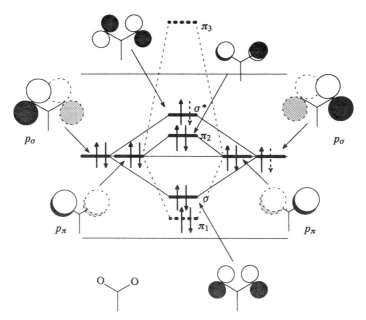

Figure C1.9. The MOs of a carboxylate anion and free radical. The dashed lines indicate the result of interaction with the central carbon $2p$ orbital. Answer to question 12.

(d) In order to understand why the Br atom is abstracted, we must realize that bromine is less electronegative than oxygen. Therefore the LUMO which is the σ^*_{BrO} orbital is polarized toward Br. Since the alkyl radical SOMO is high in energy (i.e., at α), the SOMO–LUMO interaction will be more important than any interaction with the occupied orbitals.

Q13. Reductive decyanations of 2-cyanotetrahydropyran derivatives with sodium in ammonia yield predominantly axially protonated products. The observations are consistent with the reductive decyanation proceeding via a pyramidal, axial radical which accepts a second electron to give a configurationally stable carbanion, which in turn abstracts a proton from ammonia with retention of configuration (Rychnovsky,

$R = i\text{-Pr}, \quad R' = n\text{-C}_5\text{H}_{11}, \quad M = \text{Na or Li}$

S. D.; Powers, J. P.; LePage, T. J. *J. Am. Chem. Soc.*, **1992**, *114*, 8375–8384. Provide an explanation for the axial preference of the intermediate free radical on the basis of orbital interactions. *Hint*: The title of the paper by Rychnovsky *et al.* is *Conformation and Reactivity of Anomeric Radicals.*

Q14. Cyclopentadienyl iodide (5-iodo-1,3-cyclopentadiene) reacts ca. 10 times as rapidly as cyclopentenyl iodide (3-iodocyclopentene) with tetrabutylammonium bromide under the same conditions (Breslow, R.; Canary, J. W. *J. Am. Chem. Soc.*, **1991**, *113*, 3950). Under solvolysis conditons (S_N1), the reactivity order is reversed. Provide a rationalization based on the frontier orbitals of the system. What would you predict for the analogous three-membered and seven-membered ring systems?

Q15. The **Meerwein–Pondorf–Verley** reaction involves transfer of a hydride from the oxygen-substituted carbon atom of an isopropoxide group to a carbonyl group thus effecting the reduction of the carbonyl

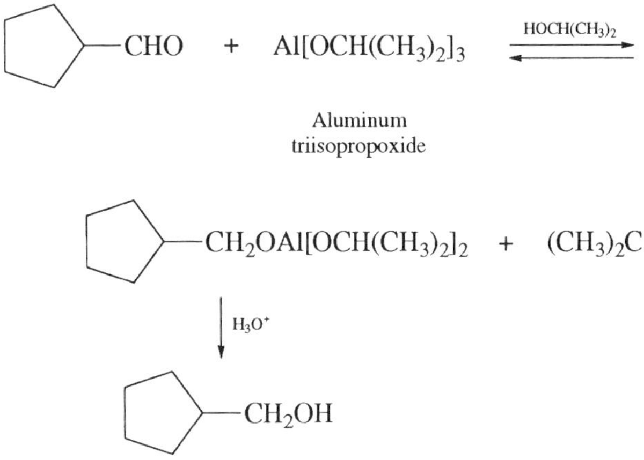

$$\text{cyclopentyl-CHO} + \text{Al[OCH(CH}_3)_2]_3 \xrightleftharpoons{\text{HOCH(CH}_3)_2}$$

Aluminum triisopropoxide

$$\text{cyclopentyl-CH}_2\text{OAl[OCH(CH}_3)_2]_2 + (\text{CH}_3)_2\text{C}=\text{O}$$

↓ H_3O^+

cyclopentyl-CH$_2$OH

compound to an alcohol. Without the aluminum triisopropoxide, the reaction does not proceed. Show the structure of the intermediate in which the hydride transfer occurs and use ideas from orbital interaction theory to discuss the factors which enhance the hydride transfer in this reaction.

ANSWER TO Q15: Three factors combine to make this reaction easy: (1) activation of the carbonyl group toward nucleophilic addition as a result of coordination to the Lewis acid (aluminum triisopropoxide), as discussed in Chapter 8; (2) activation of the secondary C—H bond as a σ donor by the presence of the very good X: substituent (—O—Al, which resembles —O$^-$), as discussed in Chapter 4; (3) *opportunity* presented by the coordination within the complex shown in Figure C1.10

Q16. An *exciplex* is a complex between two molecules, one of which has been photoexcited, i.e., A* · · · B. Show, using orbital interaction diagrams, the possible bonding in an exciplex. An *excimer* is an exciplex of the type, A* · · · A. In general, would you expect stronger or weaker bonding in an excimer? Explain. [*Of interest:* In a recent study of the Diels–Alder reaction, it was found that some enantioselectivity could be achieved by tying up one of the two prochiral (enantiotopic) faces of the dienophile, a *trans*-substituted alkene, as an

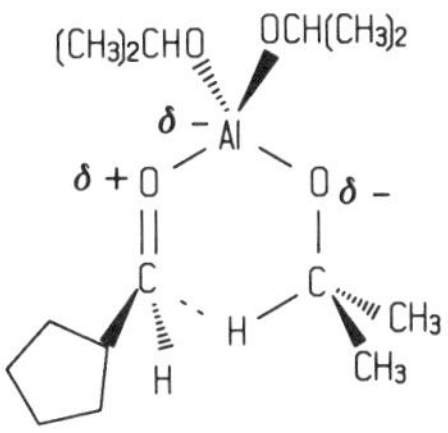

Figure C1.10. Intermediate in the Meerwein–Pondorf–Verley reaction. Answer to question 15.

exciplex with an optically active sensitizer, a twisted binaphthyl (Kim, J.-I. and Schuster, G. B. *J. Am. Chem. Soc.*, **1992**, *114*, 9309–9317).]

ANSWER TO Q16: The interaction between two ground-state molecules in close approach is shown schematically in Figure C1.11(a). The energy interaction is dominated by the four-electron two-orbital case which is repulsive. The interaction of the LUMO's which may be large, has no consequence to the energy, *unless* one of the molecules is photoexcited. In this case, shown in Figure C1.11(b), the LUMO–LUMO, as well as the HOMO–HOMO interactions become attractive. The overall energy gain depends on the energy separations and the extents of orbital overlap in the two interactions. A perfect match in nodal characteristics and energies is achieved when the two molecules are identical [Figure C1.11(c)]. Therefore, an *excimer* would be expected to be more stable in general than an *exciplex*.

Q17. Use orbital interaction theory to develop the π orbitals of the 2-oxaallyl system, $R_2C-O-CR_2$, also known as a carbonyl ylide. Show why 2-oxaallyl readily reacts with alkenes and alkynes in a 4 + 2 cycloaddition reaction (an example may be found in El-Saidi, M.; Kassam, K.; Pole, D. L.; Tadey, T.; Warkentin, J. *J. Am. Chem. Soc.*, **1992**, *114*, 8751–8752).

Q18. The photochemistry of previtamin D_3 has been intensively studied (for leading references, see Dauben, W. G.; Disanayaka, B.; Funhoff, D. J. H.; Kohler, B. E.; Schilke, D. E.; Zhou, B. *J. Am. Chem. Soc.*, **1991**, *113*, 8367, and Enas, J. D.; Shen, G.-Y. Okamura, W. H. *J. Am. Chem. Soc.*, **1991**, *113*, 3873). The thermal and photo-

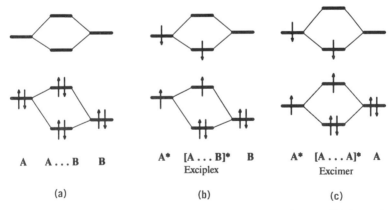

Figure C1.11. (a) Interaction of two molecules in their ground state; (b) three-electron and one-electron bonding in an exciplex; (c) bonding in an excimer. Answer to question 16.

Figure Cl.12. Photoreactions of previtamin D_3, question 18.

reactions are summarized in Figure C1.12. Discuss the various conversions using the descriptive terminology of pericyclic reactions.

Q19. Cyclobutadiene has been shown to have a rectangular geometry by competitive trapping of the two valence tautomeric 1,2-dideuteriocyclobutadienes using methyl 3-cyanoacrylate in a Diels–Alder reaction [(a) Whitman, D. W. and Carpenter, B. K. *J. Am. Chem. Soc.*, **1982,** *104*, 6473–6474. (b) Whitman, D. W. and Carpenter, B. K. *J. Am. Chem. Soc.*, **1980,** *102*, 4272–4274]. Show the products expected from the Diels–Alder reactions. Use orbital interaction theory to develop a bonding scheme for rectangular cyclobutadiene and explain why rectangular cyclobutadiene may be exceptionally reactive as a diene in Diels–Alder reactions.

REFERENCES

1. Fukui, K. *Acc. Chem. Res.*, **1971**, *4*, 57.
2. Fukui K. *Angew. Chem. Int. Ed. Engl.*, **1982**, *21*, 801.
3. (a) Woodward, R. B., and Hoffmann, R. *The Conservation of Orbital Symmetry*, Verlag Chemie, Weinheim, 1970. (b) Woodward, R. B. and Hoffmann, R. *Angew. Chem. Int. Ed. Engl.*, **1969**, *8*, 781. (c) Woodward, R. B., and Hoffmann, R. *Acc. Chem. Res.*, **1968**, *1*, 17.
4. Klopman, G. *J. Am. Chem. Soc.*, **1968**, *90*, 223.
5. Salem, L. *J. Am. Chem. Soc.*, **1968**, *90*, 543, 553.
6. Hoffmann, R. *Acc. Chem. Res.*, **1971**, *4*, 1-9.
7. Fleming, I. *Frontier Orbitals and Organic Chemical Reactions*, John Wiley & Sons, New York, NY, 1976.
8. (a) Shaik, S. S. *J. Am. Chem. Soc.*, **1981**, *103*, 3692. (b) Pross, A. and Shaik, S. S. *Acc. Chem. Res.*, **1983**, *16*, 363. (c) Shaik, S. S. *Prog. Phys. Org. Chem.*, **1985**, *15*, 197. (d) Pross, A. *Adv. Phys. Org. Chem.*, **1985**, *21*, 99.
9. (a) Halevi, E. A. *Helv. Chim. Acta*, **1975**, *58*, 2136. (b) Halevi, E. A. *Angew. Chem. Int. Ed. Engl.*, **1976**, *15*, 593. (c) Halevi, E. A. *Nouv. J. Chim.*, **1976**, *15*, 593. (d) Katriel, J. and Halevi, E. A. *Theoret. Chim. Acta*, **1975**, *40*, 1. (e) Bachler, V. and Halevi, E. A. *Theoret. Chim. Acta*, **1981**, *59*, 595. (f) Halevi, E. A. and Thiel. W. *J. Photochem.*, **1985**, *28*, 373. (g) Halevi, E. A. and Rom, R. *Israel J. Chem.*, **1989**, *29*, 311.
10. Halevi, E. A. *Orbital Symmetry and Reaction Mechanism: The OCAMS Way*, Springer-Verlag, New York, NY, 1992.
11. Dauben, W. G.; Salem, L.; Turro, N. J. *Acc. Chem. Res.*, **1975**, *8*, 41.
12. Michl, J. and Bonacic-Koutecky, V. *Electronic Aspects of Organic Photochemistry*, John Wiley & Sons, New York, NY, 1990.
13. Turro, N. J. *Modern Molecular Photochemistry*, University Science Books, Mill Valley, CA, 1991.
14. Gilbert, A. and Baggott, J. E. *Essentials of Molecular Photochemistry*, CRC Press, Inc., Boca Raton, FL, 1991.
15. Many of the ideas and examples in Chapter 1 have been extracted from notes from a course given by Kurt Mislow to a class of Princeton graduate students in 1968.
16. Jaffe, H. H.; Orchin, M. *Symmetry in Chemistry*, John Wiley & Sons, New York, NY, 1965.
17. (a) Masamune, S.; Kennedy, R. M.; Petersen, J. S.; Houk, K. N.; Wu, Y.-D. *J. Am. Chem. Soc.*, **1986**, *108*, 7404-7405. (b) Imai, T.; Tamura, T.; Yamamuro, A.; Sato, T.; Wollmann, T. A.; Kennedy, R.M.; Masamune, S. *J. Am. Chem. Soc.*, **1986**, *108*, 7402-7404.
18. Luger, P.; Buschmann, J.; McMullan, R.K.; Ruble, J.R.; Matias, P.; Jeffrey, G. A. *J. Am. Chem. Soc.*, **1986**, *108*, 7825.
19. Stair, P. C. *J. Am. Chem. Soc.*, **1982**, *104*, 4044.
20. Kuck, D. and Bögge, H. *J. Am. Chem. Soc.*, **1986**, *108*, 8107.

21. de Boer, J. A. A.; Reinhoudt, D. N.; Harkema, S.; van Hummel, G. J.; de Jong, F. *J. Am. Chem. Soc.*, **1982**, *104*, 4073.
22. Eaton, P. E. and Cole, T. W. Jr. *J. Am. Chem. Soc.*, **1964**, *86*, 3157.
23. Paquette, L. A.; Ternansky, R. J.; Balogh, D. W.; Kentgen, G. *J. Am. Chem. Soc.*, **1983**, *105*, 5446.
24. Shanzer, A.; Libman, J.; Gottlieb, H.; Frolow, F. *J. Am. Chem. Soc.*, **1982**, *104*, 4220.
25. For a review, see Buda, A. B.; Auf der Heyde, T.; Mislow, K. *Angew. Chem. Int. Ed. Engl.*, **1992**, *31*, 989–1007.
26. Zabrodsky, H.; Peleg, S.; Avnir, D. *J. Am. Chem. Soc.*, **1992**, *114*, 7843–7851.
27. Marckwald, W. *Ber.*, **1904**, *37*, 349.
28. Mislow, K. *Introduction to Stereochemistry*, W. A. Benjamin, Inc., New York, NY, 1965.
29. (a) Mislow, K. *Bull. Soc. Chim. Belg.*, **1977**, *86*, 595. (b) Mislow, K. and Siegel, J. *J. Am. Chem. Soc.*, **1984**, *106*, 3319.
30. Pirkle, W. H. and Beare, S. D. *J. Am. Chem. Soc.*, **1969**, *91*, 5150, and references therein.
31. Hofer, O. *Top. Stereochem.*, **1976**, *9*, 111.
32. Takeuchi, Y.; Itoh, N.; Note, H.; Koizumi, T.; Yamaguchi, K. *J. Am. Chem. Soc.*, **1991**, *113*, 6318.
33. Gaussian 92, Revision A.; Frisch, M. J.; Trucks, G. W.; Head-Gordon, M.; Gill, P. M. W.; Wong, M. W.; Foresman, J. B.; Johnson, B. G.; Schlegel, H. B.; Robb, M. A.; Repogle, E. S.; Gomperts, R.; Andres, J. L.; Raghavachari, K.; Binkley, J. S.; Gonzalez, C.; Martin, R. L.; Fox, D. J.; Defrees, D. J.; Baker, J.; Stewart, J. J. P.; Pople, J. A.; Gaussian, Inc., Pittsburgh, PA, 1992.
34. Szabo, A. and Ostlund, N. S. *Modern Quantum Chemistry: Introduction to Advanced Electronic Structure Theory*, Macmillan Publishing Co., Inc., New York, NY, 1982, p. 192.
35. Hariharan, P. C. and Pople, J. A. *Theoret. Chim. Acta*, **1973**, *28*, 213.
36. Allen, L. C. *J. Am. Chem. Soc.*, **1992**, *114*, 1510.
37. Allen, L. C. *J. Am. Chem. Soc.*, **1989**, *111*, 9003.
38. Reed, L. H. and Allen, L. C. *J. Phys. Chem.*, **1992**, *96*, 157.
39. (a) Pauling, L. *J. Am. Chem. Soc.*, **1932**, *54*, 3570. (b) Allred, A. L. *J. Inorg. Nucl. Chem.*, **1961**, *17*, 215.
40. Allred, A. L. and Rochow, E. G. *J. Inorg. Nucl. Chem.*, **1957**, *5*, 264.
41. For an additional discussion of electronegativities and an extensive list of references, see Luo, Y.-R. and Benson, S. W. *Acc. Chem. Res.*, **1992**, *25*, 375.
42. Allen, L. C.; Egolf, D. A.; Knight, E. T.; Liang, C. *J. Phys. Chem.*, **1990**, *94*, 5602.
43. Schlegel, H. B. *J. Comput. Chem.*, **1982**, *3*, 214.
44. Hehre, W. J.; Radom, L.; Schleyer, P. v. R.; Pople, J. A. *Ab Initio Molecular Orbital Theory*, John Wiley & Sons, New York, NY, 1986.
45. JANAF Thermochemical Tables, 2nd ed.; Natl. Stand. Ref. Data Ser. NSRDS-NBS 31. U.S. Government Printing Office, Washington, DC, 1971.
46. Langhoff, S. R. and Davidson, E. R. *Int. J. Quantum Chem.*, **1974**, *8*, 61.

47. An *ab initio* form of PMO theory has been developed: (a) Bernardi, F. and Bottoni, A. *Theor. Chim. Acta*, **1981**, *58*, 245. (b) Bernardi, F. and Bottoni, A. In *Computational Theoretical Organic Chemistry*, Csizmadia, I. G., Daudel, R., eds., Reidel Publishing Co., Dordrecht, 1981; pp. 197–231. (c) Bernardi, F.; Bottoni, A.; Fossey, J.; Sorba, J. *Tetrahedron*, **1986**, *42*, 5567–5580.

48. Hoffmann, R. *Acc. Chem. Res.*, **1971**, *4*, 1.

49. Albright, T. A.; Burdett, J. K.; Whangbo, M.-H. *Orbital Interactions in Chemistry*, Wiley–Interscience, New York, NY. 1985.

50. Wolfsberg, M. and Helmholz, L. *J. Chem. Phys.*, **1952**, *20*, 837.

51. Hoffmann, R. *J. Chem. Phys.*, **1963**, *39*, 1397.

52. Woolley, R. G. *Nouv. J. Chim.*, **1981**, *5*, 219, 227.

53. (a) Bischof, P.; Hashmall, J. A.; Heilbronner, E.; Hornung, V. *Helv. Chim. Acta*, **1969**, *52*, 1745. (b) Haselbach, E.; Heilbronner, E.; Schroder, G. *Helv. Chim. Acta*, **1971**, *54*, 153.

54. Masclet, P.; Grosjean, D.; Mouvier, G.; Dubois, J. *J. Electron Spectrosc. Relat. Phenom.*, **1973**, *2*, 225.

55. Robin, M. B. *Higher Excited States of Polyatomic Molecules*, Academic Press, New York, NY, 1975, vols. I, II, and III.

56. Kim, E. K. and Kochi, J. K. *J. Am. Chem. Soc.*, **1991**, *113*, 4962.

57. Bingham, R. C. *J. Am. Chem. Soc.*, **1975**, *97*, 6743.

58. Chen, Y.; Tschuikow-Roux, E.; Rauk, A. *J. Phys. Chem.*, **1991**, *95*, 9832.

59. Tachikawa, H. and Ogasawara, M. *J. Phys. Chem.*, **1990**, *94*, 1746.

60. Barnett, R. N.; Landman, U.; Dhar, S.; Kestner, N. R.; Jortner, J.; Nitzan, A. *J. Chem. Phys.*, **1989**, *91*, 7797.

61. Van der Waals forces are very complex and manifest themselves even at distances at which it is unreasonable to assume that orbital interactions can occur. An explanation due to London in terms of the mutual attraction of induced dipoles (dispersion forces) accounts for the long range behavior. The unoccupied–occupied orbital interactions will be the dominant component of van der Waals forces at short range. See Kauzmann, W., *Quantum Chemistry*, Academic Press, New York, NY, 1957, chapter 13, for a discussion of dispersion forces.

62. Gimarc, B. M. *Acc. Chem. Res.*, **1974**, *7*, 384.

63. Walsh, A. D. *J. Chem. Soc.*, **1953**, 2260, and papers immediately following.

64. (a) Peyerimhoff, S. D.; Buenker, R. J.; Allen, L. C. *J. Chem. Phys.*, **1966**, *45*, 734. (b) Allen, L. C. and Russell, J. D. *J. Chem. Phys.*, **1967**, *46*, 1029.

65. (a) Ritchie, C. D. *J. Am. Chem. Soc.*, **1983**, *105*, 7313–7318. (b) Ritchie, C. D. In *Nucleophilicity*, Harris, J. M. and McManus, S. P., eds, Advances in Chemistry Series, American Chemical Society, Washington, DC, 1986.

66. Armentrout, P. B. and Simons, J. *J. Am. Chem. Soc.*, **1992**, *114*, 8627–8633.

67. Shaik, S. S. *J. Org. Chem.*, **1987**, *52*, 1563–1568.

68. (a) Schmidt, H. and Schweig, A. *Tetrahedron Lett.*, **1973**, 981. (b) Schmidt, H. and Schweig, A. *Angew. Chem. Int. Ed. Engl.*, **1973**, *12*, 307.

69. (a) Benson, S. W. *J. Chem. Ed.*, **1965**, *42*, 502. (b) Kerr, J. A. *Chem. Rev.*, **1966**, *66*, 465.

70. Walsh, A. D. *Trans. Faraday Soc.*, **1949**, *45*, 179.
71. Brundle, C. R.; Robin, M. B.; Kuebler, N. A.; Basch, H. *J. Am. Chem. Soc.*, **1972**, *94*, 1451.
72. Kobayashi, T.; Miki, S.; Yoshida, Z.; Asako, Y.; Kajimoto, C. *J. Am. Chem. Soc.*, **1988**, *110*, 5622.
73. Honegger, E.; Yang, Z. Z.; Heilbronner, E.; Alder, R. W.; Moss, R. E.; Sessions, R. B. *J. Electron Spectrosc. Rel. Phenom.*, **1985**, *36*, 297.
74. Kimura, K.; Katsumata, S.; Achiba, Y.; Yamazaki, T.; Iwata, S. *Handbook of He I Photoelectron Spectra of Fundamental Organic Molecules*, Halsted, New York, NY, 1981.
75. (a) Badger, R. M. *J. Chem. Phys.*, **1934**, *2*, 128. (b) Badger, R. M. *J. Chem. Phys.*, **1935**, *3*, 710. (c) Gordy, W. *J. Chem. Phys.*, **1946**, *14*, 305.
76. (a) Mack, H. G.; Christen, D.; Oberhammer, H. *J. Mol. Struct.*, **1988**, *190*, 215. (b) Christen, D.; Gupta, O. D.; Kadel, J.; Kirchmeier, R. L.; Mack, H. G.; Oberhammer, H.; Shreeve, J. M. *J. Am. Chem. Soc.*, **1991**, *113*, 9131.
77. Røggen, I. and Dahl, T. *J. Am. Chem. Soc.*, **1992**, *114*, 511, and references therein.
78. David, S.; Eisenstein, O.; Hehre, W. J.; Salem, L.; Hoffmann, R. *J. Am. Chem. Soc.*, **1973**, *95*, 3806.
79. Laube, T. and Hollenstein, S. *J. Am. Chem. Soc.*, **1992**, *114*, 8812–8817, and references therein.
80. For a review of the structural consequences of hyperconjugation, see Radom, L. *Prog. Theor. Org. Chem.*, **1982**, *3*, 1.
81. Zurawski, B.; Ahlrichs, R.; Kutzelnigg, W. *Chem. Phys. Lett.*, **1973**, *21*, 309.
82. (a) Raghavachari, K.; Haddon, R. C.; Schleyer, P. v. R.; Schaefer, H. F. III *J. Am. Chem. Soc.*, **1983**, *105*, 5915. (b) Yoshimine, M.; McLean, A. D.; Liu, B.; Defrees, D. J. Binkley, J. S. *J. Am. Chem. Soc.*, **1983**, *105*, 6185.
83. (a) Hawthorne, M. F. *Acc. Chem. Res.*, **1968**, *1*, 281. (b) Lipscomb, W. N. *Acc. Chem. Res.*, **1973**, *6*, 257.
84. (a) Whitman, D. W. and Carpenter, B. K. *J. Am. Chem. Soc.*, **1982**, *104*, 6473–6474. (b) Carpenter, B. K. *J. Am. Chem. Soc.*, **1983**, *105*, 1700–1701. (c) Whitman, D. W. and Carpenter, B. K. *J. Am. Chem. Soc.*, **1980**, *102*, 4272–4274.
85. Borden, W. T.; Davidson, E. R.; Hart, P. *J. Am. Chem. Soc.*, **1978**, *100*, 388–392.
86. Van Catledge, F. A. *J. Org. Chem.*, **1980**, *45*, 4801.
87. Mullay, J. *J. Am. Chem. Soc.*, **1985**, *107*, 7271–7275.
88. Boyd, R. J. and Edgecombe, K. E. *J. Am. Chem. Soc.*, **1988**, *110*, 4182–4186.
89. Meot-Ner (Mautner), M. and Sieck, L. W. *J. Am. Chem. Soc.*, **1991**, *113*, 4448.
90. Lias, S. G.; Liebman, J. F.; Levin, R. D. *J. Phys. Chem. Ref. Data*, **1984**, *13*, 695 (proton affinities of 800 compounds).
91. Bailey, W. F.; Nurmi, T. T.; Patricia, J. J.; Wang, W. *J. Am. Chem. Soc.*, **1987**, *109*, 2442–2448.
92. Bailey, W. F.; Khanolkar, A. D.; Gavaskar, K. V. *J. Am. Chem. Soc.*, **1992**, *114*, 8053–8060.
93. Bailey, W. F.; Khanolkar, A. D.; Gavaskar, K. V.; Ovaska, T. V.; Rossi, K.; Thiel, Y.; Wiberg, K. B. *J. Am. Chem. Soc.*, **1991**, *113*, 5720.

94. For a review of electrophilic additions to olefins and acetylenes, see Fahey, R. C. *Topics in Stereochemistry*, **1968**, *3*, 237–342.
95. Rauk, A., and Barriel, J. M. *Chem. Phys.*, **1977**, *25*, 409.
96. Shea, K. J. and Kim, J.-S. *J. Am. Chem. Soc.*, **1992**, *114*, 3044–3051.
97. Traetteberg, M. *Acta Chem. Scand.*, **1975**, *B29*, 29.
98. Wijsman, G. W.; de Wolf, W. H.; Bickelhaupt, F. *J. Am. Chem. Soc.*, **1992**, *114*, 9191–9192.
99. March, J. *Advanced Organic Chemistry: Reactions, Mechanisms and Structure*, 3rd ed., John Wiley & Sons, New York, NY, 1985.
100. Stams, D. A.; Thomas, T. D.; MacLaren, D. C.; Ji, D.; Morton, T. H. *J. Am. Chem. Soc.*, **1990**, *112*, 1427–1434.
101. (a) Creary, X.; Mehrsheikh-Mohammadi, M. E.; Eggers, M. D. *J. Am. Chem. Soc.*, **1987**, *109*, 2435–2442. (b) Creary, X. *Acc. Chem. Res.*, **1985**, *18*, 3–8.
102. Fornarini, S. and Muraglia, V. *J. Am. Chem. Soc.*, **1989**, *111*, 873.
103. (a) Olah, G. A.; Liang, G.; Mateescu, G. D.; Riemenschneider, J. L. *J. Am. Chem. Soc.*, **1973**, *95*, 8698. (b) Saunders, M. and Kates, M. R. *J. Am. Chem. Soc.*, **1983**, *105*, 3571.
104. For a review of gas-phase carbanion chemistry, see Squires, R. R. *Acc. Chem. Res.*, **1992**, *25*, 461–467.
105. Farnham, W. B.; Dixon, D. A.; Calabrese, J. C. *J. Am. Chem. Soc.*, **1988**, *110*, 2607–2611.
106. (a) Boese, R.; Paetzold, P.; Tapper, A.; Ziembinski, R. *Chem. Ber.*, **1989**, *122*, 1057. (b) Olmsted, M. M.; Power, P. P.; Weese, K. J.; Doedens, R. J. *J. Am. Chem. Soc.*, **1987**, *109*, 2541. (c) Locquenghien, K. H. v.; Baceiredo, A.; Boese, R.; Bertrand, G. *J. Am. Chem. Soc.*, **1991**, *113*, 5062.
107. Abell, P. I. In *Free Radicals*, Kochi, J. K., ed., John Wiley & Sons, New York, NY, 1973. vol. II, p. 96.
108. McMillen, D. F. and Golden, D. M. *Ann. Rev. Phys. Chem.*, **1982**, *33*, 493–532, and references therein.
109. Bordwell, F. G.; Cheng, J.-P.; Ji, G.-Z.; Satish, A. V.; Zhang, X. *J. Am. Chem. Soc.*, **1991**, *113*, 9790–9795, and references therein.
110. Sustmann, R. and Korth, H.-G. *Adv. Phys. Org. Chem.*, **1990**, *26*, 131–178.
111. Bordwell, F. G.; Zhang, X. M.; Alnajjar, M. S. *J. Am. Chem. Soc.*, **1992**, *114*, 7623–7629.
112. Jensen, P. and Bunker, P. R. *J. Chem. Phys.*, **1988**, *89*, 1327, and earlier references.
113. Bauschlicher Jr., C. W.; Schaefer, H. F. III; Bagus, P. S. *J. Am. Chem. Soc.*, **1977**, *99*, 7106–7110.
114. Wasserman, E.; Kuck, V. J.; Hutton, R. S.; Anderson, E. D.; Yager, W. A. *J. Chem. Phys.*, **1971**, *54*, 4120.
115. Murray, K. K.; Leopold, D. G.; Miller, T. M.; Lineberger, W. C. *J. Chem. Phys.*, **1988**, *89*, 5442.
116. (a) Koda, S. *Chem. Phys. Lett.*, **1978**, *55*, 353. (b) Koda, S. *Chem. Phys.*, **1982**, *66*, 383.
117. Kim, S.-J.; Hamilton, T. P.; Schaefer, H. F. III *J. Chem. Phys.*, **1991**, *94*, 2063.

118. Russo, N.; Sicilia, E.; Toscano, M. *J. Chem. Phys.*, **1992**, *97*, 5031–5036.
119. Kesselmeyer, M. A.; Sheridan, R. S. *J. Am. Chem. Soc.*, **1986**, *108*, 99.
120. Du, X.-M.; Fan, H.; Goodman, J. L.; Kesselmayer, M. A.; Krogh-Jespersen, K.; La Villa, J. A.; Moss, R. A.; Shen, S.; Sheridan, R. S. *J. Am. Chem. Soc.*, **1990**, *112*, 1920.
121. Arduengo, A. J. III; Harlow, R. L.; Kline, M. *J. Am. Chem. Soc.*, **1991**, *113*, 361.
122. Arduengo, A. J. III; Dias, H. V. R.; Harlow, R. L.; Kline, M. *J. Am. Chem. Soc.*, **1992**, *114*, 5530.
123. McMahon, R. J.; Abelt, C. J.; Chapman, O. L.; Johnson, J. W.; Kreil, C. L.; LeRoux, J.-P.; Mooring, A. M.; West, P. R. *J. Am. Chem. Soc.*, **1987**, *109*, 2456.
124. Maier, G.; Reisenaur, H. P.; Schwab, W.; Carsky, P.; Hess, Jr., B. A.; Schaad, L. J. *J. Am. Chem. Soc.*, **1987**, *109*, 5183.
125. Tomioka, H; Hirai, K.; Tabayashi, K.; Murata, S.; Izawa, Y.; Inagaki, S.; Okajima, T. *J. Am. Chem. Soc.*, **1990**, *112*, 7692.
126. Gosavi, R. K.; Torres, M.; Strausz, O. P. *Can. J. Chem.*, **1991**, *69*, 1630.
127. (a) Suchard, S. N. *Selected Spectroscopic Constants for Selected Heteronuclear Diatomic Molecules*, Air Force Report No. SANSO-TR-74-82, U.S. Government Printing Office, Washington, D.C., 1974. (b) Gilles, A.; Masanet, J.; Vermeil, C. *Chem. Phys. Lett.*, **1974**, *25*, 346.
128. Davis, J. H. and Goddard, W. A. III, *J. Am. Chem. Soc.*, **1977**, *99*, 7111–7121.
129. Back, T. G. and Kerr, R. G. *Can. J. Chem.*, **1982**, *60*, 2711, and references therein.
130. Rauk, A. and Alewood, P. F. *Can. J. Chem.*, **1977**, *55*, 1498, and references therein.
131. Hrovat, D. A.; Waali, E. E.; Borden, W. T. *J. Am. Chem. Soc.*, **1992**, *114*, 8698–8699.
132. Kim, S.-J.; Hamilton, T. P.; Schaefer, H. F. III, *J. Am. Chem. Soc.*, **1992**, *114*, 5349–5355.
133. Travers, M. J.; Cowles, D. C.; Clifford, E. P.; Ellison, G. B. *J. Am. Chem. Soc.*, **1992**, *114*, 8699–8701.
134. For a review of nitrenium ions, see Scott, A. P., Ph.D. Dissertation, The University of New England, Armidale, N.S.W., Australia, 1991.
135. Gibson, S. T.; Greene, J. P.; Berkowitz, J. *J. Chem. Phys.*, **1985**, *83*, 4319.
136. Ford, G. P. and Herman, P. S. *J. Am. Chem. Soc.*, **1989**, *111*, 3987.
137. (a) *Multiple Bonds and Low Coordination in Phosphorous Chemistry*, Regitz, M. and Scherer, O. J. eds.; Georg Thieme Verlag: Stuttgart, 1990 (b) Cowley, A. H. and Kemp, R. A. *Chem. Rev.*, **1985**, *85*, 367.
138. Burford, N.; Parks, T. M.; Royan, B. W.; Borecka, B.; Cameron, T. S.; Richardson, J. F.; Gabe, E. J.; Hynes, R. *J. Am. Chem. Soc.*, **1992**, *114*, 8147–8153.
139. Ford, G. P. and Scribner, J. D. *J. Am. Chem. Soc.*, **1981**, *103*, 4281.
140. (a) Li, Y.; Abramovitch, R. A.; Houk, K. N. *J. Org. Chem.*, **1989**, *54*, 2911. (b) Ohwada, T. and Shudo, K. *J. Am. Chem. Soc.*, **1989**, *111*, 34.
141. Elevation of the HOMO, n_O, of the carbonyl group above $\alpha - |\beta|$ is expected

on the basis of the orbital interactions and the SHMO energy of the $2p$ orbital of a monocoordinated oxygen atom (Table 5.1, Chapter 5). This is consistent with increased basicity of the carbonyl oxygen relative to the π bond of ethylene, but is not consistent with the observed ionization potentials of carbonyls and structurally related alkenes. For instance, the I.P. of formaldehyde is 10.88 eV and that of ethylene is 10.51 eV. It may be that SHMO expectations are in error here, and that the greater basicity of the carbonyl oxygen is simply a reflection of the greater stability of the protonated carbonyl intemediate as compared to the protonated alkene, as discussed in Chapter 7.

142. Rauk, A. unpublished results.
143. For a study of hydride reductions, see Wu, Y.-D.; Tucker, J. A.; Houk, K. N. *J. Am. Chem. Soc.*, **1991**, *113*, 5018, and references therein.
144. Liebman, J. F. and Pollack, R. M. *J. Org. Chem.*, **1973**, *38*, 3444.
145. Hoz, S. *J. Org. Chem.*, **1982**, *47*, 3545.
146. Shustov, G. V. *Acad. Sci. U.S.S.R. Proceedings Chemistry*, **1985**, *280*, 80.
147. (a) Cieplak, A. S. *J. Am. Chem. Soc.*, **1981**, *103*, 4540-4552. (b) Cieplak, A. S.; Tait, B. D.; Johnson, C. R. *J. Am. Chem. Soc.*, **1989**, *111*, 8447-8462.
148. (a) Wu, Y.-D. and Houk, K. N. *J. Am. Chem. Soc.*, **1987**, *109*, 908-910. (b) Houk, K. N. and Wu, Y.-D. In *Stereochemistry of Organic and Bioorganic Transformations*, Bartmann, W. and Sharpless, K. B., eds., VCH, Weinheim, Germany, 1987, pp. 247-260.
149. (a) Klein, J. *Tetrahedron* **1974**, *30*, 3349-3353. (b) Klein, J. *Tetrahedron Lett.*, **1973**, 4307-4310.
150. Chérest, M.; Felkin, H.; Prudent, N. *Tetrahedron Lett.*, **1968**, 2199-2204. (b) Chérest, M. and Felkin, H. *Tetrahedron Lett.*, **1968**, 2205-2208.
151. (a) Anh, N. T. and Eistenstein, O. *Nouv. J. Chim.*, **1977**, *1*, 61-70. (b) Anh, N. T. *Top. Curr. Chem.*, **1980**, *88*, 145-162.
152. Li, Y.; Garrell, R. L.; Houk, K. N. *J. Am. Chem. Soc.*, **1991**, *113*, 5895.
153. Involvement of the N lone pair in "resonance" with the carbonyl group has been a subject of some controversy. For a review, see Bennet, A. J.; Somayaji, V.; Brown, R. S.; Santarsiero, B. D. *J. Am. Chem. Soc.*, **1991**, *113*, 7563.
154. Jackman, L. M. In *Dynamic NMR Spectroscopy*, Jackman, L. M. and Cotton, F. A., eds.; Academic Press, New York, NY, 1975, p. 203.
155. For a discussion of the mechanism of acid and base catalyzed hydrolysis of amides, see Brown, R. S.; Bennet, A. J.; Slebocka-Tilk, H. *Acc. Chem. Res.*, **1992**, *25*, 481-488.
156. Wang, Q.-P.; Bennet, A. J.; Brown, R. S.; Santarsiero, B. D. *J. Am. Chem. Soc.*, **1991**, *113*, 5757.
157. For a review of nonplanar and strained amides, see Greenberg, A. In *Structure and Reactivity*, Liebman, J. F. and Greenberg, A., eds.; VCH Publishers, Inc.; New York, NY, 1988, pp. 138-179.
158. Wiberg, K. B.; Hadad, C. M.; Rablen, P. R.; Cioslowski, J. *J. Am. Chem. Soc.*, **1992**, *114*, 8644-8654.
159. Richard, J. P.; Amyes, T. L.; Vontor, T. *J. Am. Chem. Soc.*, **1991**, *113*, 5871, and references therein.
160. Shaik, S. S.; Schlegel, H. B.; Wolfe, S. *Theoretical Aspects of Physical Organic Chemistry: The S_N2 Mechanism*, Wiley-Interscience: New York, NY, 1992.

161. For leading references, see (a) Cyr, D. M.; Posey, L. A.; Bishea, G. A.; Han, C.-C.; Johnson, M. A. *J. Am. Chem. Soc.*, **1991**, *113*, 9697. (b) Wilbur, J. L. and Brauman, J. I. *J. Am. Chem. Soc.*, **1991**, *113*, 9699.

162. Wladkowski, B. D.; Lim, K. F.; Allen, W. D.; Brauman, J. I. *J. Am. Chem. Soc.*, **1992**, *114*, 9136-9153.

163. Abboud, J.-L.; Notario, R.; Bertrán, J.; Taft, R. W. *J. Am. Chem. Soc.*, **1991**, *113*, 4738, and references therein.

164. (a) Edwards, J. O. and Pearson, R. G. *J. Am. Chem. Soc.*, **1962**, *84*, 16. (b) Grekov, A. P. and Veselov, V. Ya. *Usp. Khim.*, **1978**, *47*, 1200.

165. (a) Shaik, S. S. and Pross, A. *J. Am. Chem. Soc.*, **1982**, *104*, 2708. (b) Shaik, S. S. *Nouv. J. Chim.*, **1982**, *6*, 159.

166. Shaik, S. S. *J. Am. Chem. Soc.*, **1983**, *105*, 4358-4367.

167. Shaik, S. S. *J. Am. Chem. Soc.*, **1984**, *106*, 1227.

168. Pimentel, G. C. and McClellan, A. L. *The Hydrogen Bond*, W. H. Freeman and Co., San Francisco, CA, 1960.

169. Herbine, P.; Hu, T. A.; Johnson, G.; Dyke, T. R. *J. Chem. Phys.*, **1990**, *93*, 5485, and references therein.

170. As of this writing there is some controversy about the gas-phase structure of the ammonia dimer (Baum, R. M. *Chem. & Eng. News*, **1992**, October 19, 20-23). The traditional view of the dimer, containing a "linear" hydrogen bond, which is supported by most theoretical and experimental studies, has been questioned by Klemperer and co-workers on the basis of microwave measurements (Klemperer, W.; Nelson, D. D. Jr., Fraser, G. T. *Science*, **1987**, *238*, 1670).

171. Curtiss, L. A.; Frurip, D. J.; Blander, M. *J. Chem. Phys.*, **1979**, *71*, 2703.

172. Van Duijneveldt-van de Rijdt, J. G. C. M. and van Duijneveldt, F. B. *J. Chem. Phys.*, **1992**, *97*, 5019-5030.

173. Perrin, C. L. and Thoburn, J. D. *J. Am. Chem. Soc.*, **1992**, *114*, 8559-8565.

174. Gronert, S. *J. Am. Chem. Soc.*, **1991**, *113*, 6041.

175. Sunner, J. A.; Hirao, K.; Kebarle, P. *J. Phys. Chem.*, **1989**, *93*, 4010.

176. (a) Kirchen, R. P.; Ranganayakulu, K.; Rauk, A.; Singh, B. P.; Sorensen, T. S. *J. Am. Chem. Soc.*, **1981**, *103*, 588. (b) Kirchen, R. P.; Okazawa, N.; Ranganayakulu, K.; Rauk, A.; Sorensen, T. S. *J. Am. Chem. Soc.*, **1981**, *103*, 597. (c) Sorensen, T. S. and Whitworth, S. M. *J. Am. Chem. Soc.*, **1990**, *112*, 8135.

177. Dorigo, A. E.; McCarrick, M. A.; Loncharich, R. J.; Houk, K. N. *J. Am. Chem. Soc.*, **1990**, *112*, 7508-7514. Erratum: *J. Am. Chem. Soc.*, **1991**, *113*, 4368.

178. (a) Tsang, W. *Int. J. Chem. Kinet.*, **1978**, *10*, 821. (b) Tsang, W. *J. Am. Chem. Soc.*, **1985**, *107*, 2872.

179. (a) Russell, J. J.; Seetula, J. A.; Senkan, S. M.; Gutman, D. *Int. J. Chem. Kinet.*, **1988**, *20*, 759. (b) Russell, J. J.; Seetula, J. A.; Timonen, R. S.; Gutman, D.; Nava, D. F. *J. Am. Chem. Soc.*, **1988**, *110*, 3084. (c) Russell, J. J.; Seetula, J. A.; Gutman, D. *J. Am. Chem. Soc.*, **1988**, *110*, 3092. (d) Seetula, J. A.; Russell, J. J.; Gutman, D. *J. Am. Chem. Soc.*, **1990**, *112*, 1347.

180. Diederich, F.; Rubin, Y.; Knobler, C. B.; Whetten, R. L.; Schriver, K. E.; Houk, K. N.; Li, Y. *Science*, **1989**, *245*, 1088.

181. (a) Krätschmer, W.; Lamb, L. D.; Fostiropoulos, K.; Huffman, D. R. *Nature*,

1990, *347*, 354. (b) Taylor, R.; Hare, J. P.; Abdul-Sada, A. K.; Kroto, H. W. *J. Chem. Soc. Chem. Commun.*, **1990**, 1423.

182. Parasuk, V.; Almlöf, J.; Feyereisen, M. W. *J. Am. Chem. Soc.*, **1991**, *113*, 1049.
183. For a special edition on Buckminsterfullerenes, see *Acc. Chem. Res.*, **1992**, *25*, 98–175.
184. Haddon, R. C.; Brus, L. E.; Raghavachari, K. *Chem. Phys. Lett.*, **1986**, *125*, 459.
185. Cremer, D.; Reichel, F.; Kraka, E. *J. Am. Chem. Soc.*, **1991**, *113*, 9459.
186. For a review of secondary orbital interactions, see Ginsburg, D. *Tetrahedron*, **1983**, *39*, 2095–2135.
187. Houk, K. N.; Lin, Y.-T.; Brown, F. K. *J. Am. Chem. Soc.*, **1986**, *108*, 554, and references therein.
188. Bernardi, F.; Bottoni, A.; Robb, M. A.; Field, M. J.; Hillier, I. H.; Guest, M. F. *J. Chem. Soc. Chem. Commun.*, **1985**, 1051.
189. Doering, W. von E.; Toscano, V. G.; Beasley, G. H. *Tetrahedron*, **1971**, *27*, 5299–5306.
190. Morokuma, K.; Borden, W. T.; Hrovat, D. A. *J. Am. Chem. Soc.*, **1988**, *110*, 4474.
191. Maluendes, S. A. and Dupuis, M. *J. Chem. Phys.*, **1990**, *93*, 5902.
192. Lee, E.; Shin, I.-J.; Kim, T.-S. *J. Am. Chem. Soc.*, **1990**, *112*, 260.
193. Luger, P.; Buschmann, J.; McMullan, R. K.; Ruble, J. R.; Matias, P.; Jeffrey, G. A. *J. Am. Chem. Soc.*, **1986**, *108*, 7825.
194. Meier, B. H. and Earl, W. L. *J. Am. Chem. Soc.*, **1985**, *107*, 5553.
195. Padwa, A. *1,3-Dipolar Cycloaddition Chemistry*, John Wiley & Sons, New York, NY, 1984; vols. 1 and 2.
196. Samuilov, Y. D. and Konovalov, A. I. *Russ. Chem. Rev.*, **1984**, *53*, 332.
197. Houk, K. N.; Sims, J.; Watts, C. R.; Luskus, L. J. *J. Am. Chem. Soc.*, **1973**, *95*, 7301.
198. Houk, K. N.; Sims, J.; Duke R. E., Jr.; Strozier, R. W.; George, J. K. *J. Am. Chem. Soc.*, **1973**, *95*, 7287.
199. Bridson-Jones, F. S.; Buckley, G. D.; Cross, L. H.; Driver, A. P. *J. Chem. Soc.*, **1951**, 2999.
200. Herzberg, G. *Molecular Spectra and Molecular Structure III. Electronic Spectra and Electronic Structure of Polyatomic Molecules*, D. Van Nostrand Co., Toronto, Canada, 1967.
201. Fowler, J. E.; Alberts, I. L.; Schaefer, H. F. III, *J. Am. Chem. Soc.*, **1991**, *113*, 4768, and references therein.
202. Yamazaki, H.; Cvetanovic, R. J.; Irwin, R. S. *J. Am. Chem. Soc.*, **1976**, *98*, 2198.
203. Ikezawa, H.; Kutal, C.; Yasufuku, K.; Yamazaki, H. *J. Am. Chem. Soc.*, **1986**, *108*, 1589–1594, and references therein.
204. (a) Clark, K. B. and Leigh, W. J. *J. Am. Chem. Soc.*, **1987**, *109*, 6086. (b) Leigh, W. J.; Zheng, K.; Clark, K. B. *Can. J. Chem.*, **1990**, *68*, 1988.
205. Leigh, W. J. and Zheng, K. *J. Am. Chem. Soc.*, **1991**, *113*, 4019; Erratum *J. Am. Chem. Soc.*, **1992**, *114*, 796.

206. Bernardi, F.; Olivucci, M.; Ragazos, I. N.; Robb, M. A. *J. Am. Chem. Soc.*, **1992**, *114*, 2752–2754.

207. Trulson, M. O.; Dollinger, G. D.; Mathies, R. A. *J. Am. Chem. Soc.*, **1987**, *109*, 586–587.

208. Martin, H.-D.; Urbanek, T.; Pfohler, P.; Walsh, R. *J. Chem. Soc. Chem. Commun.*, **1985**, 964–965.

209. (a) Hoffmann, R. *J. Am. Chem. Soc.*, **1968**, *90*, 1475. (b) Fueno, T.; Nagase, S.; Tatsumi, K.; Yamaguchi, K. *Theor. Chim. Acta*, **1972**, *26*, 43.

210. (a) Houk, K. N.; Rondan, N. G.; Mareda, J. *J. Am. Chem. Soc.*, **1984**, *106*, 4291. (b) Houk, K. N.; Rondan, N. G. *J. Am. Chem. Soc.*, **1984**, *106*, 4293.

211. Dobson, R. C.; Hayes, D. M.; Hoffmann, R. *J. Am. Chem. Soc.*, **1971**, *93*, 6188.

212. Hoffmann, R.; Gleiter, R.; Mallory, F. B. *J. Am. Chem. Soc.*, **1970**, *92*, 1460.

213. See, for example, Zaklika, K. A.; Thayer, A. L.; Schaap, A. P. *J. Am. Chem. Soc.*, **1978**, *100*, 4916–4918.

214. Dewar, M. J. S. and Kirschner, S. *J. Am. Chem. Soc.*, **1974**, *96*, 7578–7579, and references therein.

215. Calvert, J. G. and Pitts J. N., Jr. *Photochemistry*, John Wiley & Sons, New York, NY, 1966, p. 280.

216. Frisch, M. J. *Gaussian 92 User's Guide/Programmer's Reference*, Gaussian, Inc., 4415 Fifth Avenue, Pittsburgh, PA, 15213, pp. 25, 71–74, 1990. Gaussian 92 is a registered trademark of GAUSSIAN, INC. The material in Appendix A is copyrighted by GAUSSIAN, INC. and is used with permission.

INDEX

Absorption, of photon, 215
  time scale, 216
Acetaldehyde, 122
  properties, 144
Acetoacetone, pK_a, 163
Acetone, 259
  properties, 144
Acetonitrile, pK_a, 163
Acetophenones:
  α-chloro-o-methyl, 273
  α-(o-tolyl), 274
  photocyclization, 273
Acetylene:
  pK_a, 263
  point group, 4
  van der Waals complex with O_3, 276
Achiral object, 1
Acid anhydrides, hydrolysis, 146
Acid dissociation constant, related to BDE, 133
Acid halides, hydrolysis, 146
Acidity:
  and bond dissociation energy, 133
  interaction diagram, 90
  pK_a:
    acetoacetone, 163
    acetone, 163
    acetonitrile, 163
    acetylene, 163, 263
    cyclopentadiene, 163
    dimethylsulfone, 163
    diphenylmethane, 163
    ethane, 163, 263
    ethylene, 163, 263
    hydrogen fluoride, 162
    methyl acetate, 163
    nitromethane, 163
    propene, 163
    toluene, 163
Acids, reduction, 146
Aconitic acid, 10
Acrolein, properties, 144
Activation energy, negative, 169
Adamantanones, reduction, stereoselection, 260
2-Adamantyl cation, 261
Adenine, 161
Aldol reaction, 122
Alkenes:
  electrophilic addition, 121
  and ionization potential, 124
  nucleophilic attack, 120
Alkyl halides:
  comparison of, 155
  reactions, 262
Alkynes, 125
Allene, point group, 6
Allotropes of C, 171
Allred-Rochow electronegativity, 44
Allyl:
  SHMO, 108, 109, 110
  anion, 110, 201
    HOMO, 122

295

Allyl (*Continued*):
  cation, 110, 201, 256
    acidity relative to CH_3^+, 251
    LUMO, 123, 251
  radical, 110
Alpha, Hückel:
  definition, 108, 114
  effect of coordination, 114
  and hybridization, 117
  as reference energy, 113
Alternant π systems, 110
Aluminum trichloride, 152, 168
Aluminum triisopropoxide, 281
Amide group, interaction diagram, 148, 258
Amides:
  hydrolysis, 146
  rotation barrier, 148
Amine oxides, 251
  bonding, 252
Amines, oxidation by H_2O_2, 275
2-Aminoethanol, 263
Aminomethyl cation, 239
  interaction diagram, 240
Aminonitrene, singlet-triplet separation, 137
Ammonia:
  complexes with halogens, 102, 243
  dimer structure, 292
  geometry, 47
  hydrogen bonding, 162
  I. P., 101
  MOs of, 43
  normal modes, 48
  orbital energies, 42
  vs. phosphine, 242
  point group, 5
  total energy, 41
Anchimeric assistance, 102
Aniline, 173, 178
  SHMO, 173
Anisochronous, 11
Anisole, conformation, 263
Anomeric effect, 75, 103, 277
Anomeric radicals, 280
Antarafacial reaction:
  of π system, 185, 186
  of σ bond, 190
Antisymmetrizer operator, 23
Anthraquinone, 1-methyl, photoenolization, 274
Aromaticity:
  in homotropenylium cation, 171
  4n + 2 rule, 170, 171
  in σ bonded arrays, 171
Arsenium ions, 139
Azapentadiene, 179

Azapentadienyl, SHMO, 176
Azides, alkyl, 137, 194, 196
Aziridines, 196, 201
Azo compounds, 137
Azoxy, 194, 197
Azulene, preparation, 276

Basicity:
  and interaction diagrams, 90
  table of pK_b values, 144
Basis function:
  cartesian gaussian, 35
  polarization, 36–37
  primitive gaussian, 34
  Slater type, 35
Basis functions,
Basis set, 31, 34
  split valence, 36
  STO-3G, 36
  6-31G*, 36, 223
BDE (bond dissociation energy), 133
  related to acid dissociation constant, 133
Benzaldehyde, 173, 174, 178, 179
  SHMO, 173
Benzene:
  basicity, 117
  complex with:
    $CH_2CH_2^+$, 264
    NO^+, 264, 265
  point group, 6
  SHMO, 110, 111, 170
  bond order, 113
Benzenes:
  electrophilic substitution, 172
  substituent effects, 173
  X: substituent, 173
  Z substituent, 174
  C substituent, 175
  nucleophilic substitution, 177
  substituent effects, 178
  Z substituent, 178
  by proton abstraction, 179
Benzophenone, 272
  2,4-di-*t*-butyl, 272, 273
Benzpinacol, 272
Benzvalene, 206
Benzyne, 179, 180, 181, 256, 264
  IR spectrum, 256
  MW spectrum, 256
  orbital interaction, 180
Beta, Hückel:
  definition, 108, 114
  effect of coordination, 114
  and hybridization, 117
  as unit of energy, 113

BF$_3$, 276
BF$_3$ affinity, table of, 144
BH$_3$NH$_3$, σ bond, 243
Bicyclo[2.2.1]hept-2-ene. *See* Norbornene
Bicyclo[2.1.1]hex-2-ene, rearrangement of, 268
Bicyclo[3.1.0]hex-2-ene, 268
Bicyclo[3.1.0]hex-3-en-2-one, 267
Bicyclo[3.1.0]hexan-3-one, 267
Bishomocyclopropenyl cation, 257
Bond:
  σ, as donors and acceptors, 101
  three electron, 242, 254
Bond dissociation energy, (BDE), 98, 133
Bond length, and force constant, 101
Bond lengths, SCF, 47
Bond order, in SHMO theory, 112, 113
Bond Polarity Index, 45, 46
Bond strength, and bond length, 101
Boranes, 104
Boric acid, point group, 5
Born-Oppenheimer, 19, 206
  hypersurface, 46, 213
Boron trifluoride, 152. *See also* BF$_3$
Bound state, 213, 214
BrCl, reaction with alkenes, 243
Brillouin's Theorem, 51
Bromonium cation, 129
Brucine, 10
Buckminsterfullerene (C$_{60}$), 171
  point group, 6
Bullvalene, 4, 192, 193, 194
  point group, 4
  rearrangement, 192
1,3-Butadiene, 200, 202
  in Diels-Alder reaction, 190–191
  SHMO, 108, 109
Butadienes, 187, 202, 205
2-Butene, 204
  (Z)-, I. P., 72, 99
Butenone, 144
t-Butyl carbanion, 129
t-Butyl cation, 276–277
2-Butyne, 124

"C" substituents, 124
C$_{18}$, 171
C$_{60}$, 171
  point group, 6
Cannizzaro, 103, 168
Captodative effect, 257
Carbanion:
  t-butyl, 129
  ethyl, 129
  halomalonate, 262
  methyl, 129
  2-propyl, 129
  α-sulfonyl, 257–258
Carbanions, 129
  C substituted, 130, 131
  X: substituted, 129, 130
  Z substituted, 130
Carbene:
  chloro methoxy, ClCOCH$_3$, 136
  cycloheptatrienylidene, 254
  1,3-di-1-adamantylimidazol-2-ylidene, 136
  dibromo, CBr$_2$, 136
  dichloro, CCl$_2$, 136
  difluoro, CF$_2$, 136
  diiodo, CI$_2$, 136
  dimethoxy, C(OCH$_3$)$_2$, 136
  ethynyl, 136
  fluoro, CHF, 135
  fluoro methoxy, FCOCH$_3$, 136
  formyl, 136
  methoxy methyl, CH$_3$COCH$_3$, 256
  methylene, CH$_2$, 134, 135
    complex with water, 256
    geometry of, 134
    triplet-singlet separation, 134
  phenyl, 136, 254
    bonding, 255
Carbenes, 134
  addition to olefin, 187, 210, 211
  C substituted, 135, 136
  cheletropic reaction, 202
  reaction with carbonyl, 259
  X: substituted, 135
  Z substituted, 135, 136
Carbocation:
  2-adamantyl, 261
  aminomethyl, 239
    interaction diagram, 240
  bishomocyclopropenyl, 257
  cyclopropylcarbinyl, 129, 254
  ethyl, 104, 129, 169
  ethylbenzenium, 257
  norbornyl, 104, 129
  triphenyl, 128, 262
  tropylium, 256
Carbocations, 126
  C substituted, 127
  cyclic H-bridged, 168, 169
  intermolecular reaction, 128
  intramolecular reaction, 128
  in SN1 reaction, 151, 152
  X: substituted, 126
  Z substituted, 127
Carbon, *s* to *p* promotion, 117

Carbonyl compounds. *See also* Ketones
  α-cleavage, 222
  electronic states, 208–209, 220
  Norrish Type II, 218–220
  Norrish Type I, 221–222
  reaction with carbenes, 259
Carbonyl group, 86
  effect of substituents on stability, 149, 150
  orbital interaction diagram, 87, 143
  Lewis acid catalysis, 242
  mass spectrum, 90, 91
  SHMO, 142, 143, 258
  protonated, 143
  electrophilic attack on, 142
  nucleophilic attack on, 143, 146
  electronic states, 208–209, 220
Carbonyl imine, 197
Carbonyl oxide, 197
Carbonyl ylide, 197, 282
Carboranes, 104
Carboxylate anion, 278, 279
Carboxylate free radical, 278, 279
CH bond activation:
  by alkyl halide, 164
  by C and Z substituents, 163, 164, 166, 167
CH_2:
  geometry of, 135
  triplet-singlet separation, 133
Charge:
  distribution and orbital interactions, 71
  in SHMO theory, 112
  population analysis, net, 40, 112
Chemiluminescence, 212
Cheletropic reaction, 184, 187, 202
  stereochemistry, 187
Chiral object, 1
  solvent, 11
Chloro methoxy carbene, 136
Chloroform, point group, 5
Cieplak theory, 147, 261
Citric acid, 10
Cl_2, complex with ammonia, 102, 243
ClBr, reaction with alkenes, 243
ClF, complex with ammonia, 102, 243
Coefficients, 31
Component analysis, 188, 190
  rule for, 190
Configuration, electronic, 209, 217
  symmetry of, 218
Configuration energy, 43
Configuration interaction, 49
Conrotatory, 185
Constitutional isomers, 7
Contracted functions, 36

Cope rearrangement, 184, 192
  orbital interaction, 193
Core hamiltonian, 58
Correlation diagrams. *See also* Orbital correlation diagrams, State correlation diagrams
  general principles, 198
Correlation error, 48
Coulomb integral, α, 113
Coulomb's law, 19
Coulomb repulsion, and Hückel theory, 134
18-Crown-6, 4
Cubane:
  from cyclooctatetraene, 206
  point group, 4, 6
Curtius rearrangement, 138
2-Cyanofuran, SHMO, 248
Cycloaddition reaction, 182–184
  component analysis, 189
  orbital correlation diagram, 199–200
  stereochemistry, 183
Cycloalkanols, 257
Cyclobutadiene, 119, 283
  barrier, 112
  tautomerism, 283
  SHMO, 110, 111, 170
  tetramethyl, 270
Cyclobutane, 199, 200
  orbital correlation diagram, 203
  point group, 6
Cyclobutene, photochemistry of, 201
Cyclobutenes, 201
  pericyclic reactions, 186, 206
Cyclobutyl anion, perfluoro-1-methyl-1-, 255
1,4-Cycloheptadiene, 192, 193, 194
Cycloheptatriene, to tropylium, 262
Cycloheptatrienes, rearrangement of, 267
Cycloheptatrienyl, SHMO, 170
Cycloheptatrienylidene, 254
  bonding, 255
1,3-Cyclohexadiene, 206
  photolysis, 206, 271
  ring opening, 271
1,3-Cyclohexadienes, 268
1,4-Cyclohexadiene, 99
  interaction diagram, 100
  MOs, 100
  photolysis of, 275–276
Cyclohexadienones, photochemistry of, 272
Cyclohexane, point group, 6
Cyclohexanones, 104
  nucleophilic attack on, 146–147
Cyclohexene, 99, 203
  I. P., 73

Cyclooctatetraene, 206
  point group, 6
Cyclooctatrienone, 266–267
Cyclooctene
  cis, I. P., 124
  trans:
    I. P., 204
    geometry of, 205
Cyclopentadiene, 119, 269
  pK_a, 163, 263
Cyclopentadienone, SHMO, 248
Cyclopentadienyl, 254
  SHMO, 110, 111, 170
Cyclopentadienyl anion, point group, 6
Cyclopentadienyl iodide, S_N2 vs. S_N1, 280
Cyclophane, 235
Cyclopropane:
  electronic states, 212
  bonding, 98
  from carbene reactions, 135, 187
  hybridization, 13
  by Michael reaction, 262
  point group, 6
Cyclopropenyl, SHMO, 110, 111, 170
Cyclopropyl aldehyde, 259
Cyclopropyl anion, 201
Cyclopropyl cation, 201
Cyclopropyl cations, 268, 271
Cyclopropyl group, 99, 254
Cyclopropylcarbinyl cation, 129
Cycloreversion, 183
  active components, 185
Cytosine, 161

DABCO:
  1,4-diazabicyclo[2.2.2]octane, 100, 243
  HOMO, 100, 243
  I. P., 100
Dative bond, 74, 75, 243
Dauben-Salem-Turro analysis, 217
  orbital interaction diagram, 217
Davidson correction, 51
Density, matrix, 34
Destabilization, 67
Dewar benzene, hexamethyl-, 235, 270
Dewar benzene, 268, 270
  to benzene, 270
Diastereomers, 8
Diastereoselectivity, 182
Diastereotopic, 8, 234
1,4-Diazabicyclo[2.2.2]octane, 239. *See also* DABCO
Diazenium dications, 274
1,1-Diazines, 137, 138

Diazoalkanes, 194, 196
Diborane, point group, 5
Dibromocarbene, CBr_2, 136
1,2-Dibromoethane, conformation, 243
2,4-Di-*t*-butylbenzophenone, photolysis of, 272
Dichlorocarbene, CCl_2, 136
1,2-Dichloroethane:
  conformation, 243
  point group, 4
1,2-Dichloroethene-(E), point group, 4
Dielectric constant, 81
Diels-Alder reaction, 108, 181, 183, 186, 190, 194, 199
  aromatic transition state, 171
  correlation diagram, 200, 204
  examples of, 265–266, 269–270
  orbital interaction, 191–192
  with cyclobutadiene, 284
  reverse demand, 191
  via triplex, 281
Difluorocarbene, CF_2, 136
1,2-Difluoroethane, conformation, 243
Diiodocarbene, CI_2, 136
Dimethoxycarbene, $C(OCH_3)_2$, 136
Dimethyl ether, properties, 194
Dimethyl fumarate, 132, 133
2,5-Dimethylborolane, 258
Dimethylaminofulvene, 276
2,5-Dimethylcyclopentanone, 236
Dimethylsulfone, pK_a, 163
1,3-Dioxacyclopentanes, 259
1,4-dioxane, 277
Dioxetane:
  correlation diagrams, 211
  pyrolysis, 212
Diphenylmethane, pK_a, 163
Dipolar Cycloadditions, 1,3-dipolar, 194–197
1,3-Dipoles, SHMO orbitals, energies, 195
Dipole moment, 40
Diradical states, 217
Disrotatory, 185
Dissociative state, 214
Dissymmetric, 4
1,5-Dithiacyclooctane:
  bonding, 241
  cation, 242
  dication, 242
Divinylcyclopropane, 192
  rearrangement, 192, 194
Dodecahedrane, 4
  point group, 4, 6

E1, 129, 164, 165
E1cb, 102, 165, 180

E2, 164, 165
 stereoelectronic, 164, 182, 263
Electrocyclic reaction, 182, 184, 199, 201
 component analysis, 189
 orbital correlation diagram, 202
 stereochemistry, 185, 201
Electron transfer, 74
Electronegativity:
 spectroscopic, 43
 table of, 44
Electronic states:
 carbonyl group, 208
 from MOs, 208
Electrophilicity, and interaction diagram, 90
Electrostatic energy, 81
Elements, symmetry, 2–3
Enantiomers, 8
Enantiotopic, 8, 234
 NMR, 11
Endergonic, 173
Energy:
 changes due to orbital interaction, 71
 classical, 19
 configuration, 43
 electronic, in terms of orbital energies, 37, 38
 expectation value, 23
 Hartree-Fock, 27, 38
 one electron, 25
 orbital, 37
 total, 37, 41
  ammonia, 41
  hydrogen fluoride, 41
  methane, 41
  water, 41
 units, 41
Energy Index, 45
Enol:
 orbitals, 116
 SHMO, 114
Enolate:
 orbitals, 116
 SHMO, 114, 122, 130, 131
Enzyme, 80
Epoxides, 259, 275. See also Oxiranes
Esters:
 hydrolysis, 146
 reduction, 146
Ethane:
 1,2-dihalo, conformations, 243
 pK_a, 163, 263
 point group, 6
 protonated, 168, 169
Ethanolamine, 263
 orbital interactions, 263

Ethene. See Ethylene
Ethyl carbanion, 129
Ethyl cation, structure, 169
Ethyl chloride, E2 vs. S_N2 by *ab initio*, 165
Ethylbenzenium cation, 129, 256, 257
Ethylene, 238
 interaction diagram, 109
 I. P., 99, 291
 nucleophilic attack, 120
 in SHMO theory, 108, 109
 pK_a, 163, 263
 point group, 5
Ethyne. See Acetylene
Ethynyl carbene, 136
Excimer, 281–282
Exciplex, 281–282
Excited States, from CI, 51
Expectation value, 21
 orbital energy, 60
Extended Hückel Theory, 68

F_2, complex with ammonia, 102, 243
Feist's ester, 267
Fermions, electrons, 17
FHF-, 156, 162
Fluoboric acid, 262
Fluorescence, 215, 216
 lifetime, 216
Fluoride ion, in S_N2 and E2, 165
Fluoro carbene, CHF, 135
Fluoro methoxy carbene, 136
Fluoromethyl formate, structure, 244
Fock:
 equations, 30
 matrix, 31
 operator, 30
  RHF form of, 58
FOOF, structure, 242
Formaldehyde:
 I. P., 291
 point group, 5
 properties, 144
Formamide, 242
Formate, fluoromethyl, 244
Formyl carbene, 136
Free radical. See Radical
Frequencies, 47
 of ammonia, 48
Frontier orbitals, 89
Fukui method, 183
Fullerene, 171
Fulvene, 276
 6-dimethylamino, 276
 SHMO calculation, 250

Furan:
   2-cyano-, SHMO, 248
   SHMO, 248

Gauche Effect, 243
GAUSSIAN 90, 51
Gaussian functions, 35
   primitive, 36
GAUSSIAN 92, 19, 56
   input, 223
Gaussian XX, Z-matrix, 223
Geometric isomers, 8
Geometry, optimization, 46
Gimarc's rules, 85
Group, definition, 2
Groups:
   diastereotopic, 8, 9
   enantiotopic, 8, 9
   homotopic, 8, 9
Group orbitals, 81. *See also* Orbitals
   of dicoordinated atom, 83, 84
   of methyl, 85
   of methylene, 84
   of monocoordinated atom, 83
   of tricoordinated atom, 84
   of zero-coordinated atom, 82
Guanine, 161
Guanidinium ion, 126

Halogens, complexes with ammonia, 102, 243
Hamiltonian operator:
   core, 20
   electronic, 20
   matrix, 51
Hammond principle, 126
Hartree, unit of energy, 41
Hartree product, 22
Hartree-Fock:
   approximation, 23
   determinantal wavefunction, 38
   energy, 27
   limitations, 47, 91–93
   restricted (RHF), 38
      determinant, 38
      orbital energy, 38
      total energy, 27, 39, 58
   theory, 38
   unrestricted (UHF), 38
HCl, H bonded to oxirane, 262
Heptatrienyl, SHMO, 176
Hessian matrix, 46

Heteroatoms, in SHMO theory, 114, 115
Heteroaromatic, 177
Heterolytic cleavage, 152
Heterotopic, 9
1,5-Hexadiene, 192
   Cope rearrangement, 192
1,5-Hexadien-3-ones, photolysis of, 273
Hexamethyl (Dewar benzene), 235
Hexatriene, 202, 205, 256
   barrier from cyclohexadiene, 271
   pericyclic reaction, 202
   by photolysis, 206, 271
   SHMO, 255
HF, in aqueous solution, 162
Hofmann rearrangement, 138
Homoaromaticity, 171
Homoconjugation, spiropentadiene, 253
Homotopic, 9, 234
   NMR, 11
Homotropenylium cation, 171
Hückel MO Theory, 58, 68
   simple (SHMO), 106
Hückel 4n + 2 rule, 170, 171
Hückel program, SHMO, 228
Hund's Rule, 134
Hunsdiecker reaction, 278–279
Hybridization, 12, 13, 37
   and SHMO α and β, 117
Hydrazine, structure, 240
Hydrazines, 154, 240
Hydrazinium dichloride, 277
Hydride bridge, 103, 167, 168, 169
Hydride abstraction, 166
   examples, 168
Hydrogen atom:
   ionization potential, 81
   radius, 81
Hydrogen bonding, 75, 102
   ethanolamine, 263
   in ammonia, 292
Hydrogen bonds, 160
   interaction diagram, 161
   in nucleic acids, 161
   in proteins, 161
Hydrogen disulfide, 242
Hydrogen fluoride:
   geometry, 47
   orbital energies, 42
   total energy, 41
Hydrogen peroxide, 242
   mechanism of amine oxidation, 275
   point group, 4
Hydroxylamines, 154
Hyperconjugation, 104, 244, 254
Hypervalency, 156

Imine:
  carbonyl, 197
  nitrosyl, 197
  ylide, 197
Imines, nitrile, 196
Indanones, by photolysis, 273
Independent electron approximation, 58
Integral:
  Coulomb, 26, 27
  exchange, 26, 27
  kinetic energy, 34
  interaction, 60
  nuclear electron attraction, 34
  overlap, 33, 61
  two electron repulsion, 26, 34
Interaction:
  as a function of separation, 78
  generalizations for intermolecular, 70, 77
Interaction diagram, 59, 72
  construction of, 86-89
  of ethylene, 109
  interpretation of, 89-91
  two-orbital:
    four-electron, 72
    one-electron, 76
    three-electron, 73
    two-electron, 74
    zero-electron, 76
  of norbornadiene, 73
Interaction integral, 60
Interaction matrix elements, 68
  first row elements, 94
  halides, 97
Internal conversion, 219, 220
  lifetime, 216
Intersystem crossing, 211, 215, 219, 220
  lifetime, 216
Ionization potential:
  acetaldehyde, 144
  acetone, 144
  (Z)-2-butene, 72, 99
  butenone, 144
  cyclooctene, 124
  cyclohexene, 72
  DABCO, 100
  dimethyl ether, 144
  ethylene, 99, 291
  formaldehyde, 144, 291
  hydrogen atom, 81
  Koopmann's theorem, 40
  methanol, 144
  methyl acetate, 144
  nitrous oxide, 196
  norbornadiene, 72

  norbornene, 72
  oxetane, 144
  table of, 144
  tetrahydrofuran, 144
  trimethylamine, 100
  water, 144
Isomers, definition, 6
Isocyanates, 138

Jablonski diagram, 215-216

Ketocarbenes, 194
Ketones, *See also* carbonyl compounds
  metal reduction of, 260
  photochemistry, 271-274
Kinetic energy integrals, 34
Koopmann's theorem, 40

Lagrangian multipliers, 28, 29
Lanthanide shift reagents, 11
Laplacian operator, 20
LEGO, Molecular, 270
Lewis acids and bases, 118
Limitations:
  Hartree-Fock theory, 47-48, 91-93
  orbital correlation diagrams, 206
  orbital interaction diagrams, 91-93
Lithium aluminum hydride, 259
Localized orbital. *See* Orbital
Lossen rearrangement, 138
Lowdin orthogonalization, 33
Lowry-Bronsted, 97, 102, 118, 160, 166
Lumisterol$_3$, 283

McLafferty rearrangement, 90, 91
Markovnikov's Rule, 257
Mass spectrum, of carbonyls, 91
Matrix:
  density, 33
  Fock, 32
  Hamiltonian, 51
  Hessian, 46
  interaction, 68
  overlap, 31, 33
  Z, 224
MCPBA, *meta*-chloroperbenzoic acid, 124
Meerwein-Pondorf-Verley, 103, 280-281
Methane:
  geometry, 47
  orbital energies, 42
  point group, 6
  total energy, 41
Methanol:
  point group, 4
  properties, 144

Methoxybenzene, conformation, 264
Methyl acetate, 144, 259
Methyl acrylate, 144
Methyl carbanion, 129
Methyl cation, 126, 251
Methyl iodide, 154
Methyl methoxy carbene, 253
Methyl radical, 134, 169
Methylamine, 239
1-Methylanthraquinone, photoenolization, 274
Methylene, CH_2, complex with water, 256
Methylene chloride, point group, 5
Methylstyrene, 253
MO (molecular orbital), 30
    ammonia, 43
    energies, first row hydrides, 42
    energy, 38
Molecular LEGO, 270
Møller-Plesset, 52, 55
Molozonide, 197
Momentum, quantum mechanical, 20
MTPA, 12
Mulliken population analysis, 40

N,N-dimethylaminocarbinyl radical, 254
N,N-Dimethylformamide, rotation barrier, 148
N-Bromosuccinimide, 1-e reduction of, 243
N-F bonds, unusual length-force constant relationship, 101
Naphthalene, point group, 5
Negative hyperconjugation, 103, 254
Neighboring group effect, 103
Net charge:
    Mulliken, 40
    in SHMO theory, 112
Nitrene:
    NH, electronic configuration, 137
    phenyl, electronic state, 138
Nitrenes, 136, 137
    dialkylamino, 137, 138
    substituent effects, 138
Nitrenium ion:
    NH_2^+, singlet-triplet separation, 139
    methyl, CH_3NH^+, 139
    dimethyl, $(CH_3)_2N^+$, 139
    phenyl, $C_6H_5NH^+$, 141
Nitrenium ions:
    cyclic, 139
    dimethyl, 139
    methyl, 139
    substituent effects, 140, 141

Nitrile:
    imines, 196
    oxides, 196
    ylides, 196
Nitro compounds, 197
Nitromethane, pK_a, 163
Nitrone, 194, 197
Nitrosyl:
    imine, 197
    oxide, 197
Nitrous oxide, N_2O, I. P., 196
N,N-dimethylaminocarbinyl radical, 254
Norbornadiene, 72, 99
    interaction diagram, 73
    I. P., 72, 73
    photochemistry, 204
Norbornadienone, 202, 203
Norbornene, I. P., 72
2-Norbornyl cation, 129
    structure, 104
Norrish type I, 91, 221–222
    Dauben-Salem-Turro analysis, 222
    electronic configuration, 221
    frontier orbitals, 221
Norrish type II, 91, 218–220
    Dauben-Salem-Turro analysis, 220
    electronic states of products, 219
Nucleophilic substitution, at saturated carbon, 151
Nucleophilicity
    α-effect, 154
    and interaction diagrams, 90

OCAMS, 199
Olefins. *See* Alkenes
Operator:
    antisymmetrizer, 23, 24
    Coulomb, 21
    dipole moment, 40
    exchange, 29
    expectation value of, 21
    Fock, 30
    Hamiltonian, 20
    Laplacian, 20
    momentum, 20
    permutation, 22
Optimization of geometry, 46
Orbital:
    energy, 38
    group, 15
        methyl, 85
        methylene, 84
    localized, 15, 16
    molecular, 17, 30

Orbital (*Continued*):
  π compared to σ, 69
  proper, 15, 16
  Slater-type (STO), 34–36
Orbital correlation diagrams:
  benzvalene to benzene, 207
  carbene to olefins, 210
  cheletropic, 203
  cycloaddition, 199, 200, 204
  dioxetane decomposition, 211
  electrocyclic, 202, 205
  limitations of, 206
  photochemistry from, 202
  and symmetry, 198–199
  Woodward-Hoffmann, 199
Orthonormal, 22
Overlap:
  consequences of non-zero, 66, 68–69, 70
  destabilization due to, 63
  effect on polarization, 69
  as a function of separation, 78
Overlap matrix, 31, 33
Overlap integral, 61
Overlap Integrals, and orbital interactions, 68–69
2-Oxaallyl, 197, 282
Oxaloacetic acid, 10
Oxetane, properties, 144
Oxidation potential, related to BDE, 133
Oxirane, H bonded to HCl, 262
Oxiranes, 196, 201
Oxy-Cope rearrangement, 192, 193, 194
Ozone, 197
  van der Waals complex with C_2H_2, 276
Ozonolysis, 197

Pauling electronegativity, 44
1,3-Pentadiene, 266
Pentadienone:
  Diels-Alder reaction, 248
  SHMO, 245, 246
Pentadienyl:
  amino, SHMO, 174
  formyl, SHMO, 175
  SHMO, 108
  SHMO orbitals, 109
  vinyl, SHMO, 176
Pentadienyl anions, in aromatic substitution, 178
Pentadienyl cations, in aromatic substitution, 172, 174–176
Pericyclic reaction, 182, 184. *See also* cheletropic reaction, cycloaddition reaction, electrocyclic reaction, sigmatropic rearrangement
Peroxides, 154
Peroxyacids, mechanism of olefin epoxidation, 275
Perturbative MO theory, 57
Phenyl carbene, 136, 254
Phenylnitrene, electronic states, 138
Phenylnitrenium ion, 141
Phosphenium ions, 139
Phosphide (PH_2-), in E2 and S_N2, 165
Phosphine, 242
Phosphine group, 149
Phosphonium, 130
Phosphorescence, 212, 215, 216
Photochemistry, 213
  from orbital correlations, 202
Photocyclization, arylacetophenones, 274
Photoenolization, 1-methylanthraquinone, 274
Photoexcitation, 213
Photoreactions, of previtamin D_3, 283
pK_a. *See* Acidity
pK_b, table of, 144
Point group, 2
  classification by type, 3
Polarization:
  effect of overlap, 69
  of one basis function by another, 37
Population, atomic, 40
Population analysis:
  Mulliken, 40
  in SHMO theory, 112–113
Potential energy hypersurface, 213
Previtamin D_3, 282–283
Principal axis, 2
Prochiral, definition, 10
Propellane, [1.1.1], 244
Propene, 122
  pK_a, 163
Proper orbital. *See* Orbital
2-Propyl carbanion, 129
Proton abstraction reaction, 162
Proton affinity, 119
  table of, 144
Protovitamin D_3, 283
Pyridine, 176
  SHMO, 177
Pyridine chlorine radical, 254
Pyridine-N-oxide, SHMO, 177
Pyridinium cation, 176
  SHMO, 177
Pyrrole, 176
  SHMO, 177

Pyruvate decarboxylase, from brewer's yeast, 274

Quadratic formula, 64, 67
Quadricyclane, from norbornadiene, 204, 205
Quenching, 220

Radical:
   C substituted, 132
   carbon, 131
   carboxyl, $RCO_2\cdot$, 278–279
   electrophilic, 76, 131, 169
   H atom abstraction, 169
   N,N-dimethylaminocarbinyl, 254
   nucleophilic, 74, 131, 169
   X substituted, 131, 132
   Z substituted, 131, 132
Radical stabilization energy, (RSE), 133, 134
Rayleigh-Schrödinger perturbation theory, 52, 69
Resonance integral, β, 113
RHF. *See* Hartree-Fock
Rotational motion, 214
RSE, (radical stabilization energy), 133, 134

SCF (self-consistent field), 32
Schoenflies notation, 3
Schrödinger equation, 20
Sensitization, 220
Sesquinorbornatriene, 235
SHMO program, 228
SHMO calculations:
   2-cyanofuran, 248
   allyl, 109, 251
   aminopentadienyl, 174
   aniline, 173
   azapentadienyl, 174
   benzaldehyde, 173
   benzene, 111, 166, 170
   butadiene, 109
   carbonyl group, 258
   cyclobutadiene, 111, 166, 170
   cycloheptatrienyl, 170
   cyclopentadienone, 249
   cyclopentadienyl, 111, 170
   cyclopropenyl, 111, 170
   enol system, 116
   enolate anion, 116, 131
   ethylene, 109
   formylpentadienyl, 175
   fulvene, 250
   furan, 248
   heptatrienyl, 176
   hexatriene, 255

pentadienone, 245, 246
pentadienyl, 109
pyridine, 177
pyridine-N-oxide, 177
pyridinium cation, 177
pyrrole, 177
styrene, 173
vinylpentadienyl, 176
SHMO theory:
   assumptions, 106, 108
   charge and bond order in, 112–113
   cyclic π systems, 170
   heteroatoms in, 114
Sigma bonds:
   between first row atoms, 94, 95
   carbon to halogen, 96
   cleavage, 96
   dissociation energies, 98
   heterolytic cleavage, 96
   homolytic cleavage, 97
   interactions between, 99
   as π acceptors, 102
   as π donors, 103
   as σ acceptors, 101
   as σ donors, 103
Sigmatropic rearrangement, 182, 184, 187, 199
   component analysis for, 189
   stereochemistry, 188
Singlet, 207, 209
   state, 217, 219
Slater determinant, 22, 57, 238
Slater orbital, 34
S_N1 mechanism, 151, 152
S_N2 mechanism, 102, 146
   geometry of approach, 153
   at hydrogen, 162
   nucleophilicity, 154
   leaving group ability, 154
   stereoelectronic, 182
   substituent effects, 156
   transition state, 155, 157
   VBCM description, 157–159
S_N2 reaction, of cyclopentadienyl iodide, 280
Snoutene, 234
   rearrangement of, 266–267
Sodium borohydride, 259
Solvated electron, 78
Sphere, point group, 6
Spin-spin coupling constants, 13
Spiroconjugation, 253
Spiropentadiene, 253
State correlation diagrams, 206
   carbene to olefin, 210

State correlation diagrams (*Continued*)
  dioxetane decomposition, 211
  rules for, 209–210
Stationary points, 46, 213
Stereoisomerism, classification, 7, 8
Steric interaction, 72
STO, 34
Styrene, 173, 175, 178
  1-methyl-, 252
  SHMO, 173
Substituent types, 121, 122
  C, effect of:
    alkenes, 123
    carbanions, 130, 131
    carbenes, 135
    carbocations, 127
    radicals, 132
    nitrenes, 138
    nitrenium ions, 140, 141
  X:, effect of:
    alkenes, 121, 122
    carbanions, 129, 130
    carbenes, 135
    carbocations, 126, 127
    radicals, 131
    nitrenes, 138
    nitrenium ions, 139, 140
  Z, effect of:
    alkenes, 123
    carbanions, 130
    carbenes, 135
    carbocations, 127
    radicals, 131, 132
    nitrenes, 137, 138
    nitrenium ions, 140
Sulfones, cyclic, H/D exchange, 257–258
Sulfonium, 130
Suprafacial reaction:
  of π system, 185, 186
  of σ bond, 190
Symmetric orthogonalization, 33
Symmetry:
  applications, 2
  of atomic orbitals, 14
  consequences, 1
  of electronic states, 218
  elements, 3
    interrelationships, 3
  and orbital interactions, 71
  structural parameters, 12
Symmetry adapted orbitals, 15–16

Tachysterol$_3$, 283
Tetracyanoethylene, 253
Tetrahydrofuran, properties, 144

Tetrahydrofurans, from allylic acetals, 275
Tetramethyleneethane, 253
Tetrazenes, 137, 138
Thiacyclopentenes, 187
Thiacyclopropane. *See* Thiirane
Thiirane, 240
  cation, 240
Thiirane dioxide, 202
Thiiranes, 201
Thiophene-1,1-dioxide, 276
Three-electron bond, 242, 254
Thymine, 161
Transition structure, definition, 46
  location of, 227
Tricyclo[3.1.0.0^{2,4}]hexane, 9
Triethylamine, 243
Trimethylamine, 243
Trifluoromethyl group, 116, 254
Trifluoromethyl radical, 134
Trimethylamine, 243
  I. P., 100
Triphenyl carbinol, 262
Triphenyl cation, 262
Triplet, 111, 207, 209
  states, 217
Trisethylenediamine complexes, point group, 5
Tropylium cation, 158, 256, 262
Twistane, 4
  point group, 5

Unbound surface, 213
Urea, 276

Van der Waals attraction, 78, 79, 287
Van der Waals complex, O$_3$ and C$_2$H$_2$, 276
Variation method, 21
VBCM, 92
  S$_N$2, 157–159
Vertical excitation, 215
Vibrational cascade, 215, 216
Vibrational levels, 214
Vinyl acetate, 132, 133
Vitamin D$_3$, 283

Wagner-Meerwein rearrangement, 104, 128, 168
Walsh orbitals, 85
  cyclopropane, 98
  spiropentadiene, 253
Water, properties, 144
  complex with CH$_2$, 256
  geometry, 47
  orbital energies, 42
  total energy, 41

Water dimer, 162
Wavefunction
    determinantal, 22
    many electron, 21, 22
    one electron. *See* Orbital
Wolfsberg-Helmholtz, 68
Woodward-Hoffmann correlation, 199

Y-conjugation, 276–277
Ylides, 130
    carbonyl, 197, 282

imine, 196, 197
nitrile, 196, 197

X substituents, 122

Zero overlap approximation, 62, 64
Z-matrix, 224
Z substituents, 123
Zwitterionic states, 217
Zwitterion, 259